POWER OF DEVELOPMENT

Post-colonial, post-modern and feminist thinking have focused on the power structures embedded in global development, challenging the ways in which development is conceived and practised and questioning its meaning.

These essays explore development discourse as an interwoven set of languages and practices, analysing the texts of development without abandoning the power-laden local and international context out of which they arose and to which they speak.

By conceptualizing development as a discourse, the authors argue that it cannot simply be reduced to the structures and logic of economics; development has its own logic, internal coherence and effects. Three main questions are addressed. How and why does the language of development change over time? What is the role of the spatial in the language and practices of development? Is it possible to imagine a world in which development has no redeeming features or power?

Combining analyses of development discourse with concrete examples of how that discourse is constructed and operates in particular times and places, the contributors stake out the terrain for a grounded development studies in a post-marxian world.

Jonathan Crush is a Professor of Geography at Queen's University, Canada.

POWER OF DEVELOPMENT

Jonathan Crush

London and New York

First published 1995
by Routledge
11 New Fetter Lane, London EC4P 4EE

Simultaneously published in the USA and Canada
by Routledge
29 West 35th Street, New York, NY 10001

Typeset in Garamond by
Ponting–Green Publishing Services, Chesham, Bucks
Printed in Great Britain by
T.J. Press Ltd, Padstow, Cornwall

British Library Cataloguing in Publication Data
A catalogue record for this book is available from
the British Library

Library of Congress Cataloguing in Publication Data
A catalogue record for this book has been requested

ISBN 0–415–11176–5
0–415–11177–3 (pbk)

CONTENTS

CONTENTS

LIST OF ILLUSTRATIONS

LIST OF CONTRIBUTORS

W.M. Adams is Lecturer in Geography at the University of Cambridge. His publications include *Green Development: Environment and Sustainability in the Third World* (Routledge, 1990) and *Wasting the Rain: Rivers, People and Planning in Africa* (University of Minnesota Press, 1994). His research focuses on problems of environment and development in Africa, particularly water resource development, and indigenous irrigation and wetland use.

Michael Cowen is Reader in Economics at the City of London Polytechnic. He has published extensively on agrarian issues in East Africa and British colonial policy and is co-author, with Robert Shenton, of *Doctrines of Development* (Routledge, 1995).

Jonathan Crush is Professor of Geography at Queen's University, Kingston. He is the author of *The Struggle for Swazi Labour* (McGill-Queen's Press, 1987) and *South Africa's Labor Empire* (Westview Press, 1991) and co-editor of *Liquor and Labor in Southern Africa* (Ohio University Press, 1992), *Crossing Boundaries* (IDASA/IDRC, 1995) and *White Farms* (Heinemann, 1995). He is co-director of the IDRC South African Migrant Labour Project.

Arturo Escobar is Associate Professor of Anthropology at the University of Massachusetts, Amherst. He is the author of *Encountering Development: The Making and Unmaking of the Third World* (Princeton University Press, 1994). His research interests include the anthropology of development, social movements, and science and technology.

Kenneth Hewitt is Professor of Geography and the Cold Regions Research Centre at Wilfrid Laurier University, Waterloo, Ontario. He has authored or co-authored four books, including *Regions of Risk: Hazards, Vulnerability and Disaster* (Longman, forthcoming), and edited a further two books including *Interpretations of Calamity* (Allen and Unwin, 1983). His current interests include hazards and disasters research, high mountain environments and war risks for civil societies and habitats.

Fiona Mackenzie is Associate Professor of Geography at Carleton University, Ottawa. She has written widely on rural landholding in Kenya, gender and the environment, and farming in eastern Ontario. She is the co-editor of *Development from Within* (Routledge, 1991) and author of a forthcoming book on rural transformation in Kenya.

Kate Manzo is Lecturer in Political Science at the Australian National University. She is the author of several articles on development discourse and two books: *Domination, Resistance and Social Change in South Africa: The Local Effects of Global Power* (Praeger, 1992) and *National Families: Race, Empire, Scripture* (Rienner, forthcoming).

T.G. McGee is Professor in the Department of Geography at the University of British Columbia. He is the author of *The Southeast Asian City* (Praeger, 1967), *The Urbanization Process in the Third World* (Bell, 1974) and *Hawkers in Hong Kong* (Centre of Asian Studies, 1974), and co-author of *Theatres of Accumulation* (Methuen, 1985) and *The Extended Metropolis* (University of Hawaii Press, 1991). He is currently carrying out research on urbanization in Southeast Asia.

Timothy Mitchell is Associate Professor of Politics at New York University and the author of *Colonising Egypt* (Cambridge University Press, 1990). He is currently working on the invention of the modern idea of the economy in the 1930s and 1940s.

Jane L. Parpart is Professor of History, Women's Studies and International Development Studies at Dalhousie University. She is the author of *Labor and Capital on the African Copperbelt* (Temple University Press, 1983) and co-editor of *Patriarchy and Class* (Westview Press, 1988), *Women and the State in Africa* (Rienner, 1989), *Women, Employment and the Family in the International Division of Labour* (Macmillan, 1990) and, most recently, *Feminism/Postmodernism/Development* (Routledge, 1985). Her research and teaching interests include gender and development theory, and gender and the construction of middle-class identity in southern Africa.

Doug J. Porter of the Division of Society and Environment, Research School of Pacific and Asian Studies, Australian National University, has worked with rural development agencies in east and southern Africa and South East Asia for the past fifteen years. He is the co-author of *Development in Practice: Paved with Good Intentions* (Routledge, 1991) and *Economic Change and Rural Transformation in Vietnam* (Westview Press, 1995). His recent research and development work has concentrated on the impact of macro-economic change on marginal households in rural Indochina.

Robert Shenton is Professor of History at Queen's University, Kingston. He is author of *The Development of Capitalism in Northern Nigeria* (Heinemann, 1986) and co-author of *Doctrines of Development* (Routledge, 1995), as well as articles on development theory, rural transformation and Nigerian history. He is currently supervisor of an IDRC project on environment and resource management in Uganda.

Nanda Shrestha is Associate Professor in the School of Business and Industry at Florida A&M University. He is the author of *Landlessness and Migration in Nepal* (Westview Press, 1991) and a forthcoming book entitled *I Am a Development Victim* (Westview Press). He has published numerous articles on development issues and is currently working on a project on the political ecology of land encroachment funded by the National Science Foundation.

Chris Tapscott is the former Director of the Namibian Institute for Social and Economic Research at the University of Namibia, Windhoek. He has written extensively on development issues in South Africa and Namibia and on the political transition in Namibia.

Michael Watts is Director of the Institute of International Studies at the University of California, Berkeley. He is author of *Silent Violence* (University of California Press, 1983) and *Reworking Modernity: Capitalism and Symbolic Discontent* (Rutgers University Press, 1992) and editor of *State, Oil and Agriculture in Nigeria* (University of California Press, 1987) and *Living Under Contract* (University of Wisconsin Press, 1994). He has published widely on agrarian change in Africa, and the agrarian question in comparative perspective. He is currently working on agricultural change in the wake of NAFTA in California, and a comparative study of decollectivization in formerly socialist agricultures.

Gavin Williams is Fellow of St Peter's College and Lecturer at the University of Oxford. He is author of *State and Society in Nigeria* (Afrografika, 1980) and *The Nigerian Civil War* (Open University, 1982) and co-editor of *Sociology and Development* (Tavistock, 1974), *Nigeria: Economy and Society* (Collings, 1976), *Rural Development in Tropical Africa* (St Martin's Press, 1981), and *Sociology of Developing Societies: Sub-Saharan Africa* (Monthly Review Press, 1982). He has also published numerous articles on the politics and agricultural policies of Nigeria, and on the World Bank in Africa.

PREFACE

In South Africa, the language and practice of development have long been hopelessly tainted by their association with segregation and apartheid. During its years in exile the African National Congress (ANC), though resolutely modernist in most of its ideas, consciously tried to avoid this language in articulating its vision of an alternative South Africa. The uncomfortable radicalism of the ANC, and its close ties to the former Eastern bloc countries, proved too much for most international development agencies and Western governments. Until Nelson Mandela was released from prison in 1990, Chief Mangosuthu (Gatsha) Buthelezi of the Inkatha Freedom Party was usually far more welcome than the ANC in Western capitals. He, at least, spoke 'the right language.'

During the agonizing interregnum between Mandela's release in 1990 and his ascent to the Presidency in May 1994, South Africa was wracked by violence, disorder and a deepening sense of pessimism that any kind of humane future was possible. In the midst of the mayhem and gloom, an extraordinary transformation began to take place. Western development agencies and their legions of 'experts' poured into South Africa to set up shop. Unsure of whether South Africa was indeed a 'developing country,' and lacking any basic knowledge of this complex society, they handed out development consultancies to former anti-apartheid activists who gratefully received what they were given. The World Bank (for International Reconstruction and Development) rapidly positioned itself as an adviser to the government-in-waiting. In a few short months, the Bank (using a rhetoric of consultation and benign assistance) was having an inordinate influence over the agenda for agricultural policy and post-apartheid land reform in the country. Another agency, the International Development Research Centre of Canada, sponsored a series of missions to produce the knowledge, which the 'Mass Democratic Movement' supposedly lacked, on basic development issues of urban policy, local government, industrial strategy, science and technology, and macroeconomics. These reports offered a series of influential blueprints for ANC policy and future foreign aid to South Africa.

Meanwhile, institutions and individuals with close ties to the apartheid regime underwent a rapid rehabilitation. The Southern African Development Bank (an underwriter of the apartheid state's bantustan policy since its establishment in the late 1970s) reconstituted itself as the premier agency of rural development in the country and was quickly welcomed into the boardrooms of banks in Europe and North America. This new-found legitimacy had nothing to do with any fundamental shift in the personnel, policies or pronouncements of the Bank. Indeed, there was precious little need to change the language at all. For over a decade the Bank had been speaking in words not of its own making. The technocratic, depoliticized language of international development had already found a good home halfway between Johannesburg and Pretoria.

The discourse of development flowed easily across the political divide between the old South Africa and the new. The discourse increasingly began to frame the protracted negotiations between the ANC and the incumbent National Party government, staking out areas of common interest and agreement. When the parties agreed on a federal dispensation and mapped South Africa's nine new provinces, for example, they replicated the nine 'development regions' demarcated by the apartheid government in the early 1980s. The ANC's public policy language was quickly cleansed of the marxian rhetoric which was so central during the later years in exile, replaced by the more 'pragmatic' language of 'reconstruction and development.' When the new ANC government announced its first five-year plan in early 1994, it was suitably titled *The Reconstruction and Development Programme* (RDP). All political parties, including the National Party and the right-wing Freedom Front, indicated their support for the plan. A prominent left-wing trade unionist, now a minister in the new government, was charged with overseeing the implementation of the Plan. The transformation seemed complete. But how and why had it happened?

Clearly, this is not the first time that South Africa has fallen under the soothing power of development at a time of social disorder and crisis. It had happened in the 1930s, in the 1960s, and again in the late 1970s and early 1980s. What was it about this discourse that allowed it to re-emerge untainted and reinvent itself every time? And how could it now have captured the public face of South Africa's greatest political transformation ever? The ANC was not the only organization, nor South Africa the only country, to undergo this transformation in the terms of public discourse in the last few years. Neighbours Mozambique and Namibia have recently experienced something similar, as have Eastern Europe, the former Soviet Union, and Central America. Namibia was the last in a long line of decolonizing African countries in the post-1945 era to articulate its post-colonial aspirations in the language of development. According to many accounts, this was the era when development had actually 'begun,' when the regime of colonialism was replaced by the regime of development. Yet its purchase in South Africa clearly went back much further than that.

Conventional academic development theory (itself in a profound state of disarray) offers limited guidance about how to situate South Africa's transition. There are few answers that still do not eventually lead to the conclusion that development always emanates from the West (or the North), and is an expression either of Western progress and enlightenment or projected economic power and domination. To argue that the ANC is simply being 'captured' by the forces of international capitalism and its vast development apparatus seems to me to be both simplistic and misleading.

A newer wave of scattered scholarship more influenced by post-structural thought appears to offer important alternative insights. In attempting to conceptualize development as a *discourse*, as an interwoven set of languages and practices, these writers see it also as a modernist regime of knowledge and disciplinary power. As such it cannot simply be reduced to the out-working of deeper economic logics and structures but has its own logic, internal coherence and effects. This approach is also particularly attuned to the language of development itself, pushing towards an analysis focused on the texts of development without abandoning the power-laden local and international context out of which they arise and to which they speak.

Writing in this vein is growing rapidly though not much of it is yet readily available or easily accessible to students. There seemed a need therefore for a volume of essays which would make these new directions more accessible to the student and, perhaps simultaneously, provoke further engagement with this perspective. As an edited sample of some of the new work on the writing of development, this book cannot pretend to explore the power of the discourses of development in a particularly systematic or comprehensive way. What it can do is signpost fragments of a much wider discursive field and highlight areas for further consideration and analysis.

A number of the chapters in this volume were originally presented in three panels on 'Discourse and Development' at the annual meeting of the Canadian Association of Geographers at Queen's University in 1991. In selecting which papers to include in this volume, and which other authors to invite to contribute to the book, I was motivated by four main considerations: first, I wanted to combine more abstract, conceptual analyses and reviews of development discourse with concrete examples of how that discourse is constructed and operates in particular times and places. Second, development studies has always been recognized for its interdisciplinarity. This is a very useful tradition to continue. If there are more geographers than anyone else that is because of my third concern; a desire to reassert the fundamentally spatial character of the languages, disciplines and practices of development and anti-development. Finally, I wanted to include chapters by authors who were engaging with post-structural ideas from a personal intellectual history of looking at development from other, more materialistic, perspectives. Many of the authors, for example, have written earlier books on development within a political economy tradition. This is not a book which systematically sets out to explore the interface between Marxism and post-structuralism in development studies. But it is useful to approach the question of the writing of development from a background in less literary traditions of analysis. The result may be some rather uneasy tensions but it does at least help to keep one's feet on the ground.

The essays inevitably cover a broad field and a multiplicity of themes. In an effort to give this volume internal coherence I have included, as an introductory chapter, my own reading of some of the common themes and used this as a basis for grouping the chapters themselves. Inevitably the chapters themselves engage in much greater depth with these issues and more besides. In addition, the grouping is somewhat arbitrary since many of the chapters span more than one of the themes.

I am very grateful to the various authors who agreed to provide chapters for this volume and then waited, with considerable patience, as it took shape. My thanks also to Tristan Palmer and Patricia Stankiewicz at Routledge for similar forbearance. I am particularly grateful to Wilmot James for his insights into the current transition in South Africa and to Bob Shenton for similar advice on the history of development. I am also indebted to several other colleagues at Queen's University for their ideas on these and related questions: Alan Jeeves, Bruce Berman, Bob Stock, John Holmes and Colin Leys. My thanks to Miriam Grant for helping to organize the original conference panels in 1991. Linda Crush, Sally Peberdy and Meagen Freer provided invaluable editorial assistance. Finally, I am very grateful to the Social Sciences and Humanities Research Council of Canada for its financial support of this project.

ABBREVIATIONS

ANC	African National Congress
AZAPO	Azanian People's Organization
BCM	Black Consciousness Movement
BENBO	Bureau for Economic Research re Bantu Development
BENSO	Bureau for Economic Research, Co-operation and Development
CLACSO	Latin American Social Science Council
GAD	Gender and Development
GATT	General Agreement on Tariffs and Trade
GNP	Gross national product
HSRC	Human Sciences Research Council
IDNHR	International Decade for Natural Hazard Reduction
IDRC	International Development Research Centre (Canada)
IMF	International Monetary Fund
IUCN	International Union for Conservation of Nature and Natural Resources
IRSAC	Institut pour la Recherche Scientifique en Afrique Central
KLC	Kenya Land Commission
MAB	Man and Biosphere Programme
NGO	Non-governmental organization
NP	National Party
NSM	New Social Movement
OAU	Organization of African Unity
ORSTOM	Office de Recherche Scientifique et Technique d'Outre Mer
PAR	Participatory action research
PC	Peace Corps
RDP	Reconstruction and Development Programme
SADCC	Southern African Development Co-ordination Conference
UNEP	United Nations Environment Programme
USAID	United States Agency for International Development
WAD	Women and Development
WCS	World Conservation Strategy

WID   Women in Development
WWF   Worldwide Fund for Nature (formerly World Wildlife Fund)
ZDSDP Zamboango del Sur Development Project

INTRODUCTION
Imagining development

Jonathan Crush

Development occupies the centre of an incredibly powerful semantic constellation . . . at the same time, very few words are as feeble, as fragile and as incapable of giving substance and meaning to thought and behaviour.

(Esteva 1992: 8)

In March 1895, on his return from a tour of duty as Commissioner in British Central Africa (Malawi), Sir Harry Johnston spoke to the Royal Geographical Society about the changing character of British rule in Central Africa. Johnston drew a vivid picture for his audience of two contrasting landscapes:

The Lower Shire district . . . was a marshy country, with only one European occasionally residing at a half-formed station, and with a native population scarcely exceeding one thousand. The country had become almost uninhabited through the raids of certain Mokololo chiefs and some powerful tribes north of the Zambesi. . . . In the Mlanje District there was practically chaos. The chiefs of the aggressive Yao tribes . . . had taken complete possession of this rich district, the few European planters were menaced in their lives and property, and the only mission station had to be abandoned. . . . In short, throughout all this country there was absolutely no security for life and property for natives, and not over-much for the Europeans. . . . Everything had got to be commenced; there was no proper postal service, there were no customs-houses, no roads suitable for wheeled traffic, very little labour in the coffee plantations; the forests of the land were being steadily destroyed year by year by bush fires, and the navigation of the Shire River was entirely at the mercy of evil-minded slave traders.

(Johnston 1895: 194–6)

A mere three years later, according to Johnston, the visible landscape had been dramatically transformed under the benign influence of British rule:

An increasing number of natives are able to read and write, and, above

1

all, are trained to respect and to value a settled and civilized government. ... Here will be seen clean broad level roads, bordered by handsome avenues of trees, and comely red brick houses with rose-covered verandahs peeping out behind clumps of ornamental shrubs. The natives who pass along are clothed in white calico, with some gaudy touch of colour superadded. A bell is ringing to call the children to the mission school. A planter gallops past on horseback, or a missionary trots in on a fat white donkey from a visit to an outlying station. Long rows of native carriers pass in Indian file, carrying loads of European goods, or a smart-looking policeman, in black fez, black jacket and breeches marches off on some errand. You will see a post-office, a court of justice, and possibly a prison, the occupants of which, however, will be out mending roads under the superintendence of some very business-like policeman of their own colour. The most interesting feature in the neighbourhood of these settlements at the present time is the coffee-plantation, which, to a great extent, is the cause and support of our prosperity.

(Johnston 1895: 202, 211)

Johnston's was a highly stylized rendering of the reordering of space: the civilized, ordered, white, male, English landscape erases its unordered, savage, chaotic, dangerous, African predecessor. For Johnston, colonialism was about gaining control of disorderly territory and setting loose the redemptive powers of development. The African landscape is rewritten, figuratively and literally, to reflect the subsumption of one reality by another. Africans are incorporated into this landscape as garbed agents of a higher power. Their bodies and behaviours testify to the new order. The text smooths out incongruities and inconsistencies, and erases all oppositional voices and spaces of dissent. Africans become objects for the application of power rather than subjects experiencing and responding to the exercise of that power. This is the power of development: the power to transform old worlds, the power to imagine new ones.

Johnston's audience knew exactly what these encouraging scenes meant. He was, after all, talking about the 'cause and support of our prosperity.' In contrast, as Michael Cowen and Robert Shenton point out in the first chapter of this volume, open almost any contemporary development text and all is confusion. Both the meaning and the purpose of development look rather like the Lower Shire in 1892; at best 'marshy,' more often 'practically chaos.' And yet, as an arena of study and practice, one of the basic impulses of those who write development is a desire to define, categorize and bring order to a heterogeneous and constantly multiplying field of meaning. In a recent spate of development dictionaries we sense an urgent, even desperate, attempt to stabilize development and bring order out of ambiguity (Eatwell *et al.* 1989; Welsh and Butorin 1990; Fry and Martin 1991; Sachs 1992; Hadjor 1993).

These dictionaries merely confirm that development is a most elusive concept. Perhaps, as Sachs (1992: 1–5) suggests, it ought to be banned. But first it would be necessary to say what exactly should be banished. Thus, in the very call for banishment, Sachs implicitly suggests that it is possible to arrive at an unequivocal definition.

This book does not attempt to provide a more precise definition of 'development' and none of the contributors were asked to offer one. Nor is it about 'development theory' – the self-designated academic field which attempts to verbally model 'real world' processes of development – and its recurrent internal crises and impasses (Booth 1985; Binder 1986; Edwards 1989; Hunt 1989; Mathur 1989; Sutton 1989; Corbridge 1990; Hettne 1990; Slater 1990; Manzo 1991; Kay 1993; Schuurman 1993; Leys, forthcoming). Much continues to be written on the theme of what development is (or should be), what it does (or fails to do) and how it can be better implemented (Toye 1987; Kothari 1988; Norgaard 1992; Alvares 1992a; Pottier 1992; Hobart 1993; Moser 1993). Rather than asking what development is, or is not, or how it can be more accurately defined, better 'theorized,' or sustainably practised, the authors in this volume are generally more interested in a different kind of question. Here the primary focus is on the texts and words of development – on the ways that development is written, narrated and spoken; on the vocabularies deployed in development texts to construct the world as an unruly terrain requiring management and intervention; on their stylized and repetitive form and content, their spatial imagery and symbolism, their use (and abuse) of history, their modes of establishing expertise and authority and silencing alternative voices; on the forms of knowledge that development produces and assumes; and on the power relations it underwrites and reproduces.

The *discourse* of development, the forms in which it makes its arguments and establishes its authority, the manner in which it constructs the world, are usually seen as self-evident and unworthy of attention. This book's primary intention is to try and make the self-evident problematical. The concern with this issue is influenced by similar concerns in other disciplines and fields. Three connections, in particular, should be mentioned: first, there is the 'textual turn' in the social sciences and humanities which has focused attention on the conventions of writing and representation by which Western disciplines and institutions 'make sense' of the world (see, for example, Said 1983; McCloskey 1985, 1990; Clifford and Marcus 1986; White 1987; Atkinson 1990; Crush 1991; Barnes and Duncan 1992; Campbell 1992; Dalby 1992; Preston and Simpson-Housley 1994). Second, there is the impact of post-modern, post-colonial and feminist thought which have converged upon the truth claims of modernism and shown how the production of Western knowledge is inseparable from the exercise of Western power (for example, Said 1978, 1993; Minh-Ha 1989; Spivak 1990b; Young, R. 1990; Mohanty *et al.*

3

1991; Ahmad 1992; Norris 1992; Godlewska and Smith 1994). And third, there is the growing struggle within postcolonial thought to loosen the power of Western knowledge and reassert the value of alternative experiences and ways of knowing (for example, Fanon 1968; Thiong'o 1986; Spivak 1987, 1990b; Stauffer 1990; Nandy 1991; Long and Long 1992; Momsen and Kinnaird 1993; Appiah 1992; Breckenridge and van der Veer 1993; Corbridge 1993; Sardar, Nandy and Davies 1993; Bhabha 1994; Crush 1994).

Perhaps, it might be objected, to subject development to such an inquisition is simply another form of faddish intellectualism destined, like all the others, to bloom and fade. Certainly it is true that the work and words of development will continue on pretty well regardless. However, this form of analysis does, I believe, offer new ways of understanding what development is and does, and why it seems so difficult to think beyond it. The idea that the texts of development might be analysed as a form of writing is not altogether new (Escobar 1984, 1988, 1994; Horesh 1985; Wood 1985; Apter 1987; Ferguson 1990; Apffel Marglin and Marglin 1990; Manzo 1991; Parajuli 1991; Pieterse 1991; Slater 1992a, 1992b). But what, it might be asked, is the point of literary pursuits in such a non-literary domain? The developer will say that there is no time for such esoterica. Surely the practical challenges of development are so pressing that we can scarcely afford to bother with this kind of armchair contemplation? By bringing together a selection of the work of scholars who are currently grappling with these issues, and trying to make it accessible to an interdisciplinary audience of students of development studies, this book will hopefully further the debate around the issue of whether it is possible to extricate ourselves from the development morass.

As most of us are aware, development rarely seems to 'work' – or at least with the consequences intended or the outcomes predicted. Why then, if it is so unworkable, does it not only persist but seem continuously to be expanding its reach and scope? Could it be that development does in fact work very well? It is just that what it says it is doing, and what we believe it to be doing, are simply not what is actually happening. And if this is so, then perhaps we need to understand not only why the language of development can be so evasive, even misleading, but also why so many people in so many parts of the world seem to need to believe it and have done so for so long.

Language is fundamental to the way in which we order, understand, intervene and justify those interventions into the natural and social world. Admittedly, most writing on development is prosaic in the extreme – leaden, jargon-ridden, hackneyed and exclusionary. In addition, the structure and form of the development text is highly stylized and repetitive. Nevertheless, for all their pedantry and pretension, the texts of development are, of necessity, also written in a representational language – a language of metaphor, image, allusion, fantasy, and rhetoric. These imagined worlds of development writing and speaking often appear to bear very little resemb-

4

lance to any commonsense reality. To find out about a country, one usually does not read its development plan. In a textual field so laden with evasion, misrepresentation, dissimulation and just plain humbug, language often seems to be profoundly misleading or, at best, have only limited referential value. How then does it have such staying power?

The texts of development have always been avowedly strategic and tactical – promoting, licensing and justifying certain interventions and practices, delegitimizing and excluding others. An interest in how the texts of development write and represent the world is therefore, by extension, an interest in how they interact with the strategies and tactics of their authors and of those who lend them authority. What is expertise, after all? And why is there so much of it inside what James Ferguson (1990) aptly calls 'the development machine'? Why does expertise license certain forms of speech and not others? What do the texts of development not say? What do they suppress? Who do they silence – and why?

In identifying an object for analysis this book focuses first on the texts of development and only secondarily on its projects and practices. In generic terms, the objects of analysis are the reports, plans, analyses, evaluations, assessments, consultancies, papers, books, policies, speeches, discussions, debates, presentations and conversations that circulate within and through the apparatus of agencies and institutions of the development machine. The authors of these texts include the legions of planners, practitioners, consultants, experts, scholars, advocates, theorists and critics in the employ of or associated with this institutional and disciplinary nexus. Their names and individual identities are generally not that important, so stylized are their texts, though like any disciplinary field, development has its authority figures whose ideas prompt genuflection and ritual obeisance by others.

In arguing that more attention should be paid to the language of development, we need simultaneously to resist the submersion of the world by the words of development. Though development is fundamentally textual it is also fundamentally irreducible to a set of textual images and representations. Even as they explore facets of the rhetoric and language of development, the essays in this volume implicitly reject the conceit that language is all there is. The primary purpose of the development text (like most others) is to convince, to persuade, that this (and not that) is the way the world actually is and ought to be amended. But ideas about development do not arise in a social, institutional or literary vacuum. They are rather assembled within a vast hierarchical apparatus of knowledge production and consumption sometimes known, with metaphorical precision, as the 'development industry.' This industry is itself implicated in the operation of networks of power and domination that, in the twentieth century, have come to encompass the entire globe. As Claude Alvares (1992b: 230) points out, 'knowledge is power, but power is also knowledge. Power decides what is knowledge and what is not knowledge.' A contextual reading of the literature of development therefore has a great deal to say about the apparatuses of power and domination within which those texts emerge, circulate and are consumed. The aim in this kind of approach is literary analysis as prelude to *critique*. As Said (1983: 221) has noted 'the fascinated description of exercised power is never a substitute for trying to change power relationships within society.'

Languages are never self-referential but are instead constructed within 'social fields of force, power and privilege' (Polier and Roseberry 1989). The challenge, therefore, is both to situate the texts of development in their historical and social context, and to decode 'the subtleties of contextual presences in texts' (Cunningham 1994: 45). Many of the authors in this volume come out of a political economy tradition that argues that politics and economics have a real existence that is not reducible to the texts that describe and represent them. Textual analysis is a dangerous activity if it succeeds in supplanting political engagement with poetical reflection, in 'reducing life to language and obliterating the relations of power, exploitation and inequality that order society and history' (Palmer 1990).

Development discourse promotes and justifies very real interventions and practices with very real (though invariably unintended) consequences. To incarcerate or confine these (often catastrophic) effects within the text is to embark on a dangerous 'descent into discourse' (Palmer 1990). In this volume, poetics and politics are generally envisioned as discrete, though interwoven, strands of social life. In this way, conceptual space is made for an exploration of the links between the discursive and the non-discursive; between the words, the practices and the institutional expressions of development; between the relations of power and domination that order the world and the words and images that represent those worlds.

Development discourse is constituted and reproduced within a set of material relationships, activities and powers – social, cultural and geopolitical. To comprehend the real power of development we cannot ignore either the immediate institutional or the broader historical and geographical context within which its texts are produced. The immediate context is provided by 'the development machine.' This machine is global in its reach, encompassing departments and bureaucracies in colonial and post-colonial states throughout the world, Western aid agencies, multilateral organizations, the sprawling global network of NGOs, experts and private consultants, private sector organizations such as banks and companies that marshall the rhetoric of development, and the plethora of development studies programmes in institutes of learning worldwide.

As Arturo Escobar (this volume) suggests, development can be seen as an apparatus 'that links forms of knowledge about the Third World with the deployment of forms of power and intervention, resulting in the mapping and production of Third World societies.' Development is thus fundamentally about mapping and making, about the spatial reach of power and the control and management of other peoples, territories, environments, and places. In their chapter, Cowen and Shenton argue that development at its birth involved the crafting of a set of managerial strategies (what they call trusteeship) to cope with the disruptions of social disorder within Europe and, later, the colonial and post-colonial worlds. But, as they imply, development is not simply a closed system of 'arrogant interventionism' (Sachs 1992: 2) – an unproblematical set of instruments and justifications for the application of strategic Western power and domination and the subjugation of the dismissively labelled 'Third World.'

Power, as Said (1983: 221) suggests, is analogous 'neither to a spider's web without the spider nor to a smoothly functioning flow diagram; a great deal of power remains in such coarse items as the relationships between rulers and ruled, wealth and privilege, monopolies of coercion, and the central state apparatus.' Power in the context of development is power *exercised*, power *over*. It has origins, objects, purposes, consequences, agents, and, *contra* Foucault, much of this seems to lie quite patently within the realm of the economic and the political. There are also 'ascertainable changes stemming from who holds power and who dominates who' (Said 1983: 221). The imaginary and practice of development are not static entities impervious to change. Development discourse, despite enormous continuity over time, also changes its language, strategies and practices. One of the reasons is its reciprocal relationship with shifts in 'who holds power and who dominates who.'

The work of Edward Said (1978, 1983, 1993) provides a useful point of departure for a volume of this nature. Said himself actually has remarkably little to say about development as a component of Orientalism. Possibly this is because he focuses more on the novelists, scholars, and travellers of empire than the prosaic managers of the imperial and post-colonial estate, amongst whom development was and is a recurrent obsession. Said provides a clear reminder of the need to situate all Western words within imperial worlds. To argue that development (like, say, the novels of Jane Austen) needs an imperial context may seem like a statement of the obvious. But the point is that within the texts of development themselves, this context is either ignored, downplayed or (as in much neo-Marxian 'development theory') made completely determining (Peet 1990).

Orientalism, in Said's (1978: 3) oft-quoted definition, is a 'systematic discourse by which Europe was able to manage – and even produce – the Orient politically, sociologically, militarily, ideologically, scientifically, and imaginatively.' This definition – with appropriate substitutions ('the West' for 'Europe,' the 'Third World' for 'the Orient') – would serve for many as a working hypothesis about the power and purpose of development. But it fails in two respects inherent in the original conception. First, Said's critics

7

point out that he has a great deal to say about the ideological, scientific and imaginative production of the Orient, but is rather less forthcoming about its economic and political production and their interrelationship (see Sprinker 1992; Breckenridge and van der Veer 1993). Though Said could hardly be accused of always privileging the text over the material context, some of his followers are not so subtle. Second, his critics charge that he draws the Orientalist web too tightly around the diffuse representational practices of the West. The result is an image of a homogenizing disciplinary power that is too tidy, too seamless, too unitary. In the case of development, it would be a mistake to view power as emanating exclusively from one space and being directed exclusively at another. Spatially, the power of development is far more diffuse, fragmented and reciprocal than this.

Development, for all its power to speak and to control the terms of speaking, has never been impervious to challenge and resistance, nor, in response, to reformulation and change. In a startling reversal, Fanon (1968) once argued that 'Europe is literally the creation of the Third World.' There is a great deal about the form and content of development that suggests that it is reactive as well as formative. As a set of ideas about the way the world works and should be ordered, understood and governed, development should also be glimpsed if not as 'the creation of the Third World,' then certainly as reflecting the responses, reactions and resistance of the people who are its object. Without the possibility of reaction and resistance, there is no place for the agents and victims of development to exert their explicit and implicit influence on the ways in which it is constructed, thought, planned and implemented. Put simply, we simply do not yet know enough about the global, regional and especially local *historical geographies* of development – as an idea, discipline, strategy or site of resistance – to say much with any certainty about its complex past.

HISTORIES OF DEVELOPMENT

In his review chapter in this volume Michael Watts identifies many of the conflicting intellectual currents flowing through the contemporary academic domain of development studies. He concludes that in order to give development back its history, we need to pursue both an archaeology and a genealogy of development. Genealogy traces the recurrence of the idea, imagery and tropes of development across a range of nineteenth- and twentieth-century contexts. Archaeology attempts to uncover how and why development emerged as a problem 'grounded in the European experience of governability, disorder and disjuncture.' Only with this two-pronged approach can we begin to comprehend the power of development to make and remake the world (see Peet and Watts 1993) .

Even a cursory glance at the basic liturgy of post-World War II development discourse – the national development plan – will demonstrate contemporary development's almost overwhelming need to reinvent or erase the past. Most plans contain a formulaic bow to the previous plan period, a technocratic assessment of its failings designed as a prelude to the conclusion that this time 'it'll go much better.' But prior histories of the object of development – the people, country, region, sector or zone – are deemed irrelevant, best left to the ivory tower academic who has, by definition, no contribution to make to today's problems and tomorrow's solutions. Because development is prospective, forward-looking, gazing towards the achievement of as yet unrealized states, there seems little point in looking back. The technocratic language of contemporary plan writing – the models, the forecasts, the projections – all laud the idea of an unmade future which can be manipulated, with the right mix of inputs and indicators, into preordained ends. The past is impervious to change, untouchable and irredeemable. It is of no interest in and of itself. Occasionally it might have 'lessons to teach,' but not very often.

Not only are the objects of development stripped of their history, but they are then reinserted into implicit (and explicit) typologies which define a priori what they are, where they've been and where, with development as guide, they can go. Perhaps the best known of these formal typologies to students of development is Rostow's 'stages of growth model.' But the basic trope – that Europe shows the rest of the world the image of its own future – is of much broader and deeper purchase. Development, as Watts argues, has rarely broken free from linearity, from organic notions of growth and teleological views of history. With the idea of an original steady state from which all evolves, 'it became possible to talk of societies being in a state of "frozen development".' Deeply embedded within development discourse, therefore, was a set of recurrent images of 'the traditional' which were fundamentally ahistorical and space-insensitive. Collectivities (groups, societies, territories, tribes, classes, communities) were assigned a set of characteristics which suggested not only a low place in the hierarchy of achievement but a terminal condition of stasis, forever becalmed until the healing winds of modernity and development began to blow.

What is the point of constructing the objects of development as existing outside, rather than as products of, the tide of modern history? Two of the chapters in this collection try to answer this question in specific contexts. In his analysis of the construction of Egypt in the development texts of the United States Agency for International Development (USAID), Timothy Mitchell argues that the Nile Valley is imagined as a site in which life has remained virtually unchanged for centuries, if not millennia. Rather than being a product of the political and economic transformations of the twentieth century, the Egyptian peasantry has always existed in its present state. 'The image of a traditional rural world' concludes Mitchell 'implies a

9

system of agriculture that is static and therefore cannot change itself.' Unable to change, and no longer able to cope with the growing imbalance between population and resources, it must be changed by the injection of technology and expertise from outside. Only then can the primordial be dragged into the twentieth century.

'Traditional society,' though motionless and misrepresented, is not often overly-romanticized in the development text. To do this would be to run the risk of implying that there is no necessity for outside intervention and management. When Harry Johnston described the state of Mlanje before British rule, therefore, the language was traumatic not romantic – the area was practically in 'chaos,' virtually uninhabited and uninhabitable, racked by internal violence and insecurity. Development – the rebuilding of the landscape and the reclothing of its benighted inhabitants – is redemptive power. Without it, order cannot be restored, improvement is impossible. Johnston's imagery is replete with another recurrent trope in development – the idea that development works on a chaotic and disorderly terrain.

The language of 'crisis' and disintegration creates a logical need for external intervention and management. Accompanying the imagery of crisis is an implicit analysis of causation – sometimes external, more often internal. The causes are mostly endogenous – tribalism, primitivism and barbarism in older versions; ethnicity, illiteracy and ignorance in more modern incarnations. The reality of broader connections and causes is not always spurned, however. Development animates the static and manages the chaotic. It has a powerful habit of using history to apportion blame to its immediate predecessors for the disorder it attempts to amend. In industrializing Europe, as Cowen and Shenton suggest, development emerged to mitigate the disorder of progress. In Mlanje it is the 'evil-minded slave-traders,' the agents of a pre-modern era, who have created the turmoil that now needs management. In the post-colonial era, the colonial inheritance (either the destructive colonial impact or the lack of a democratic culture, education, skills, expertise, and so on) can be blamed (Watts 1991b; Leys 1994). In the current era, misguided left-wing ideologies are culpable (Berman and Dutkiewicz 1993). In each case the aim, as Mitchell points out, is always to distance development from any complicity in chaos – development is always the cure, never the cause.

The chapters in this volume place slightly different emphases on development's own history. In the contemporary era, argue Cowen and Shenton in their chapter, the period of development is routinely assumed to be the span of history since 1945. They then turn this argument around by suggesting that there are really no predecessors – development was always implicated and from the first. The modern idea of development, they suggest, can be traced to where it was first invented, amidst the throes of early industrial capitalism in Europe. Development emerged to ameliorate the chaos apparently caused by progress, 'to create order out of the social disorder of rapid

10

urbanization, poverty and unemployment.' In similar vein, Watts concludes that the trope of crisis was therefore built into development 'from the very beginning.' In the writings of a number of major nineteenth-century thinkers who grappled with the notion of development as an antidote to progress, Cowen and Shenton discern all of the central ideas of contemporary development. Development discourse is thus rooted in the rise of the West, in the history of capitalism, in modernity, and the globalization of Western state institutions, disciplines, cultures and mechanisms of exploitation. But this does not mean reducing all interpretation to 'superannuated nineteenth-century conceptions of political economy' (Said 1983) or filtering them through a functionalist master-narrative in which development is a mere instrument of Western domination, drained of ambiguity, complexity and contestation.

While not disputing the deep historical origins of development, some of the other chapters in this book have slightly different readings of its archaeology and genealogy. Escobar's position is perhaps the most distant from that of Cowen and Shenton, though elsewhere (Escobar 1992d: 132) he has also argued that development is inextricably linked to 'the rise of Western modernity since the end of the 18th century.' While clearly cognizant of the need to situate development in its broader historical and imperial context, Escobar discerns a sea change in the institutions and discourses of development in the post-1945 period. Clearly there is, at the very least, a disjuncture here that needs to be explained. For Escobar, the essence of the change is that a threshold of internalization is crossed. People who were once simply the objects of development now came to see and define themselves in its terms. They began, to rework E.P. Thompson's felicitous phrase, to fight 'not against development, but about it.' Three other chapters – by Porter, Watts, and Manzo – are closer in spirit to Cowen and Shenton though they have slightly different perspectives on the origins of the imaginary of development. In his chapter, Doug Porter points to the profound effect of nineteenth-century natural science on the metaphorical language of development. Evolutionary science provided a 'clutch of biological, organic and evolutionary' images while nineteenth-century physics donated a set of images about order, stability and constraint.

While Cowen and Shenton propose a reading that is part materialist, part theological, and emphatically Western and European, Watts chooses a cultural location for development under 'the broad arch of modernity.' One strand of development is rooted in the general normalizing practices of the modern state – the effort to produce and reproduce disciplined citizens and governable subjects. Another is rooted in fundamental points of difference between modern and pre-modern societies. The desire for accumulation – so central to modern society and its notions of development – only had meaning in a world where 'primitive economies' had no desire. Thus, development was 'neither *sui generis* nor simply imposed (subsequently) on the non-developed ("uncivilized") world, but rather ... in an important way a product of the non-developed.' Development required non-development 'and to this extent the origins of modernity were not simply located in the West.' Finally, if development was a cultural reaction to progress 'generated from within the belly of capitalism,' it was also a point of connection with the non-developed realm, an ever-present reminder of a world lost and perhaps of impending doom.

Kate Manzo develops this point, both here and elsewhere (Manzo 1991), by tracking a set of modernist images first attributable to European thinkers who gazed upon the peoples of North America from the shores of Europe and constructed a set of dichotomous images contrasting the civilized European with the untutored, natural, childlike native American. Thus, it was that 'those defined solely by Europeans as inferior or "primitive" to themselves are presumed to advance in direct proportion to their acquisition of European traits, so that normal development entails becoming, figuratively, white.' Science and reason prevented European degeneration into 'a state of nature typified by brutality, poverty, evil and immanent death.' The labyrinthine task of tracing such tropes and images of development from their early modern origins through to the development machinery of the present day is very much in its infancy. These essays can make only an incremental contribution to this important task (see also Escobar 1994; Moore and Schmitz, forthcoming).

One historical method is suggested by Cowen and Shenton's chapter – a kind of comparative inventory of the rhetoric of early nineteenth- and late twentieth-century development writing. They do this in order to demonstrate not only the deep continuities in development thought and practice but to elucidate an important historical lesson – that development failed then even as it will fail now. A second method is suggested by the work of David Spurr (1993). Spurr's genealogy categorizes the recurrent tropes of colonial discourse – surveillance, aestheticization, classification, debasement, affirmation, naturalization, eroticization and appropriation – and then ransacks a wide variety of periods, places and texts for evidence of their presence.

While Cowen and Shenton reinvigorate the notion of trusteeship as a central trope in development writing, Manzo focuses on the related metaphor of guardianship. Following Nandy (1987), she argues that familiar dichotomies such as white/black, civilized/uncivilized, European/native are underpinned by a parent/child metaphor. Amongst the continuities between early and late-modern discourses of development, Manzo cites the idea of the modern West as a model of achievement, and the rest of the world as a childish derivative. The metaphor of adult and child 'continues to inform analysis of the "modern world" of development.' Manzo tracks this metaphor to nineteenth- and twentieth-century South Africa where, she argues, it constituted a fundamental metaphorical underpinning for segregation and apartheid.

Doug Porter also finds continuity and persistence in the underlying metaphors of development despite what he sees as apparent change in the 'fashion-conscious institutional language of development' since 1945. Porter solves this paradox by suggesting that there are three kinds of metaphors – organizing metaphors (those pertaining to post-1945 development), master metaphors (those which recur repeatedly independent of time and place) and metaphors of practice (those that arise in particular local contexts). The logic of Cowen and Shenton's argument is that everything is prefigured, that there are only master metaphors. But they would surely not disagree with the central role accorded by Porter to metaphors of order, stability and constraint. In his chapter, Porter traces the genealogy of several metaphors from this trilogy of types, and crucially highlights the very non-discursive implications of metaphor-making for development as practised.

In the ensuing discussion of a Philippines development project, Porter exemplifies a third method for tracking the history of development (see also Tennekoon 1988; Pigg 1992). Here the focus is primarily on the 'privileged particles' of the development process – the fragmented discourses that swirl around local projects and practices when general tropes are forced into direct engagement with the local histories and geographies of particular localities. By mapping these emergent local languages of development it is possible, suggests Porter, to work out both how universal master metaphors are mediated by the particularities of time and place, and how locality generates its own distinctive metaphors and tropes. That this is not an unproblematical task has been clearly demonstrated elsewhere by a vigorous debate between Beinart (1984) and Phimister (1986) over whether to give greater weight to universal explanation or local context in unveiling the vocabularies and practices of conservationism and development in southern Africa in the 1920s and 1930s.

In the same part of the world, Chris Tapscott's chapter in this volume shows the appeal of the comforting words of development to segregation and apartheid (see also Dubow 1989; Ashforth 1990b). Many of the central spatial and organic tropes of a broader development imaginary flowed smoothly into the apartheid strategy of separate development, rationalizing rather than challenging its basic precepts. In the 1970s, development was reinvented as part of a more general strategy to, in Stanley Greenberg's (1987) phrase, 'legitimate the illegitimate.' A vast development machine was constructed in which a depoliticized, technocratic language of development, bearing all the old familiar trademarks, circulated. The failure of this project to buy consent and maintain order was all too apparent by the mid-1980s. It is ironic, but hardly surprising, that the new South African government is reinventing development for a third time to manage the ravages of past policies legitimated by development. The programme for reconstruction and development may be new but the purpose and the images conjured up harken back to a time long distant.

13

JONATHAN CRUSH

GEOGRAPHIES OF DEVELOPMENT

Development discourse can do without its history but not its geography for, without geography, it would lack a great deal of its conviction and coherence. Spatial and organic images and metaphors have always been used to define what development is and does. The language of development constantly visualizes landscape, territory, area, location, distance, boundary and situation (Slater 1993). Similarly, analogies from the natural world are used to picture the process through which development occurs (McCloskey 1990). Development writing constantly delineates and divides territory by means of a relentless dualistic logic. The binary oppositions between developed (territories that have) and the undeveloped (territories that lack) created by this cartographic exercise are very familiar. But development also needs geography to link these binary oppositions, a task performed through the language of spatial dispersion and diffusion.

In this way, it is possible to visualize how dominant, superior parcels of space can (and will) supersede their inferior other. In order to map this process, the static language of spatial demarcation needs the dynamism of historical narrative. As Emery Roe (1991) has recently suggested, it is sometimes helpful to see development as a form of story telling. Put this way, the idea of development as a narrative with stage, plot, characters, coherence, morality and an outcome has its appeal (White 1987). Roe concludes that by tinkering with the plot, more realistic narratives are possible and better development practice may result. That may be so, but what is of more interest in the context of this volume are the analytical possibilities opened up by viewing development as a form of writing amenable to narrative analysis in which geography is both stage and actor.

One of the primary elements in the development narrative is a setting of the geographical stage. Open almost any academic or development text dealing with the African country of Lesotho, for example, and you will find that it begins with the same textual ritual. 'Lesotho' we are always informed, 'is a small landlocked African country completely surrounded by South Africa.' Since anyone interested enough to pick up a learned text on the country probably already knows where it is, this incantation is hardly necessary to impart information. Is it therefore meaningless? Or is it an opening gambit by minds too hidebound to think of an original entrance? The significance of the ritual probably lies in the much broader cartographic anxiety that adheres to the imaginary of development (Porter 1991). Indeed, one can often be forgiven for thinking that the country has no context at all – its boundaries mark the limits of its world.

What is happening, as Mitchell argues in his Egyptian case study, is the marking of boundaries, the designation of a nation-state as a 'free-standing unit, lined up in physical space alongside a series of similar units.' The consequences of this convention are twofold: first, it creates an illusion that the nation-state is a functional unit rather than the product of a larger constellation of forces. In the case of Lesotho, it is surely important to establish that the country is *completely surrounded* since this has important implications for its prospects of development. The problem is that in many of these same texts this location is subsequently ignored as a factor constraining or enabling the process of 'development.' Secondly, described as a self-contained, bounded object, the country is constructed as something apart from the discourse that describes it. Lesotho, Egypt, 'the developing country,' are all laid out as mapped objects of development, those who bring development are not in any sense part of that object's prior history and geography.

In demarcating, dividing and sealing territories as objects of outside intervention, development simultaneously assigns each territory a characteristic morphology. Sometimes, as in many development plans, geography is a largely inert spatial inventory of physical and social facts. But the language of development also brings a powerful set of landscape images into play. When Johnston described the changes wrought by three brief years of British rule, he visualized a transforming power literally remaking the landscape. The landscape before development was at best only 'half-formed,' but more accurately a blank landscape, a landscape of absences: 'uninhabited (with) no customs-houses, no roads suitable for wheeled traffic, very little labour in the coffee plantations.'

Onto this empty scene come 'clean broad level roads, bordered by handsome avenues of trees, and comely red brick houses with rose-covered verandahs ... a post-office, a court of justice, [and] the coffee-plantation, which ... is the cause and support of our prosperity.' The new landscape is a vital and living testimonial to the power that made it. Landscape description, the spreading out of a country or territory as a picture to be gazed upon from above, provides a powerful means of visualizing what it is that development does. As Gavin Williams argues in his chapter, development discourse represents whole countries or regions in 'standardized forms' as objects of development. This tendency finds fruition in the simplistic reaggregation of demarcated units into homogenous swathes of territory that span the globe – the 'developing world,' the 'Third World,' the 'South.' These global spaces are inhabited by generic populations, with generic characteristics and generic landscapes either requiring transformation or in the process of being transformed.

If the human landscape is both object of development and a testimonial to its power, so too is its physical counterpart. In Johnston's virtually uninhabited pre-colonial landscape, natural forests are destroyed by bush fires set by the evil-minded. This image is given analytical substance by several of the chapters in this volume. In her study of the silencing power of environmental discourse in colonial Kenya, for example, Fiona Mackenzie argues that Africans are constructed as 'unscientific exploiters' of the environment. Their knowledge of the local environment can then be disqualified as pre-modern and 'unscientific.' The peasant farmer, undifferentiated and ungendered, is established as the object in need of exogenous agricultural science and 'expertise.' In South Africa, the language of environmental mismanagement was also central to the idea of 'separate development' described by Tapscott. Betterment was premised on the notion that African cultivation and pastoral practices despoiled the environment. Only scientific management could redeem the environment and re-educate the despoilers. This notion resonates into the present. Development itself is never the disease, only the cure. It proceeds, Escobar suggests, by creating abnormalities which it can then treat or reform. Development discourse has a remarkable capacity for forgiving its own mistakes and reinventing itself as the remedy for the ills it causes. One of the primary mechanisms for this periodic reinvention is the appropriation of the language and imagery of other, related, modernist discourses.

In the nineteenth century, as a number of chapters in this volume point out, Christian theology and the natural sciences provided a rich well from which to draw metaphorical inspiration. In the secular late twentieth century environmental science continues to offer useful possibilities. Bill Adams, in his chapter, argues that the fashionable idea of 'sustainable development' needs to be located within 'Northern environmentalism' rather than the genealogy of development *per se*. Imported into development, the ideas and images of environmentalism are 'encoded invisibly . . . within the simplistic problem-solving spreadsheets' of development. In particular, reformist and technocratic images and strategies have worked their way into the idea of sustainability in development. Adams would, I think, agree with Cowen and Shenton's claim that this is certainly not something new for, as he argues, the environmental imagery of colonial environmental science and conservationism had earlier found its way into development, where it resides still. The metaphorical power of sustainability in contemporary development, however, lies in its promise of 'escape from the environmentally destructive record' of past practice. Like other (re)inventions this one too, argues Adams, will fail to be much more than a transient label on a set of power relations which are much deeper and more durable than the words used to describe them.

Ken Hewitt comes at the issue of writing the environment within the 'viewpoints of power' of Western hazards research. 'Natural' hazards are discussed within these managerial texts neither as the predicaments or crises of capitalist modernity nor as failures of policy and management. Rather, hazards are constructed as problems due to external factors beyond managerial control – natural extremes, impersonal forces of demography, accident and error. Hazards are situated, metaphorically, at the frontier, part of the 'unfinished business of modernization.' They are explained by 'extraordinary events to be combated by extraordinary measures.' Similarly, in development writing, 'natural hazards' (and even the environment more generally) are seen as being outside and in an adversarial relationship to development. Ecology, climate, soils, water – the physical geography of the landscape – is temperamental and threatening, punishing mismanagement by its indigenous inhabitants, but ultimately amenable to the soothing touch of development.

Geography, argues Hewitt, is also quite central to the strategic thinking and inner logic of the dominant discourse of hazards. Maps of natural agents and their relative intensity and frequencies define the incidence and basic pattern of risk for natural disasters, cordoning off areas of disorder and disorganization. The 'bad geography' of hazards discourse is not confined, however, to that discourse. Similar geographies are imagined in such diverse arenas (with common roots) as colonial literature, Orientalism, travel writing and development. These are the 'the master texts of dominant views,' crafted by 'atlas-gazers and intelligence-gatherer's visions,' gazing down from the lofty heights of The Centre, the metropolis, the dominant states and institutions. But, as Hewitt goes on to argue, bad geographies are not simply mapping exercises. What the map makes invisible is just as interesting as what it includes for this says a great deal about those who compile the maps (Harley 1992; Pickles 1992). That the fruits of development practice may flow *from* rather than *to* the groups and areas 'targeted' is certainly not part of 'the map of development.' The 'interests of power' demand a rather different geography of development.

The final chapter in this volume, adopting an explicitly geographical purview, is Williams's analysis of the narrative strategies of recent World Bank reports on population and the environment. Williams argues that a basic rhetorical strategy of these texts is argument by 'common sense.' Rather than problematizing the association between population growth, land scarcity, environmental degradation and food shortage, the relationships are assumed to be axiomatic. The Bank's generalized analysis of African demography, he argues, 'ignores the complex and varied historical processes which have shaped the rise, fall, and age- and gender-distributions of populations, and their patterns of settlement and migration . . . history is replaced by stylized transition . . . geography is simply ignored.' Williams contests the stylized transitions and blank geographies with the litmus test of basic fact. Given the transparent superficiality and erroneousness of so much that passes for factual analysis, why, asks Williams, is there a depressing sameness and persistence to World Bank discourse? The answer is provided by Ferguson (1990) who has suggested that what is happening is not 'staggeringly bad scholarship' but something entirely different. The accuracy or plausibility of the argument to those who do not have to believe in it is irrelevant to those who do.

Ferguson's (1990) distinction between 'development discourse' and academic discourse on development is a useful one in this context, though the division is far from absolute. Another project, currently in progress, is exploring precisely this interface (Cooper and Packard 1992; Packard 1994). In this volume, Terry McGee's more contemplative chapter looks at the producers of geographical development texts on Asia. McGee charts the progress of his own personal enlightenment as a geographer caught within the conventions of representation of a discipline – geography – that was not only a child of empire (Livingstone 1992; Gregory 1994; Godlewska and Smith 1994) but has never perceived a need to break with that past. The changing geographical representation of Asia is based on many of the same spatial tropes and images to be found within development discourse. Self-reflexivity now unfortunately tends to be viewed more as a means of establishing authority than visualizing how alternative worlds might be imagined and made (Geertz 1988a; van Maanen 1988). McGee, with character-istic honesty, tries to chart a personal and general route around the spatial dualism and teleological models that underlie western representations (both academic and non-academic) of the Asian city.

The commonsense histories and bad geographies of the development agency's text are constructions which license some forms of intervention (their own) and delegitimize others. As long as the interventions persist, so do the constructions irrespective of how 'right' or 'wrong' they may seem to everyone else. As MacCarney (1991) suggests, in an analysis of the interior of the World Bank, the images and ideas can change quite independent of any engagement with what is happening on the ground. Using the example of World Bank low-cost housing strategy, she argues that the internal bureau-cracies within organizations such as the World Bank provide their own momentum to the rhetoric of development. Thus it is, that perfectly workable (and often quite effective) policies, even judged by the Bank's own stated aims, can be marginalized by the culture of careerism and competition within the organization itself. The same is undoubtedly true of the language and images through which those policies are spoken and justified.

OTHER DEVELOPMENTS

Is there a way of writing (speaking or thinking) beyond the language of development? Can its hold on the imagination of both the powerful and the powerless be transcended? Can we get round, what Watts calls, the 'develop-ment gridlock'? Can, as Escobar puts it, the idea of 'catching up' with the West be drained of its appeal? Any contemporary volume of development-related essays can no longer afford to ignore these questions. One of the most damaging criticisms levelled against Said's (1978) notion of Orientalism is that it provides no basis for understanding how that discourse can be overcome. This book also, by definition, cannot stand outside the phenomenon being analysed. The text itself is made possible by the languages of development and, in a sense, it contributes to their perpetuation. To imagine that the Western scholar can gaze on development from above as a distanced and impartial observer, and formulate alternative ways of thinking and writing, is simply a conceit. To claim or adopt such a position is simply to replicate a basic rhetorical strategy of development itself. What we can do, as a first step, is to examine critically the rival claims of those who say that the language of development can, or is, being transcended.

To assert, like Esteva (1987: 135), that 'development stinks' is all very well, but it is not that helpful if we have no idea about how the odour will be erased. The authors in this volume are by no means agreed on whether the language of development (and its associated practices) is here to stay or whether it should, or could, be transcended. A number of them take issue with recent arguments that suggest that this is not only necessary but possible. Anti-capitalist discourses such as dependency and underdevelopment theory, for example, are sometimes represented as proffering a radically different discourse of anti-development. If this is so, no-one bothered to tell the developer. Even political movements that once drew spiritual inspiration from this 'alternative' discourse have increasingly found the development imaginary a far more appropriate and concrete vehicle for articulating their aspirations. The extraordinary metamorphosis of the African National Congress is only one case in point.

Watts, drawing on the work of various post-colonial scholars, suggests that the quest for an *alternative* development is in some sense misplaced. The radical anti-capitalism of the 1970s, which asserted that autonomy and delinking were the key to 'development,' looks decidedly threadbare in the face of a counter-critique that they are as guilty of 'Eurocentric universalism' as those they criticize. Cowen and Shenton are even harsher in their judgement of the proponents of (an)other development. Development, they argue, is criticized as Eurocentric, but how could it be anything else? That accepted, the idea of 'autonomous development,' 'development from within' or 'development from below' as *alternatives* to development is nonsensical. And, in any case, the argument that there are real alternatives between externally-managed and internally-generated 'development' is simply a reprise of an age-old image.

Deploying Derrida's concept of logocentrism, Manzo proceeds to argue that romantic images of indigenous societies and their authentic knowledges do not push beyond modern relations of domination and threaten to reinscribe them in their most violent form. Hence, 'efforts in the post-colonial world to reinvent a pre-colonial Eden that never existed in fact, have been no less violent in their scripting of identity than those that practise domination in the name of development.' This trap – the reinscription of modernist dualisms – is also inherent in any claim that there can be pristine counter-hegemonic discourses of anti-development which are implacably opposed

19

and totally untainted by the language of development itself. Here Foucault's notion of the 'tactical polyvalence of discourses' seems particularly useful. He argues (Foucault 1990: 100–1) that we should not imagine a world of dominant and dominated, or accepted and excluded, discourses. We should think instead of a 'complex and unstable process whereby discourse can be both an instrument and an effect of power, but also a hindrance, a stumbling-block, a point of resistance and a starting point for an opposing strategy.'

Watts argues that alternative discursive strategies and frameworks have always constituted a 'pronounced undertow' within development discourse. There have always been oppositions and contestations in both centre and periphery which have structured, in complex ways, the very imagination of development itself. This would suggest that development discourse is not hermetically sealed, impervious to challenge or reformulation in the face of contest. As Adams argues, for example, contemporary environmentalism (and its manifestation within the arena of development) is riven with contradictions and conflict between dominant and avowedly counter-hegemonic discourses. From this perspective, perhaps the most interesting task is to decipher exactly how deeply development (and the discourses that claim to reject it) are implicated within one another.

In this context, Watts puts a certain faith in the capacity of populism to articulate discontent and imagine alternative worlds although, as he points out, much of the language of *fin de siècle* populism is in some sense still contained within (and even a necessary part of) modernist development discourse. Manzo and Escobar are both much more optimistic about the possibility of articulating truly alternative visions in a post-modern world. Manzo finds, in the writings of dependency theory, liberation theology, feminist ecology and participatory action research, a strong counter-modernist impulse. Counter-modernism, through its rigorous questioning and relentless critique, begins, she suggests, to provide the basis for thinking (and writing) beyond development. Her chosen example is unusual: the Black Consciousness Movement (BCM) in South Africa. Manzo shows how the BCM vision echoed that of other counter-modernisms in Latin America. Whether this means that it truly transcended the development imaginary of modernism is a point for debate.

Arturo Escobar is more sceptical of the dependency theorists, arguing that they 'still functioned within the same discursive space of development' as those they criticized. He advances, instead, the claims of new social movements as the medium through which alternative discourses *to* (rather than *of*) development are being articulated in the contemporary world. In his essay in this volume, he continues to make the case for social movements as the best hope for 'a more radical imagining of alternative futures.' NSM (New Social Movement) discourses are, by definition, polyvalent, local, dispersed and fragmented. To attempt to generalize across this heterogeneity, to bring rigid classificatory order to this diversity, or to suggest that there are dominant tropes and images that are common to all, would be to compromise their status as anti-developmental. Escobar is prepared, nonetheless, to venture the following generalization about NSM discourse: it strives for 'analyses based not on structures but on social actors; the promotion of democratic, egalitarian and participatory styles of politics; and the search not for grand structural transformations but rather for the construction of identities and greater autonomy through modifications in everyday practices and beliefs.' Thus, social movements constitute a *potential* terrain in which 'the weakening of development and the displacement of certain categories of modernity . . . can be defined and explored.' Where the NSMs already err though is in representing themselves as a total break with the past. There are, as Escobar suggests, important continuities (not least, one supposes, their capacity to reinvent history as well as development itself as the complete antithesis of everything they stand for).

Escobar concludes that 'inordinate care' must be taken to safeguard the fragile discourses of the NSMs from the appropriating appetite of development. Notions of 'sustainable development,' 'grassroots development,' 'women and development,' and so on perhaps exemplify the dangers most clearly. Why this should be so is a question that Jane Parpart addresses in her chapter on 'women in development' (WID). Parpart makes two basic arguments: first, that conventions of representation lodged deep within colonial discourse flowed easily into post-World War II development discourse. Where women were 'seen' at all, they were simply one more obstacle to modernization and progress. Second, she argues, the 'discovery' of women's voices might well have presaged a radical new challenge to the whole imaginary of development itself. Instead, claims Parpart (following Mohanty), Western feminism devised a new set of tropes in which women became 'benighted, overburdened beasts, helplessly entangled in the tentacles of regressive Third World patriarchy.' The imaginary of poverty, powerlessness and vulnerability was readily captured by development discourse. The very rubric of appropriation – women in development – carried the message that women's lives were now to be bounded by the power of development. They, like the colonial estate and post-colonial territory, would be managed by outside expertise. For Parpart, the way forward is through a reconstructed post-modern feminism which 'recognizes the connection between knowledge, language and power, and seeks to understand local knowledge(s), both as sites of resistance and power.'

Like Parpart, Mackenzie believes that the recovery of unheard voices and subjugated knowledges, as an act of critical scholarship, may undermine the power of development (see also Scott 1994). The place to start, perhaps, is by asking what development has meant for those spaces and peoples who it defines as its object. There is a large social science literature which tries to answer this question, primarily by examining the material and social impacts of development strategy and practice. By and large, there are three types of answer to the question: development has had a very negative impact; people would have been a lot worse off without it; or some benefit while the majority do not. All of these answers represent the recipients of development – either as victims or beneficiaries – as homogenized, voiceless subjects of outside forces. But those defined in development discourse as the subjects of development are also active agents who contest, resist and divert the will of the developer in greater or lesser ways.

Writings on protest and rebellion in the colonial and post-colonial periods, to which a number of the contributors in this volume have elsewhere contributed, have begun to unravel the ways in which development discourse and practice have been received, internalized and/or resisted on the ground. For the student of development discourse, there are at least two fruitful ways forward from this point. One, it seems to me, is to move backwards. We do not need, in other words, only to search in the present for visions of a future beyond development. The current obsession with Western representation of 'the other' is a field of rapidly diminishing returns. There are still large chapters of the story of development that need to be written and told in this mould. But these stories should be told, and heard, in concert with other stories – stories of what development meant for those whose visible and hidden lives it transformed. These stories at the very least provide 'a hindrance' and 'a stumbling-block' to the discursive power of development. But they might also constitute 'a point of resistance and a starting point for an opposing strategy.' This is not a task which this particular book set out to achieve. However, it concludes with a short autobiography which represents one way this might be done. The power of development is the power to generalize, homogenize, objectify. One way to contest this homogenizing power, albeit in an incremental way, could be through the articulation of individual biographies and autobiographies of the development experience. Nanda Shrestha – in a chapter which can profitably be read in tandem with Pigg (1992) – casts a retrospective eye over a personal trajectory of disillusionment with what development has 'done' to Nepal. Yet, here lies the irony, for if that disillusionment had come sooner (as presumably it did for many of his confreres) then he would not now be in a position to stand outside it and cast a critical gaze over its social and profoundly personal effects. Shrestha also confirms – in his discussion of the meaning of *bikas* – the importance of Escobar's point about the 'internalization' of development. In the curious mix of modernity and pre-modernity that is *bikas* are traces of the truth that development is not only internalized but rescripted by those it most affects. In 'the hidden transcripts' and everyday resistances of the weak (Scott 1985, 1990; Beinart and Bundy 1987; Haynes and Prakash 1992; Kirby 1994) the power of development to remake the world according to the word is relentlessly contested. In that sense, Esteva's epigraph is correct. When confronted with the power of the ordinary, development discourse (as Sir Harry Johnston attests) is forced to assume the most fantastical forms. That is actually when it is at its most transparent, fragile and feeble.

Part I

HISTORIES OF DEVELOPMENT

1

THE INVENTION OF DEVELOPMENT

Michael Cowen and Robert Shenton[1]

A mankind which no longer knows want will begin to have an inkling
of the delusory, futile nature of all the arrangements hitherto made to
escape want, which used wealth to reproduce want on a larger scale.

(Adorno 1993 [1951]: 156–7)

INTRODUCTION

'Development' has been called the central organizing concept of our time.
The United Nations has its development agencies, and the World Bank takes
development as part of its official name – the International Bank for
Reconstruction and Development. Hundreds and thousands of people are in
development's employ and billions are spent each year in its pursuit. It
would be difficult to find a single nation-state in the North which does not
have its departments or ministries of local, regional and international development. Nor can any Third World nation expect to be taken seriously without
the development label prominently displayed on some part of its govern-
mental anatomy.

What is development? This question is often posed in the initial stages of
books or courses on 'development studies,' but is rarely coherently answered.
Rather the reader or student is told that development means different things
to different people. The authors or lecturers then move rapidly on to a
discussion of 'the development debate.' As they proceed, the issue of what,
precisely, development is grows even murkier. While the student may initially
be informed that development is about the betterment of human kind through
the alleviation of poverty and the realization of human potential, he or she
quickly discovers that there also exists 'good' and 'bad' development, as well
as 'under' and 'over' development. Development can be 'autonomous,'
'appropriate,' 'gender-conscious,' 'sustainable,' or the opposite of all these

1 Robert Shenton wishes to acknowledge research support from the Canadian Social Science
and Humanities Research Council. This is a much-truncated version of Chapter One of our
Doctrines of Development, Routledge, forthcoming.

and much else besides. Which words are used seems to depend upon the views and policies of those in positions of authority in universities, national governments and international agencies. Even critics of development policies affirm, in the very act of criticism, development's existence.

Development thus defies definition, although not for a want of definitions on offer. One recent development studies text, notably entitled *Managing Development* (Staudt 1991), advances 'seven of hundreds of definitions of development.' Development is construed as 'a process of enlarging people's choices'; of enhancing 'participatory democratic processes' and the 'ability of people to have a say in the decisions that shape their lives'; of providing 'human beings with the opportunity to develop their fullest potential' and of enabling the poor, women, and 'free independent peasants' to organize for themselves and work together. Simultaneously, development is defined as the means to 'carry out a nation's development goals' and to promote economic growth, equity and national self-reliance (Staudt 1991: 28–9).

Here, quite typically, the distinction between development as an action and development as a goal of action is conflated with another important distinction. When, for example, someone mentions the 'development of capitalism' we take development to be an immanent and objective process. But when the same person says that it is desirable that state policy should achieve 'sustainable development,' we are now told that there is a subjective course of action that can be undertaken in the name of development. This distinction, between development as an immanent process and an intentional practice, has been central to European theological and philosophical debate. In contemporary development texts it is often lost.

Staudt (1991) assumes that development can happen as the result of decision and choice. Yet, the question of how actions taken in the name of development relate to any preconceived end of development is unanswered. Furthermore, because development is both means and goal, the final outcome is routinely assumed to be present at the onset of the process of development. Thus, Staudt argues that the goal of development is to enlarge choice. Yet, for choice to be exercised, let alone enlarged, there must be desire and capacity to choose as well as knowledge of possible choices. Choice is as much a precondition for development as its result. Further confusion arises when Staudt ignores another distinction between, on the one hand, a state policy of development and, on the other, attempts to empower people through or indeed against the state in the name of development. Empowerment is merely another name for development and thus embodies all its difficulties.

During the nineteenth century, those who saw themselves as developed, believed that they could act to determine the process of development for others deemed less-developed. The development problem was thus resolved by the doctrine of 'trusteeship,' a doctrine which became central to the historical project of European empire. Now, in the late twentieth century, the 'entrusting' of the means of development to 'the developed' has no conviction. As a doctrine, trusteeship is condemned as Eurocentrism, a vestige of the post-1945 attempt to improve the living standards of poor colonies and nations through external state administration. Development as trusteeship is taken to have no meaning for Third World countries. It has had its day and failed as an idea and a practice.

The period of development is now routinely assumed to be the span of imperial and post-colonial history since 1945. The subject of development is the imperial state (before and after political dismemberment); its object is colonial and Third World peoples. The recent *Development Dictionary* (Sachs 1992) informs us that 'underdevelopment began' on 20 January 1949, the day on which Truman called for a 'bold new program' for the improvement, growth and development of underdeveloped areas (Esteva 1992: 7). Five of the twenty essays in the *Dictionary* start with Truman's address and he is quoted at least ten times in the volume. Students are thus regularly instructed in the shallowness of development's history. A good deal of the confusion surrounding development arises from this sort of exercise in historical truncation. This essay argues, on the contrary, that development is a state practice rooted in the nineteenth century.

By truncating development's historical domain, we lose the crucial sense in which it emerged in the nineteenth century as a counterpoint to 'progress.' Development emerged to ameliorate the perceived chaos caused by progress. In many texts, the ideas of development and progress are seamlessly stitched together. There is little sense of a dynamic relationship between the two concepts (Harris 1989: 4–11; Thomas 1992: 7). In this chapter, we argue that the modern idea of development is necessarily Eurocentric because it was in Europe that development was first meant to create order out of the social disorder of rapid urbanization, poverty and unemployment. The story is of development's failure – a failure which occurred long before its supposed mid-twentieth-century birth.

THE EUROPEAN SETTING

The doctrine of development was already old before Truman's invocation. For example, Bourdillon, the British Governor of Nigeria, addressing the Royal Empire Society in London in 1937, stated that 'the exploitation theory ... is dead and the development theory has taken its place' (quoted in Cowen and Shenton 1991: 165). Nearly a century before, Robert Chambers (1969 [1844]: 360) wrote in his widely read anticipation of Darwin, *The Natural History of the Vestiges of Creation*, that 'the inorganic has one final comprehensive law, GRAVITATION. The organic, the other great department of mundane things, rests in like manner on one law and that is, DEVELOPMENT.'

By the early twentieth century, the idea of development was already well established in British thought. So was that of underdevelopment. Responding to Joseph Chamberlain's injunction to develop Britain's imperial estates in Africa, the Liberal Prime Minister, Campbell-Bannerman, put forward a project of his own, 'to develop our underdeveloped estates in this country; to colonise to our own country' (quoted in Cowen and Shenton 1991: 147). Much earlier, John Henry (Cardinal) Newman, a contemporary of Robert Chambers, writing in a field totally remote from the concerns of late twentieth-century developers, had begun to define the modern idea of under-development. The concept emerged as part of a theory of development in Newman's 1845 'Essay on the Development of Christian Doctrine' (Newman, 1992[1845]), which, the historian Acton said, 'did more than any other book of the time to make the English "think historically," to watch the process as well as the result' (quoted in Chadwick 1957: ix). Although concerned with the development of Christian doctrine, Newman's book had a far wider influence. Pattison, a fellow cleric, wrote to Newman in 1878 that it was:

A remarkable thing that you should have first state the idea – and the word – development, as the key to the history of church doctrine, and since then it has gradually become the dominant idea of all history, biology, physics, and in short has metamorphosed our view of every science, and of all knowledge.

(quoted in Chadwick 1957: ix–x)

Newman's argument encompassed both development and underdevelopment. Emphasizing that history moved through 'true' developments, Newman believed that there were doctrines and practices which did not evolve in a manner that remained faithful to the originating concept. He called this 'corruption' and used it to distinguish between the true development and the perversion of religious doctrine. Newman's idea of corruption conveys much of the sense of decay and decomposition, of disarticulation and disintegration that is essential to the modern meaning of underdevelopment. In the history of Western thought, the idea of corruption as part of a theory of change long predates Newman. As Robert Nisbet (1969) has noted, it developed in classical times and was based on the life cycle of all living things. The growing and maturing organism deposits a seed to recreate life amidst its own decay and destruction. Applied to the history of the state, the metaphor predicted that periods of state building would lead to periods of 'disorder' and 'ruin' which were, in turn, the prerequisites for renewed political construction. Successful statecraft and the art of politics, as in the work of Machiavelli, aimed to prolong maturity and forestall degeneration. It could do no more. This older meaning of development expressed a dual character of change. Positive, constructive change emerged from negative moments of destruction and decay. Purposive human intervention could ameliorate and forestall but not prevent the destruction which was intrinsic to an ordered, determined and inevitable cyclical process.

The classical theory of cyclical change remained dominant until it was challenged by modern ideas of progress that began to emerge in work of thinkers from at least Fontenelle to Hegel more than a century later and that gradually but, not wholly, supplanted the concept of inevitable degeneration. The idea of progress also had Christian origins stemming from the doctrine of divine revelation in which Providence through history maps out a design in advance of human efforts (Brunner 1948; Baillie 1950; Wager 1967). Enlightenment thinkers constructed secular variants of this idea, giving autonomy to human purpose and proposing the prospect of unlimited improvement through unaided human effort.

It has become commonplace for those attempting to legitimize the supposed modern sub-discipline, development economics, to rummage through the writings of the Scottish Enlightenment, especially those of Adam Smith. These eighteenth-century writers are supposed to have come up with the first theory of development in their idea of human economic activity evolving through a series of stages, commencing with hunting and fishing, progressing through pastoralism and settled agriculture and culminating in commerce and manufacturing. As one commentator (Meek 1976: 255 quoted in Skinner 1982: 91) has suggested, 'the four stages, at any rate at the outset of its career, usually took the form of a theory of development embodying the idea of some 'natural' or 'normal' movement through a succession of different modes of subsistence.'

The attempt by modern would-be developers to find the origins of their development practice in the Scottish Enlightenment requires a very selective reading of texts. Smith, like Locke, was fundamentally concerned with the Hobbesian problem of how social and political order might be maintained. In addressing the old idea of the inevitable corruption of the 'body economic' Smith simultaneously, and perhaps unwittingly, held out the possibility of progress as a linear unfolding of the universal potential for human improvement which need not be recurrent, finite or reversible. In his *Inquiry into the Nature and Causes of the Wealth of Nations* (Smith 1937[1776]), commonly seen as an eighteenth-century precursor of the modern 'development debate,' Smith is credited with rejecting the classical view, that national wealth and power were subject to inevitable cycles of advance, decline and stagnation, replacing it with a 'new description of rich and poor countries interlocked in a system of free trade reflecting the realities of a changing world. He could see the possibilities of progress both for rich and poor countries offered by a system of natural liberty in foreign trade' (Hont 1983: 302). Yet, Smith's optimism was immediately challenged by contemporaries who observed the political disorder that followed the French Revolution and the social disorder that accompanied the birth of industrial capitalism. In these grim facts, they found reasons to question the idea of boundless human improvement.

Although Thomas Malthus' *Essay on the Principle of Population* (1986[1798]) is usually presented as an anodyne theory concerning the limits to the growth of population imposed by agriculture, his book, published in the wake of one of the first crises of industrial capitalism, presented primarily a moral argument. He used his theory of population growth to argue against the possibility of limitless social perfectibility. Malthus attacked Smith's optimistic views about the possibility of unlimited human improvement. In particular, he challenged the idea that increases in society's wealth would necessarily bring improvement or happiness to every part of it. He argued that Smith had 'not stopped to take notice of those instances where the wealth of a society may increase without having any tendency to increase the comforts of the labouring part of it.' For Malthus, an increase in society's wealth would only mean a consequent growth in population which was 'the great obstacle in the way of any extraordinary improvement in society and was 'of a nature that we can never hope to overcome' (Malthus 1986[1798]: 183–4, 198–9).

For those who lived through the early decades of the nineteenth century, Malthus' grim predictions seemed to be coming true. E.J. Hobsbawm (1968: 58–9) captures the spirit of the times when he observes that no period of British history was as tense, or as politically and socially disturbed, as the 1830s and early 1840s. According to Hobsbawm, much of the tension of the period was the result of the working class despairing that they had not enough to eat and manufacturers despairing because they genuinely believed that the prevailing political and fiscal arrangements were slowly throttling the economy. These 'troubled times' gave birth to Luddites, Radicals, trade unionists, utopian socialists, Democrats and Chartists (Hobsbawm 1988: 13; Hobsbawm 1968: 55). They also gave rise to development. In France in the 1840s, the same sense of imminent turmoil was apparent (Sewell 1980). The French economy was more backward than the British but it was plunged, by industrial capitalism, into even greater social and political turmoil. In 1848, the *Journal des travaillers* declared that 'unemployment is the most hideous sore of current social organisation' (quoted in Sewell 1980: 249). The modern meaning of development (imbued with an overt sense of design) emerged to confront this 'hideous sore.' This new meaning was in direct opposition to the idea of progress as a 'natural' process without intentionality.

THE DESIGN OF DEVELOPMENT

The Saint-Simonians, writing at the end of the 1820s, were the original positivists. Their ideas were nurtured during the rise of industrial capitalism and, like Adam Smith, they posed the problem of creating order in a society undergoing radical transformation. Their analysis and solutions differed markedly from Smith's. The Saint-Simonians argued that humanity was a collective entity that had grown from generation to generation according to its own law of 'progressive development' (Iggers 1972[1892]: 28). Progressive development demanded 'a progressive amelioration of the moral, physical, and intellectual condition of the human race.' By this process everyone would realize that individual prosperity was inseparable from the prosperity and growth of all (Iggers 1972: 79).

The Saint-Simonians divided human history into 'organic' and 'critical' epochs. Organic epochs were characterized by harmony, widerning associations and common goals. The Middle Ages was one such epoch. Religious ideas were paramount and the economy was regulated by feudal corporations. By contrast, 'critical' epochs of human history were 'filled with disorder; they destroyed former social relations, and everywhere tend(ed) towards egoism.' Although harmful, they were necessary and indispensable, destroying 'antiquated forms' and facilitating the emergence of 'better forms' (Iggers, 1972: 28).

The thinkers of the Enlightenment and the French Revolution had succeeded in destroying the basis of the old 'organic epoch' but were incapable of visualizing a new era. This was because they remained wedded to ideas which put a premium upon self-interested individual action. This prolonged the social agony and delayed the arrival of a new organic epoch in which men would once again be 'associated,' though on a new footing. The new basis of human association, would be industrialism, guided by intellect, and governed by 'sympathy.' Unlike many of their conservative contemporaries, the Saint-Simonians applauded industrialization but argued that egoism produced irrational and destructive industrial practices: 'The industrialist is little concerned with the interests of society,' they argued, 'his family, instruments of production, and the personal fortune he strives to attain, are his mankind, his universe, and his God' (Iggers 1972[1829]: 12).

According to the Saint-Simonians, 'laissez faire' was the source of disorder. Amidst the 'throngs' who could testify to the disasters of this principle was the man 'who lives by his hands' who could never applaud the introduction of steam power to his trade. If, as the Saint-Simonians willingly acknowledged, steam was necessary and potentially beneficial for all, this was little consolation to the 'thousands of famished men' whose lives were dislocated by technological change. Moreover, laissez faire was continually threatened by overproduction as enterprise and capital rushed to profitable sectors of the economy. The outcome was 'overcrowding' and a 'death struggle' with a few triumphant and the majority completely ruined. 'Numberless catastrophes' and terrifying 'commercial crises' soon followed. Hard-working men were ruined and people began to believe that in order to succeed, 'more seems to be needed than honesty and hard work' (Iggers 1972[1829]: 13–15).

The Saint-Simonian critique of self-interest, the market and unregulated competition was echoed in England by movements of 'physiocratic anticommercialism' and 'communitarian political economy.' These groups sought solace in a romanticized agrarian past or an unhinged millenarian future (Thompson 1988).

The Saint-Simonians, in contrast, tried to impose constructive order upon the industrial disorder of the present. Their designation as 'scientific socialists' indicates the extent to which they gave agency and purpose to the development process. No longer was development something that occurred during a period of history; it was the means whereby the present epoch might be transformed into another superior order through the actions of those who were *entrusted* with the future of society.

A THEORY OF TRUSTEESHIP

For the Saint-Simonians, the remedy for disorder lay with those who had the capacity to utilize land, labour and capital in the interests of society as a whole. Property was the major obstacle to this programme. 'Idle owners' entrusted the instruments of production to the 'hands of a skilful worker' and reaped the profits. The Saint-Simonians argued that this evil could only be overcome if property was placed in the hands of 'trustees' chosen on the basis of their ability to decide where and how society's resources should be invested. These trustees were none other than banks and bankers. The banks would be fitted for trusteeship through the creation of a 'general system of banks' headed by a central government bank. They would be the 'depository of all the riches, of the total fund of production, and all the instruments of work.' The destructive effects of progress would be tamed through reform of the banking system and the personal morality of the banker (Iggers 1972: 103–10).

Auguste Comte completed the invention of development which his Saint-Simonian colleagues had begun. Comte saw 'progress as the development of Order under the influence of Love' (Comte 1875[1851]: 264). Progress had to be made compatible with order. The two could be reconciled through the understanding and application of the science of history or 'sociology' which embodied the laws of 'social evolution' which, in turn, had two aspects: 'the *development*, which brings after it the *improvement*' (Lenzer 1983: 234).

Improvement had been stilted because of the failure to reconcile progress with order. To resolve this problem, Comtean positivism recast the problem of progress and social order. Comte argued that humanity was subject to laws of 'development' analogous to those pertaining in the natural world that could be discovered and understood. Humans, however, could comprehend the fact and possibility of order. To reconcile order with progress, humans needed to comprehend the applicability of these laws. Progress was relentless and inconstant but it had to be given the consistency and morality of order. Development was the means by which progress would be subsumed by order (Lenzer 1983: 329, 341–2).

Comtean positivism was to be the true path of developmental knowledge, a mature and altruistic mode of thought. In proclaiming altruism as the guiding force of the positivist age, Comte challenged the Scottish Enlightenment by arguing that 'No calculations of self-interest' could 'rival this social instinct' (Comte 1875[1851]: 12). Anticipating at least one stream of thought on 'women and development,' Comte argued that an essential precondition of the triumph of 'social sympathy' was the fulfilment of the 'social mission of woman.' Women had innate qualities, particularly 'the tendency to place social above personal feeling', which made them 'undoubtedly superior to man' (Lenzer 1983: 373). Only through this innate quality, itself rooted in maternity, could mankind's morality and 'progressive development' be assured. All of this could happen once those who had the knowledge of 'sociology' were in a position to guide development. The high priests of positivism would join the Saint-Simonian bankers as the trustees of collective development.

UNDERDEVELOPMENT

For Newman, true development was the counterpoint to corruption. Once Newman's understanding was extended from theology to political economy, corruption became synonymous with underdevelopment. The old idea of corruption as a distinct temporal stage in a cycle of birth, growth, maturity, and decay, was replaced by a view which saw underdevelopment as simultaneously part of development itself. Modern underdevelopment theories, as well as their critics, all derive from this view in one way or another.

Underdevelopment theorists argue that industrial progress and the emergence of a proletariat at the 'centre' is the only 'true' development. In contrast, capitalism is seen as incapable of true development in the 'periphery' (Frank 1983: 186).

As Gavin Kitching (1982: 143) has written, Friedrich List's *National System of Political Economy* (1991[1885][1856]) was the prototype for all subsequent defences of effective industrial protectionism. In 1825, List emigrated to the United States. There, he observed that the passage from 'the condition of the mere hunter to the rearing of cattle – from that to agriculture, and from the latter to manufacturers and commerce' that took untold centuries in Europe 'goes on' in North America 'before one's eyes' (List 1991[1885]: xxix). Progress was telescoped through the constructive, intentional intervention of the state. The state had to act because the unfolding of immanent progress (as promised by the Scottish Enlightenment) was impossible in a world of British commercial and industrial supremacy. Smith's focus on private enterprise was, for List, besotted by a disorganized individualism. Like the Saint-Simonians, who influenced him, List argued that the self-interested individual was feeble and destitute alone. But, unlike the early Saint-Simonians, he deflected development doctrine to the ends of the nation-state (List 1991[1885]: 136–7, 165, 171–2).

List distinguished between a subjective, technical division of labour in which individuals contribute to a single activity, such as the making of pins, and an objective division of labour in which an individual simultaneously contributes to a number of different activities, all toward 'one common object.' Since it possessed a common end, the nation was the proper organizing agency for individual work. List wanted to furnish the economical education of the nation. Movement from an agrarian to an industrial economy rested upon such economical education, something which could only be accomplished by state activism. List argued that the state needed to take up the task of constructive development. Nature intended that industry, cultivation, riches and power should not be the exclusive possession of any single nation. Only when all had elaborated their productive powers to their fullest extent *in the bosom of the nation* would it be possible to realize the Saint-Simonian dream of reconciling progress with order. This would occur through the harmonious universal association of nations. In the interim, the fact of war separated the agriculturalists of one nation from the manufacturers of another, while commercial revulsion by one nation prompted another to react to the selfish policy of the first. No spurious balance of trade could reconcile the disharmony between nations. For harmony, it was first necessary to reconcile humanity through national policies designed to augment their inhabitants' productive powers (List 1991[1885]: 119–308).

Without state-directed development, the future of agrarian nations would be bleak: 'A nation exchanging its agricultural products for articles of foreign manufacture is like an individual with but one arm who invokes the assistance of a foreign arm for his support.' A 'foreign arm' could never take the place of its own missing limb because its movements were 'subject to the caprice of a foreign head' (List 1856: 241). List described the loss of productive force of his native Germany in sentiments which would later be repeatedly echoed by dependency theorists:

> Germany would scarcely have more to supply this English world with than children's toys, wooden clocks, and philological writings, and, sometimes also an auxiliary corps, who might sacrifice themselves to pine away in the deserts of Asia or Africa, for the sake of extending the manufacturing and commercial supremacy, the literature and language of England.
>
> (List 1991[1885]: 160–2, 131)

Only through constructive development by the state could the loss of productive force, through idleness or emigration, be checked.

List argued that nations had unequal productive potential but that through the policy of 'economical development' all could activate their fullest potential. Then came the caveat: his precepts were not to be followed by that part of the human race existing in the 'savage states' of the 'torrid zone.' He argued that countries of the 'torrid zone' would make 'a very fatal mistake,' if they tried to become manufacturing countries. They would progress more rapidly 'in riches and civilization' if they continued to exchange their agricultural products for the manufactured goods of the temperate zone. List turned this caveat to an imperial advantage. It was a similar idea that, a century later, generated the controversy over the possibility of capitalist development in the Third World (List 1856: 75, 112).

While the tropical territories were naturally unsuited to manufacturing, they could still assist the process elsewhere. England had shown the way by augmenting the savings of its landlords and farmers with money profits from overseas colonization which were invested in manufacturing in Britain. Germany, too, must acquire colonies to break the English domination of the world. An imperial policy would, it was true, make tropical countries 'sink thus into *dependence* upon those of the temperate zone,' but dependency would be mitigated by the competition between temperate zone nations. Competition between industrialized imperial nations would 'ensure a full supply of manufactures at low prices' and thus 'prevent any one nation from taking advantages by its superiority over the weaker nations of the torrid zone.' A beneficial economic process would be complemented by a political mission within an imperial arena (List 1856: 75–8, 199).

Alongside the extension of trade would go 'the mission of political institutions to civilize barbarian nationalities, to enlarge those which are small, to strengthen those which are weak, and above all, to secure their existence and duration' (List 1856: 263). Agrarian countries of the empire would be 'civilized' by free trade:

> The economical education of a country of inferior intelligence and culture, or one thinly populated, relatively to the extent and fertility of its territory is effected more certainly by free trade, with more advanced, richer, and more industrious nations.
>
> (List 1856: 77–8)

If free trade proved insufficient to the civilizing mission then the surplus population of the temperate zone could be exported to assist in 'torrid zone' development (List 1856:199). In such circumstances, emigration would augment rather than sap the productive force of the industrialized nation. Ironically then, it was through the paleo-dependency theorist List that the positivist idea of constructive development (articulated by the Saint-Simonians as a remedy to surplus population created by progress in Europe) became a theory of imperial development. More ironic still is the contribution made by the British theorist of liberal democracy, John Stuart Mill.

POSITIVISM IN BRITAIN

Mill owed a great intellectual debt to the Saint-Simonians and especially Comte. In later years Mill tried to hide this debt but he did not change the ideas which he earlier acknowledged had been lifted from Comte. Reflecting on the Saint-Simonians, Mill wrote that they were 'writers by whom, more than any others, a new mode of political thinking was brought home to me.' Comte, in particular, 'seemed to give a scientific shape' to Mill's thought (Mill 1989[1873]: 131–2).

After he returned to England in 1831, Mill issued his *Spirit of the Age* (1942[1831]), which strikingly echoed Saint-Simonian thought in its analysis of the crisis of British society. Mill wrote that the affairs of mankind were always in one or the other of two states; the *natural* state and the *transitional* state. Again echoing the Saint-Simonians, Mill argued that in the natural state, the material interests of the community were managed by those with the greatest capacity for management. By contrast, the transitional state occurred when 'worldly power, and the greatest existing capacity for world affairs' are severed, when there were no established doctrines, and when the world of opinions was a 'mere chaos.' A transitional state would continue until a moral and social revolution replaced worldly power and moral influence in the hands of the most competent. Then society would be 'once more in its natural state' and able to 'resume its normal progress, at the point where it was stopped before by the social system which it has shivered' (Mill 1942[1831]: 35–7).

For Mill, as for the Saint-Simonians and Comte, the present age was one of transition: 'the old order of things, had become unsuited to the state of Society and of the human mind.' Referring to government, Mill cried out for 'not merely a new machine, but a machine constructed in another manner' (Mill 1942[1831]: 6–7). Yet if Mill agreed with the Saint-Simonians about the need for social transformation, he eschewed their solution for swiftly directed social change. Mill could not agree with a theory that made improvement conditional on individual genius. Generalized knowledge, not the genius of a secular priesthood, was the condition for development (Mill 1942[1831]: 8–10). The necessary preconditions were increased education and a radical extension of liberty. Education had played a central role in Comte's idea of 'development' but only in 'critical epochs' as a means of challenging and helping to destroy old ideas. There was no role for education in Mill's sense in the 'organic epoch.' Indeed, it was antagonistic to the epoch to be ushered in by Comtean positivism. However, Mill agreed with the positivist doctrine that progress and order needed to be reconciled. The problem preoccupies Book VI of Mill's *Logic* (1974[1843]), the publication of which marked his arrival as an influential intellectual figure.

To grasp the significance of Mill's reworking of positivism it is crucial to see what he believed education, in the widest sense, to be. Education encompassed issues as diverse as electoral and land reform, birth control and equality for women, as well as the rights of labour. All were simultaneously prerequisites for, and classrooms of, the education that would make 'the new machine' possible. The social machine could only be constructed through developing the minds of human beings. And minds could only be developed under the condition of liberty, a condition characterized by choice (Mill 1974[1843]: 833, 845, 861, 869, chs 3, 4, 5 *passim*).

Thus, the problem of choice in development was posed long before it became associated with colonial and Third World trusteeship. Mill, writing not as a late twentieth-century cultural anthropologist of Africa but as a mid-nineteenth-century commentator on England, observed that while:

customs be both good as customs and suitable . . . yet to conform to custom, merely *as* custom, does not educate or develop ... any of the qualities which are distinctive endowment of a human being. The human faculties of perception, judgement, discriminative feeling, mental activity, and even moral preference, are exercised only in making choice. The mental and moral, like the muscular powers, are improved only by being used.

And then continued:

He who lets the world, or his own portion of it, choose his plan of life for him, has no need of any other faculty than the ape-like one of imitation. He who chooses his plan for himself, employs all of his faculties. He must use observation to see, reasoning and judgement to foresee, activity to gather materials for decision, discrimination to decide, and when he has decided, firmness and self-control to hold to his deliberate decision. . . . It is possible that he might be guided in some good path, and kept out of harm's way without any of these things. But what will be his comparative worth as a human being?

(Mill 1965[1859]: 307)[2]

Mill emphasizes the necessity of choice. Humans should 'use observation to see, reasoning and judgement to foresee, activity to gather materials for decision, discrimination to decide, and when (they have) decided, firmness and self-control to hold to (the) deliberate decision' (Mill 1965[1859]: 307). Choice implied the capacity to choose. Choice was a condition for development, and distinguished humans from apes. Individuality was the result of choice. In Mill's words it was 'the same thing with development,' the only thing that could produce well-developed human beings (Mill 1965[1859]: 312). Only 'well-developed human beings' were fit to enter the end-point of development – Mill's 'stationary state.'

A stationary state is often seen as the antithesis of progress. However, this confuses progress and development. For Mill, the stationary state was the condition where material progress (prompted by necessity in which minds were 'engrossed by the art of getting on') would cease to increase wealth and produce instead a legitimate effect – that of abridging labour. The 'Art of getting on' would be replaced by the 'Art of Living.' Development would do the work of education and make it possible to understand how the drive for material progress could be made consistent with an ordered life (Mill 1985[1848]: 115–17).

Progress not only needed to coexist with order but each was necessary for

2 It would be unfair to Mill not to reproduce the following footnote from his *Logic*: 'The pronoun *he* is the only one available to express all human beings, none having yet been invented to serve the purpose of designating them generally, without distinguishing them by a characteristic so little worthy of being made the main distinction as that of sex. This is more than a defect of language; tending greatly to prolong the almost universal habit, of thinking and speaking of one-half the human species as the whole' (Mill 1974[1843]: 837).

the other. Development also required a political system consistent with this goal. In Mill's words:

A party of order or stability, and a party of progress or reform, are both necessary elements of a healthy state of political life. Each of these modes of thinking derives its utility from the deficiencies of the other.

(Mill 1965[1859]: 297)

The contradiction between progress and order – where what was negative for one gave positive purpose to the other – would continue until development had done its work. Then a single party would have so enlarged its 'mental grasp' that it could be a party equally of order and of progress. Only then might the stationary state prevail (Mill 1965[1859]: 297).

In Mill's vision of a stationary state, development successfully acts against the chaos of progress. This stationary state has a close kinship with what today's developers call 'sustainable development' though Mill's soaring rhetoric ought to make the prosaic modern day exponents of sustainable development blush:

A world from which solitude is extirpated, is a very poor ideal. Solitude, in the sense of being often alone, is essential to any depth or meditation of character; and solitude in the presence of natural beauty and grandeur, is the cradle of thoughts and aspirations which are not only good for the individual, but which society can ill do without. Nor is there much satisfaction in contemplating a world with nothing left to the spontaneous activity of nature; with every rood of land brought into cultivation, which is capable of growing food for human beings; every flowery waste or natural pasture ploughed up, all quadrupeds or birds which are not domesticated for man's use exterminated as his rivals for food, every hedgerow or superfluous tree rooted out, and scarcely a place left where a wild shrub or flower could grow without being eradicated as a weed in the name of improved agriculture.... If the earth must lose that great portion of its pleasantness which it owes to things that the unlimited increase in wealth and population would extirpate from it, for the mere purpose of enabling it to support a large, but not a Better or Happier population, I sincerely hope, for the sake of posterity, that they will be content to be stationary, long before necessity compels them to it.

(Mill, 1985[1848]: 115–16)

The development necessary to bring about the stationary state could only occur in societies not bound by 'custom' and where tolerance and rational discussion flourished. For Mill, as for unwitting modern development theorists, development could only occur where the conditions of development were already present. Societies in which they were not present had to be guided by the trustees from societies where they were.

As an employee and theorist of the East India Company, Mill argued strenuously that India needed to be governed despotically by an incorruptible imperial cadre who exercised trusteeship in order to create the conditions under which education, choice, individuality – in a word development – might occur. Mill even opposed the transfer of administration from the East India Company to the British Crown after the 1857 mutiny on the grounds that the rule of India would be corrupted by British democracy. Although responsibility for Indian administration was transferred to the Crown, Mill counted it a personal victory that the principles of trusteeship would underpin the new administration (Mill 1990: 45, 50, 85).

CONCLUSION

Mill's direct involvement in the administration of India highlights one of the aims of our interpretation of the nineteenth-century origins of development. The words of the theorists discussed in this chapter are not instances of some disconnected, textualized 'developmental discourse.' Each theorist (alone and in combination with others) was a part of nineteenth-century developmental practice. The Saint-Simonians, for example, were the direct inspiration for the building of the Suez and Panama Canals. Comte's phrase 'Order and Progress' not only adorns the Brazilian flag but inspired a whole phase of nineteenth-century Latin American development policy. Australians and Indians alike came to see in the work of Friedrich List inspiration for the construction of their own national economies. Development was not simply thought or rhetoric; it was mid-century state practice.

For today's practitioners and students of development, the result of forgetting nineteenth-century development doctrine is all too evident. It is an attitude that Mill himself castigated, writing of it that:

> According to this doctrine we reject the sophisms and prejudices which misled the uncultivated minds of our ancestors, because we have learnt too much, and have become too wise ... We have now risen to the capacity of perceiving our true interests; and it is no longer in the power of impostors and charlatans to deceive us.
>
> (Mill 1942[1831]: 9–10)

Bjorn Hettne's *Development Theory and the Three Worlds* is one of a handful of development studies texts which pays some attention to the origins and ambiguities of development in earlier nineteenth-century European thought. After re-identifying the 'crises of development theory' in the East, West and South, Hettne proceeds quite reasonably to try and understand the 'development ideologies' in Western thought. For Hettne these ideologies and their progeny stand condemned as 'Eurocentric.' Yet, he never asks the obvious question: How could they be anything else? Hettne, like many others, has failed to grasp that the conditions which gave rise to intentional development as a redress to progress arose first in Europe (Hettne 1990: chs 1–2).

The dissemination of these Eurocentric theories to the Third World is condemned by Hettne as a form of 'academic imperialism' whose counterpoint is the rise of the dependency school's analysis of underdevelopment. But is dependency a *sui generis* response to Eurocentric thought? We have argued that the intellectual origins of what is now called underdevelopment are also fundamentally European. The problem is compounded when Hettne sets about proposing 'another development.' This will supposedly 'transcend the European model' and create a new kind of development thinking. This new model of development will be egalitarian, 'self-reliant,' 'eco-' and 'ethnodevelopmental.' Unfortunately for Hettne, when he introduces the precursors of the new thinking, we find ourselves face to face with the ghosts of Saint-Simon, Comte, Proudhon, and the Norodniks amongst others (Hettne 1990: ch. 6).

Hettne actually admits that the wellsprings of 'another development,' are to be found 'back where we started' in the developed world. He approvingly quotes Denis De Rougement who noted that:

> Europe is the continent that gave birth to the nation state, that was the first to suffer its destructive effects upon all sense of community and balance between men and nature . . . the continent which, therefore, has every reason to be first to produce the antibodies to the virus it itself generated.
>
> (Hettne 1990: 195)

Thus, in Hettne's words, development theory 'returns to Europe.' From a base in the 'new Europe,' the gospel of 'another development,' will be spread to the Third World (Hettne 1990: 195).

Hettne is certainly aware of some of the obvious questions raised by the idea of 'another development.' First, he asks why there is such interest in 'another development' in the West. The answer is that the 'collective consciousness of the industrially advanced countries is going through a transformation, against which spokesmen for the Mainstream will have a hard time finding a way out of the present impasse; a solution consistent with a worldview of automatic growth and eternal progress.' No Saint-Simonian ever said it better. Growth and progress are symptoms of De Rougement's 'virus.' Second, Hettne asks why the concepts of 'another development' (small-scale solutions, ecological concerns, popular participation, and the establishment of community) meet with greater enthusiasm in the rich countries than in the poor. The disinterest in the South is the product of corrupt Third World leadership. Small may be beautiful, but it does not mean power. Third World elites who aspire to Western standards of living 'do not intend to be fooled into some populist cul-de-sac' (Hettne 1990: 195–6, 154–5).

Given this impasse, how is another development to be brought to the masses in the Third World? The main vehicle will be an amalgam of official and non-governmental aid organizations whose task, in assuming the mantle of development, is to confront the destruction wrought by progress. In the face of a corrupt leadership, trusteeship (though none dare speak its name) will have to be exercised by the knowing and the moral on behalf of the ignorant and corrupt. The assembled ghosts of the Saint-Simonians, Comte, List, Mill and Newman would be much amused. So too would be the spirit of Marx seeing his own work consigned to the rubbish bin of history by those who scavenge through decidedly riper tips for choice intellectual morsels to be offered up as new fare. To the purveyors of an alternative and another development we may *either* reply with the words of Goethe's Mephisto,

> I am the spirit that negates all!
> And rightly so, for all that comes to be
> Deserves to perish wretchedly . . .
> (I am) part of the power that would
> Do nothing but evil
> And yet creates the good.
> (Goethe quoted in M. Berman 1982: 47)

or of Marx himself:

Modern bourgeois society, a society that has conjured up such gigantic means of production and exchange, is like the sorcerer who is no longer able to control the powers of the subterranean world that he has called up by his spells.

(Marx 1968[1848]: 66–7)

2

'A NEW DEAL IN EMOTIONS'
Theory and practice and the crisis of development

Michael Watts

[T]he modern public expands, it shatters into a multitude of fragments, speaking incommensurable private languages: the idea of modernity . . . loses much of its vividness, and its capacity to organize and give meaning to people's lives. As a result of this, we find ourselves today in the midst of a modern age that has lost touch with the roots of its own modernity.

(Berman 1982: 17)

To break through the barriers of stagnation in a backward country, to ignite the imaginations of men, and to place their energies in the service of economic development, a stronger medicine is needed than the promise of better allocation or resources or even the lower price of bread. . . . What is needed to remove the mountains of routine and prejudice is faith – faith in the words of Saint-Simon that the golden age lies not behind but ahead of mankind. . . . In a backward country the great and sudden industrialization effort requires a New Deal in emotions.

(Gerschenkron 1992[1952]: 87)

INTRODUCTION

Two striking paradoxes have attended the end of the Cold War. Both are inscribed in Francis Fukuyama's provocative and influential book *The End of History and the Last Man* which proposed, with a tip of the hat to Hegel, that the tearing down of the Berlin Wall marked the final victory of liberal democracy, 'the end point of mankind's ideological evolution' and 'the final form of human government' (Fukuyama 1992: xi). The world court had met, the jury deliberated, and the vote was in. Fukuyama, who spent his days analysing states of socialist orientation for the RAND Corporation, saw in the popular overthrow of Communism a sort of developmental *Geist* – what he curiously refers to as 'the Mechanism.' The motor

44

of history which keeps Fukuyama's Mechanism firing on all cylinders is the coupling of the science of economic development with the 'desire for recognition,' a sort of impulse for freedom (*thymos* in his lexicon). Its apotheosis is, not surprisingly, the universalization of liberal democracy. Fukuyama's grandiose, and in some sense Warrenite, conclusions that capitalism will bring everyone up to the current levels of material welfare of North Atlantic states and that liberal democracy is spreading like wildfire everywhere, do not deal terribly well with either the realities of global polarization that the World Bank Development Report scrupulously details for us every year or the small problem of devastating environmental 'externalities,' what Alain Lipietz characterized as the contemporary equivalent of the Great Plague (1989: 43). He seems equally unconcerned either by the deep fractures which plague liberal democracies everywhere (Halliday 1991; P. Anderson 1992) or by his own admission that 'social inequalities will remain even in the most perfect liberal societies' (Fukuyama 1992: 292).

Fukuyama's tract captures, nonetheless, some deep tensions which reside in the efforts to resurrect (neo)modernization theory and to preach the Olympian virtues of the World Market. The first paradox is that the collapse of western communisms in Eastern Europe and the former Soviet Union coupled with a growing market integration, has not produced the unambiguous ascent of a single model of capitalist democracy or the end of history but rather the aggressive assertion of difference in the guise of ferocious nationalisms and ethnic violence, what Ben Anderson (1992) calls 'a new world disorder.' And second, the spectacle of capitalist triumphalism and neo-liberal reform in the Third World – the so-called 'new realism' promoted by the International Monetary Fund and the World Bank – is simultaneously marked by a highly visible rejection of modernity in some quarters, that is to say the emergence of a coherent 'anti-development' discourse (Manzo 1991; Pieterse 1991; Shiva 1991; Escobar 1992a; Norgaard 1992).

As its custodians are at pains to point out, anti-developmentalism proposes not so much a development alternative as an *alternative to development*. Anti-development, reminiscent in some ways of the Fanonite assault on neo-colonialism in the 1960s (Fanon 1967), unequivocally rejects development as degenerate, ecologically maladaptive, an empty dream. 'You must be either very dumb or very rich if you fail to notice', notes Gustavo Esteva (1987: 135), 'that development stinks.' Sachs (1992: 1, 3) expresses a sentiment widely shared by critics of Eurocentric development discourses (Amin 1989; Wignaraja *et al.* 1992), when he concludes in apocalyptic terms that development 'appears as a blunder of planetary proportions,' an edifice which 'shows cracks and is starting to crumble.' At its most understated, there is a sense of malaise, a 'crisis of developmentalism' (practice) and an 'impasse' in development thinking (theory)

(Edwards 1989; Stone 1991; Schuurman 1993); at its most radical, there is a demand to abandon Western universalism or Euro-Americanism (Sachs 1992; Slater 1992a), a call for other worldviews and other visions (Kothari 1988, Parajuli 1991), and a recognition, to employ Wallerstein's (1991) felicitous phrase, that development has been an 'illusion' not a 'lodestar.' There has been no progress since 1400 according to Wallerstein; indeed the record shows an overall human decline – all downhill since Vasco da Gama.

In this chapter I shall explore the notion of an impasse in development theory and a crisis in development practice. Like Cowen and Shenton (this volume), I wish to explore these questions historically, pushing the origins of development back into the eighteenth and nineteenth centuries. In so doing I want to make two claims: the first is that the visage of crisis is, and was, built into development from the beginning, and to this extent development must be rooted in what Marshall Berman (1982: 15) calls the 'paradoxical unity' of modernity, a unity of disunity rooted in the maelstrom of 'perpetual disintegration and renewal, of struggle and contradiction, of ambiguity and anguish.' The second is that development, while fundamentally modernist in its genesis and character (infused with beliefs in progress, forward movement, the unfettered power of techno-logy and so on) is periodically reinvented, a reinvention located in what Schumpeter (1952: 83) called the 'essential fact' of capitalism, namely 'Creative Destruction.'

One can, therefore, pursue both an *archaeology* of development – to identify how and why development emerges as a late eighteenth- and nineteenth-century problem containing certain tropes grounded in the European experience of governability, disorder and disjuncture – and a *genealogy* of development in which repetitions and variations (inventions and reinventions) of these tropes operate across a range of ninteenth- and twentieth-century contexts (Foucault 1984; Spurr 1993). This latter exercise is not a simple mapping by which, say, new post-colonial elites naively adopt something called development; Indian development planning may draw upon Fabian or Keynesian notions but it is forged in a particularly Indian way drawing upon both Soviet and indigenous (Mahalanobis, Nehru) knowledges and practices of state intervention.

I argue that the current crisis of development is, therefore, intrinsic to development itself:

> In reality things do not last for any length of time; through the restlessness with which they offer themselves at any moment . . . every form immediately dissolves in the very moment when it emerges; it lives, as it were, only by being destroyed.
>
> (Simmel cited in Frisby 1992: 68)

If the project of development lives by being destroyed, then its restlessness will necessarily pose sharp questions of the Saint-Simonian belief that the golden age lies ahead. A key question might be to explore how the current impasse, the effort to reinvent development, is distinctive, a distinctiveness that I shall argue resides not in the existence of post-modern alternatives to development – as Buck-Morse (1989) says modernism and post-modernism are not chronological eras but political positions in a century-long struggle between culture and technology (see also Harvey 1993: 14) – but in the confluence around civil society (and I shall suggest certain populisms) as the way out of development gridlock. From quite different vantage points, but driven in large measure by an ideological antipathy to the state, neo-liberalism and neo-institutionalism on the one side and the new social movements and post-modernist alternatives on the other meet on the common ground of civil society calling for what I, following Gerschenkron (1992[1952]), see as 'a New Deal in emotions.'

DEVELOPMENT AS KEYWORD

While development came into the English language in the eighteenth century with its root sense of unfolding, it was readily granted a metaphorical extension by the new biology and by ideas of evolution (Williams 1976; Hirschman 1981; Rist 1990; Hosle 1991). Development has as a consequence rarely broken free from organicist notions of growth and from a close affinity with teleological views of history, science and progress in the West (Apffel Marglin and Marglin 1990; Brown 1991; Parajuli 1991; Alvares 1992a). By the nineteenth century the central thesis of developmentalism as a linear theory of progress rooted in Western capitalist hegemony was cast in stone; it became possible to talk of societies being in a state of 'frozen development.' Even alternatives to classical development thinking – such as dependency and Marxisms of various sorts – frequently shared the economism, linearity, and scientism of 'developmentalism.' Development's universalism carried the appeal of secular utopias constructed with the bricks and mortar of rationalization and enlightenment (Slater 1992a). Development was redemption, merging Enlightenment and Christian discourses: 'Providence cast as Progress. Predestination reformulated as determinism' (Pieterse 1991: 15). It came to constitute, in sum, 'an expression of modernity on a planetary scale' (Berthoud 1990: 23).

Tracing, and reconstructing, the genealogy of development is to obviously return to the nineteenth century when, as Cowen and Shenton (this volume) point out, development was in the air. In their view, development is a singularly nineteenth-century problem, a sort of theological discourse set in the disorder and disjunctures of, in their case, a Victorian sense of Progress. As Gerschenkron (1992[1952]: 125) noted, it was not for nothing that Saint-Simon devoted his last years to the formulation of a new creed, the New Christianity, to accompany his industrial and scientific *telos*. For Cowen and

Shenton (1993), trusteeship, mission and faith are the nineteenth-century touchstones of development; it is in this sense a reaction to Progress.

However, there is also a sense in which development was both an eighteenth-century problem, and also a product of what Herbert (1991: 301) calls 'an ethnographic interpretation of culture' resting upon a view of society as 'a self-enclosed system of symbolic equivalences in which the grand principle of order is exchange rather than control.' In fact by pushing deeper into the past, development can be more properly located under the broad arch of modernity. This is, of course, an enormously complex field and I only wish to touch upon three aspects here.

The first begins with Canguilhem's (1966: 182–3) observation on *normalization*: 'Between 1759, the date of the first appearance of the word *normal*, and 1834, the date of the first appearance of the word *normalized*, a normative class conquered the power to identify the function of social norms with its own uses and its own determination of content.' The ground on which such normalizing efforts have been made is typically the welfare of populations, and the relations of individuals (for example the poor) to society. The study of welfare as a 'third leg of modernity' (Rabinow 1990: 8) speaks of course to Foucault's study of discursive and non-discursive practices which he called 'bio-technico-political.' In other words, another strand of development is its relationship to the diverse norms and forms by which a developed or modern society can be understood and regulated; the efforts to produce, in short, governable subjects.

As Rabinow (1990) shows in his study of French colonialism, the norms and forms included fields of knowledge (hygienic, statistical, geographic), social technologies of pacification (welfare and planning), the creation of new forms and space (new towns, urban laboratories, architecture). In each case, the norms and forms of regulation were staffed and constituted by 'technicians of general ideas,' the inventors and practitioners of 'one subset of the practices, discourses and symbols of social modernity' (Rabinow 1990: 9). In the case of empire, Britain led the way in the nineteenth century in managing information – a new symbiosis of knowledge and power that Richards (1993) calls 'the imperial archive' – as a basis for civilizing and developing the colonies.

A second aspect is the genesis of classical political economy in which wealth and value are central, what Lasch (1991: 52) refers to as 'Adam Smith's rehabilitation of desire.' With its origins in the late eighteenth century, political economy was 'the favourite subject in England from about 1810 to about 1840' (Bagehot, cited in Herbert 1991: 74). As Christopher Herbert (1991) has shown, political economy can be understood as an effort to wrestle with the contradictory logic of the desire of accumulation as a precondition for the functioning of the capitalist machine and desire as the origin of misery and vice. Classical political economy strove to defuse these tensions and contradictions in part through 'a theory of culture based on the interlocking concepts of "value" and of integrated social systems . . . [which] arises from a state of urgent cultural instability' (Herbert 1991: 149).

Political economy, insofar as it endeavoured to provide a formal basis for development, turned, in other words, to culture; the desire for accumulation was, as John Stuart Mill noted, inconstant across societies. In this sense the desire for accumulation was obviously central to notions of development as part of a modernist project, but this desire – and indeed the study of political economy from its inception – drew upon knowledge of 'primitive economies' and cultural difference. Development as a nineteenth-century concern was, therefore, neither *sui generis* nor simply imposed (subsequently) on the non-developed ('uncivilized') world, but rather development was in an important way a product of the non-developed. Development required non-development and to this extent the origins of modernity were not simply located in the West.

Finally, development as a cultural condition has been linked historically to the absence of development ('the primitive,' the 'uncivilized') and to development's alternatives. On the one hand, a panoply of cultural rhetorics of development cross cut the nineteenth and twentieth centuries. David Spurr (1993) has documented what he calls the 'rhetoric of empire' through a number of tropes which emerge from the Western experience of colonialism. Surveillance, aestheticization, classification, debasement, affirmation, naturalization, eroticization and appropriation appear and reappear across a number of contexts of colonial (and indeed post-colonial) development. How these tropes are configured and attached to particular development discourses – say soil conservation in the 1930s, community development in the 1950s, market liberalization in the 1980s – is a key to the genealogical aspect of development and modernity.

On the other hand, development itself – the unregulated desire for accumulation – is, as Herbert noted, a product of cultural instability. The late nineteenth century was, to use Lasch's (1991) language, a *fin de siècle* obsessed by cataclysm. Henry George's sense of impending doom was certainly an attack on Herbert Spencer's social Darwinism but it also reflected a longer tradition – often populist in character – rooted in the nostalgia for a lost past, and in the burdens of 'creative destruction.' If cultural reactions to progress were generated from within the belly of capitalism, it was nonetheless a point of suture with the non-developed or uncivilized realm, a realm which spoke to a world lost, and perhaps also to impending doom. Progress produced its instabilities, as it were, from within and from without.

By tracking development historically, one can appreciate the complex origins of what came to be the unitary meaning of development that seemed to surface in the late colonial period in and around the Second World War. Development, as Castoriadis (1991: 186) notes, came to signify indefinite growth, maturity and capacity to grow without end. This social imaginary was 'consubstantial' with a group of developmental postulates: the omnipotence of technique, the asymptotic assumption relating to science, the rationality of economic mechanisms, and the presumption of social engineering as a prerequisite for growth. According to Castoriadis (1991: 186) development, its postulates and its imaginary significations now lie in ruins because 'the imaginary significations are accepted less and less within society.' One can, of course, contest Castoriadis' characterization and the profundity of the crisis he posits. But in my view this crisis was, as it were, built into development from the very beginning.

DEVELOPMENT CONTESTED: MODERNITY, COUNTER-MODERNITY AND POST-COLONIALISM

The paradoxical unity of development as a modernist project is captured vividly in two new development dictionaries. The first, a contribution to *The New Palgrave: Dictionary of Economics* series (Eatwell *et al.* 1989), sketches the definitional contours of development as economic growth which turns out to be of recent provenance emerging from the post-war 'interests and concerns in the West inspired by humanitarian considerations . . . and the Cold War' (Bell 1989: 1). The development story is of latecomers, pioneers and catching up; the economic coordination mechanisms which permit late developers to compensate for the historical accident of their lateness (Bell 1989: 2–17). Entries – the keywords of contemporary development theory – address great ideas (balanced growth, dependency, dual economies, nationalism, uneven development) and great men (Gerschenkron, Prebisch, Seers, Schumacher, Lewis, Mahalanobis). Theoretical and political differences course through the dictionary but the commonalities – the assumptions and projections that link state- and market-based models of growth in which the 'West . . . [acts] as the transcendental pivot of analytical reflection' (Slater 1992a: 312) – are never questioned. In describing the social science of development in Latin America, Sorj (1990) captures this orthodox narrative in terms of what he aptly calls 'the sociology of the non-existent' (the Third World is not developed, it wants to be developed, it cannot be because something is missing) and the 'sociology of desire' (what is missing is articulated by projecting desires on social reality which exaggerates the relative importance of certain agents, forces and institutions). The *Dictionary of Economics* details, without ever challenging, the developmental postulates outlined by Castoriadis (1991: 186) (see above).

The *Development Dictionary* edited by Wolfgang Sachs (1992), conversely, is about 'knowledge as power,' a dictionary of toxic words. Development is the problem, something to be deconstructed and abandoned; a universalizing discourse of Westernization and a set of positions occupied by 'admen, experts and educators.' The shapeless and eradicable virus of development must be exposed by tackling central concepts that 'set boundaries on the thinking of our epoch' (Sachs 1992: 4). Resonant with the legacies of Fanon,

Cesaire and Foucault, the mental space in which people dream about development is, in this view, largely occupied by Western imagery. Entries are keywords (state, science, resources, poverty, planning, needs, market) constituted as intellectual genealogies and parts of discursive fields. 'Needs' in late twentieth-century development become a discourse constituting citizens as cyborgs (Illich 1992); 'planning' speaks to the normalization of the social (Escobar 1992d); 'poverty' is a myth, a construct, an invention (Rahnema 1992); 'science' entails violence and the loss of vernacular spaces (Alvares 1992b); and resource exploitation desacralizes Nature (Shiva 1992).

Sachs and his group of largely Third World scholars, practitioners and activists are engaged in a sort of development archaeology, tracing the complex mapping procedures by which the iron cage of development was put together (Gendzier 1985; Scaff 1991, Escobar 1994). In their view development was invented in the 1940s and institutionalized (nationally and internationally) during the heyday of modernization theory. In taking Raymond Williams' admonition seriously – namely to recognize that words are never innocent, and their meaning rarely stable – the *Development Dictionary* curiously foreshortens the history of development which seems to have been the invention of President Truman. Development – as Bob Shenton (personal communication) points out – is somehow presented without a history while its alternative, often drawing on a crude nostalgia for the past, assumes mythical (one might also say mystical) proportions.

Any development dictionary must be a Foucauldian project, that is to say both archaeological and genealogical. In the same way that planning in nineteenth-century Europe opened up a realm of state intervention to manage poverty, so the institutions and practices of development attempt to create, and regulate, a realm of the social on a global scale (Dean 1991). Development disciplines, as much as the infrastructure and institutions of international governance, attempt to produce 'governable subjects' (Burchell *et al.* 1991; Rabinow 1990). A part of the armoury of development is, therefore, the measures (and practices of measurement) of development itself. La Touche (1992) traces the origins of the standard of living and the projection of global poverty as a relatively new construct encased in development plans, foreign aid reports and national accounts data building in part on the Keynesian revolution. Indeed, the particular significance of the colonial development and welfare legislation of the 1930s and 1940s was that it welded together concerns over growth with the apparatuses of welfare and social regulation (Cooper 1989; Guyer 1990). The recent debate over the politics of compiling the UNDP human development index (Kelley 1991; *Financial Times of India* 2 April 1992) and Jean Chesneaux's (1989: 64) observation that 'the unemployed worker in the slums of Caracas discovers with amazement that he enjoys a standard of living defined in terms of GDP which is worthy of envy', speak to both modernity's universalism and to the 'biopower' of development.

51

Development practice has, of course, always been contested even on its own terms. Paul Richards (1990) has pointed, for example, to the fractures in colonial agricultural policy in West Africa between the ecological populists and those who adhered to the Caribbean model of 'modern,' mechanized agriculture for which the estate provided the blueprint. The larger story here is, as Cowen and Shenton (1991) have shown, a struggle between Fabian colonialism (in its various guises and with its roots in Comtean positivism and its assumption of an identity of interest between peasant proprietorship and European liberal capitalism), and Chamberlain's industrial capitalist vision of 'imperial estates.' In the former, something called tradition ('customary land tenure' for example) could be harnessed for developmental purposes. In the post-1945 period when colonial powers, armed with big science and big capital, would lead a drive toward efficiency, modernization and planning, such traditions came to be seen as blockages to modernity (Cooper 1989; Stoler 1992).

These divergences were prefigured in late nineteenth-century India in quite contrary theories of economic development, though both in fact shared the common understanding that economic growth constituted the heart of social development. On the one side were the British officials influenced by J.S. Mill who saw India as 'advancing'; on the other were the nationalist intellectuals – Naoriji, Ranade, Joshi – who argued that India was economically under-developing (Chandra 1991). These views were relatively stable until the 1930s when a radicalization of development around socialism and state regulation produced an eruption within Indian society – most vividly between Nehru and Gandhi (Chakrabarty 1992) – over planning and industrialization. What is at stake here is the realization that understanding how development of a particular sort emerged out of the colonial crucible must be sensitive to competing strategies, uses of power and discourses, what Cooper and Stoler (1989) properly call the 'tensions of empire.' In this sense, the knowledge systems which were the disciplinary mechanisms of colonialism – the conscious way in which for example the Raj went about creating the categories in which British and Indian were to define themselves – are perhaps more complex and fractured than is sometimes realized. Orientalism attempted, among other things, to define the constituents of a certain sort of society and a certain sort of development, but as Lowe (1991: 4–6) points out it is 'heterogeneous and contradictory . . . consist[ing] of an uneven matrix of orientalist situations across . . . sites . . . [while] each of these orientalisms is internally complex and unstable.'

Development as theory and practice has also been contested in a somewhat different, and potentially more radical, manner under the guise of postmodernism (R. Young 1990; Slater 1992a). Here the reference point is probably Edward Said's (1978) *Orientalism* in which he claimed that the texts of Orientalism can not only create knowledge of other places and subjects but also the very reality they appear to describe (see Mudimbe 1988).

Orientalism is, in other words, inscribed in development history and development practice. Developmentalism, in this sense, contains a classical modernist procedure, what Derrida (1978) calls 'logocentrism': namely a disposition to impose hierarchy between places and subjects, a nostalgia for origins, and a philosophic predisposition to foundationalism which provides a standard or vantage point independent of interpretation (Manzo 1991). For Derrida, logocentrism renders even the most radical discourses complicit with that which it seeks to reject. Hence Said can claim that:

> The theories of accumulation on a world scale, or the capitalist state, or lineages of absolutism [read Wallerstein, Wolf, Braudel and Anderson, MW] depend (a) on the same displaced percipient and historicist observer who has been an orientalist or colonial traveller three generations ago; (b) they depend on a homogenizing and incorporating world historical scheme ...; (c) they block and keep down latent epistemological critiques of the institutional, cultural and disciplinary instruments linking the incorporative practice of world history with partial knowledges like Orientalism.
>
> Said (1985: 22)

Ironically, Said stops at the point at which much critical theory also cries wolf, namely a critical evaluation of itself. This lack of what R. Young (1990: 140) calls Orientalism's 'inner dissension' leads Said to recapitulate Orientalism's own structure.

The post-modern alternative (and its implicit critique of development) appears less in Said than in the literature on post-colonialism (Spivak 1987, 1990a; Bhabha 1984, 1991). Bhabha, in a heady mix of psychoanalysis, opaque prose and post-modern injunction, focuses on alterity and ambivalence in what he calls colonial discourse. In his voice, 'what is the desire of the repeated demand to modernise? ... what is to be done in a world where even where you were a solution you were a problem?' (Bhabha 1991: 203–61). He attempts to show how Western rationalities (in history, in practice) experience a dislocation when they are projected upon the colonial Other. According to Young (1990: 156) his interrogation undermines authority, prevents closure and 'shifts control away from the dominant Western paradigm of historicist narrative.' Development discourse for Bhabha would be an apparatus of recent provenance whose strategic function is the creation of a space for developing subjects who are construed in certain ways through the production of knowledges and through surveillance. Spivak, like Bhabha, sees imperialism (and by extension, development, though she rarely employs the term) as not only territorial and economic but also a project which constitutes a subject. She seeks to attack the notion of the Third World itself as a means by which a hegemonic signifier homogenizes its subject into nationalism and ethnicity. Rather she seeks to retrieve the subaltern, to make him/her visible, to pose the question how, and under what conditions, can the subaltern enunciate? And in this way, 'what goes on over here [is] defined in terms of what goes on over there' (Spivak 1990a: 84).

The post-colonialism literature has unleashed a ferocious debate which speaks directly to the writing of development history and the practice of development. Prakash (1992: 13) claims that 'critical history cannot simply document the process by which capitalism becomes dominant for that amounts to repeating the history we seek to displace.' For O'Hanlon and Washbrook (1992), Prakash accordingly makes capitalism a 'disposable fiction' and post-modernism a disposition hostile to class analysis in such a way that systems (like capitalism) can only produce sameness. For Prakash this is Eurocentric universalism of the old sort, devoid of heterogeneity, without ambivalence, without other voices. Others have suggested that post-modernism offers the Third World only intellectual escapism, critical paralysis (Portes and Kincaid 1989) and 'ambiguities and positions that are intellectually and morally unacceptable' (Sorj 1990: 115).

Post-colonialism is a highly problematical term precisely in relation to those issues (linearity, binarism) that it seeks to transcend (McClintock 1992; Shohat 1992). What post-colonialism does articulate clearly, nonetheless, are some difficult questions about writing the history of 'development,' about imperialist representations and discourses surrounding the 'Third World' and not least about the institutional practices of development itself. Mohanty's (1991) devastating critique of the concept of 'the Third World Woman' inscribed in feminist development writing – through 'discursive homogenization' (Mohanty 1991: 54) the native constructed into a narcissistic, truncated other for the Western feminist – alerts us to a paternalizing, colonial project embodied *practically* in multinational development initiatives such as Women in Development (Mueller 1987a). Insofar as one must take seriously Mohanty's injunction that deconstruction must be accompanied by reconstruction, the question of quite what a Derridian strategy of developmental alterity would look like remains, perhaps appropriately, somewhat opaque. What seems to emerge is an often uncritical celebration of difference, an emphasis on local knowledges, and on writing and self-relexivity.

THE REINVENTION AND INSTITUTIONALIZATION OF POST-WAR DEVELOPMENT

A public and intellectual concern with economic growth and social transformation is of great antiquity, traceable at least to the Greeks, but the idea of development as a specific domain of enquiry and state intervention is of relatively recent origin (Hunt 1989; Cooper and Packard 1992). In Africa, a self-conscious effort to plan comprehensively something called development only came about in the inter-war period through the British and French colonial development and welfare legislation (Coquery-Vidrovitch *et al.* 1988) in the wake of a few earlier desultory efforts at conservation and health planning (Beinart 1984; Packard 1989). Indian nationalist intellectuals did not take on national development as such until the 1930s, prompted by the planning experience of the Soviet Union, the impact of the Great Depression and the enhanced political role of the Congress Party (Chakrabarty 1992; Ludden 1992). Similarly, in the late 1940s the Latin American structuralists under Prebisch, in asserting the centrality of domestic planning, laid the foundation stones for the developmentalism of the 1960s and 1970s (Dietz and James 1990; Pantojas-Garcia 1990; Sikkink 1991). The collapse of Communism has meant that similar sorts of debates have been triggered (over shock therapy, market integration, the constitution of mixed economies) in the 'Second World'; development is no longer the monopoly of Africa or South Asia (Colclough and Manor 1991; Bierstecker 1992, 1993; Cooper and Packard 1992).

At a certain historic juncture these local and national debates were globalized in the sense that President Truman's programme of international 'fair dealing' helped produce an unprecedented explosion of international institutions, professions, organizations and disciplines whose *raison d'être* was the lodestar of development. While Kothari (1988: 143) may be right in saying that 'where colonialism left off, development took over,' it is surprising how little work has focused on the invention of institutions which produce, transmit and stabilize regimes of development 'truth' – and to this extent Kothari's judgement leaves much unsaid. Escobar (1988) tackles this question head on by proposing that the reinvention of nineteenth-century develop-ment dressed up in the garb of post-war neo-colonialism can be understood as 'professionalism' (a set of techniques and disciplinary practices through which knowledges are organized and disseminated) and 'institutionalization' (an institutional field in which discourses are produced, recorded and implemented). For the former, the building blocks are the new disciplines (such as development economics), the training of local development experts, and the production and socialization of experts – from, for example, the functionalist anthropologists who staffed the colonial research service to the development anthropologists and social soundness experts employed by USAID in the 1970s (Pletsch 1981; Robertson 1984; Rosen 1985; Escobar 1991; Karp 1991).

The intention is to identify, in some cases ethnographically, the local knowledge systems and 'epistemic communities' within the complex field of development (Haas 1989). And it is a vast institutional field extending its reach from the global regulatory agencies such as the World Bank to the panoply of small non-governmental organizations. While a great deal is already understood about the political economy of such organizations, much less is known about how they function as organs within a political economy of truth (Pottier 1992; Hobart 1993). Development experts inhabit these institutional environments as cosmopolitan intellectuals, members of a 'new tribe' (Hannerz 1990; Klitgaard 1990). They are the scribes who oversee the production and reproduction of knowledge and practices which purport to measure well-being and poverty, national growth and standards of living, who negotiate the re-entry of national economies into the world market through the science of adjustment, who attempt to 'mobilize' and 'animate' peasants in the name of basic needs. Research on how the international agricultural research stations (and specifically how the research agenda is set, fought over and legitimated) and the history of the traffic in plants from the imperial botanical gardens to current debates over intellectual property rights and germplasm are simply differing facets of Escobar's institutional field (Pain 1986; Lipton and Longhurst 1989; Juma 1990; Shiva 1991).

Historicizing development and locating discussions of growth, welfare and needs, and so on in institutional fields, opens the possibility for exploring specific sorts of development dialogues in an interpretive vein (Mitchell in this volume). Some recent scholarship self-consciously explores development as a text, or more properly a series of texts, from a semiotic or rhetorical perspective (McCloskey 1985; Wood 1985; Apthorpe 1986; Roe 1989, 1991). Development, for McCloskey (1985: 252) is a metaphor, 'limiting our thinking at the same time it makes thinking possible.' Development economics is a matter of words, rhetoric and metaphor. McCloskey's argument is presented as an attack on modernism, on economics as a hard science, seeking instead to show that economists like everyone else employ tropes to convince (and not to prove), to converse, to convey authority in specific contexts. Development can also be, according to Emery Roe (1991: 288), a story or narrative, 'less hortatory and normative than ideology' and 'more programmatic than myths.' Development is a sort of Proppian folktale; the dominant narratives are populated by villains, heroes and donors. In Roe's (1989: 287) view, development narratives must be dislodged not evidentially or by Popperian refutation but by counter-narratives which 'stop talking about everything as a constraint.' Apthorpe (1986: 387) in fact sees development policy as 'the equivalent of performance' which has 'powerfully constructed sets of escape routes . . . [like] "exigencies," "constraints" and so on' to escape from the responsibility of its claims and failures. However, Roe's analysis, like those of Apthorpe and McCloskey, is generally weak with respect to the social basis of ideas, the social and historical context in which such stories are produced and told, and how or why some stories become dominant and others are relegated to the periphery of the development 'community.' Development narratives remain *only* narratives.

A more compelling body of work locates development as a discursive field, a system of power relations which produces what Foucault (1984) calls domains of objects and rituals of truth. An archaeology of twentieth-century development would, among other things, seek to discover the institutions, social processes and economic relations on which the discursive formation of development is articulated, and to uncover how history gives shape to different sorts of discourse. Development is a series of 'situated knowledges,' part fact and part fiction, which are 'artifactual' (Haraway 1992). Arturo Escobar (1992b, 1994) is a key figure in the Foucauldian turn in development studies. Following Deleuze (1988), he identifies development discourse as a series of statements and 'visibilities' linked together as a *dispositif* or a diagram of power. The discursive field of development is, in this sense, a cartography of power and knowledge. An innovative study by Ferguson (1990) analyses a specific development plan in Lesotho in which experts represented the country as 'a nation of farmers not wage labourers ... [with] a geography but no history; with people but no classes; values but no structures; administrators but no rulers; bureaucracy but no politics' (Ferguson 1990: 66). Ferguson shows how these representations were embodied in development practice (for example livestock projects) and how the contradictions between the real and the imaginary are negotiated (i.e. by making social or political problems 'technical'). As he points out, project failure should not be equated with an absence of local effect (see also Crehan and von Oppen 1988; Pigg 1992).

If not directly framed by Parisian intellectual culture, a great deal of the new work on the ethnography of development and the anthropology of modernity recognizes that all societies create their own modernity, and hence the question becomes on what terms what Appadurai calls 'public culture' is debated and defined and who gets to participate (Appadurai 1990). The idea of 'reworking modernity' (Pred and Watts 1992) posits popular culture and memory as much as 'globalization' as central ingredients in the recipe of late twentieth-century development. Some of the efforts to deconstruct the meanings and ideologies of discourses on global change (Buttel *et al.* 1990; Buttel and Taylor 1992), or contestations over sustainability in development (Adams, this volume), and narratives of health (Packard 1989) share these broad concerns with institutions, knowledge and power (see also Tennekoon 1988; Karp 1991; Lavie 1990; Lansing 1991; Nandy 1991; Feeley-Harnik 1992; Scheper-Hughes 1992).

REVIVING CIVIL SOCIETY IN THE 1990s: RECYCLED POPULISMS AND THE CRISIS OF DEVELOPMENT

What the World Bank (1989a: 4) refers to as its 'new development strategy' for Africa was hatched in the context not only of widespread criticism of its 1980s adjustment programmes, but also against a backdrop of economic crisis and the prospect that 'Africa will be affected more than any other region by the likely slow growth in demand for primary commodities' (World Bank 1989a: 32). In their long-term diagnosis, agriculture will be the primary source of growth, transforming and expanding its productive capacity in a manner which is at once sustainable, equitable and self-reliant (World Bank 1989a: 89). Equity in this context means poverty alleviation for the poor (in Africa they are primarily rural and peasant) via asset provision and productivity enhancement. Sustainability refers to 'sound environmental management and human resource development' (World Bank 1989a: 44). And self-reliance speaks not to a delinking from the world economy but rather to 'building African capacities' (World Bank 1989a: 186). The two key themes, then, are 'an enabling environment' which includes a strong defence of devaluation and exchange rate policy ('getting the prices right'), and 'building African capacities' understood as institutions for enhancing entrepreneurial, managerial and technical capabilities (World Bank 1989a: 122). At the heart of its long-term strategy, says the Bank, is the desire to release those energies that permit *'ordinary people ... to take charge of their lives'* (World Bank 1989a: 4, my emphasis).

Talk of new strategies and the powers of ordinary people in society by the World Bank has been coeval with consolidation and hardening in development economics, expressed most concretely in the so-called confluence of analytics around institutions and transaction costs (Bates 1989; de Janvry *et al.* 1992). During the 1960s and 1970s there was a broad shift from market to state-centred alternatives with minor attention to the role of civil society. The 1980s brought debt, retrenchment and austerity; short-term crisis management, stabilization and the so-called neo-liberal counter-revolution in development theory were their handmaidens. Inflation, lack of investable funds and external conditionalities imposed by global regulatory agencies rendered alternative development strategies, as much as statist development initiatives, largely irrelevant. Concurrent with the regime of austerity, however, was a reintegration of development economics back into the mainstream as such, signalled by theoretical developments in transaction costs, contracts and the design of institutions in situations of incomplete or failing markets (Bardhan 1989; Stiglitz 1989). This rehabilitation was echoed in the 'new developmentalism' which sought, in its analyses of the newly industrializing states, to link neo-Weberian concerns with institutions and organizations to political economy (Vandergeest and Buttel 1988; Haggard 1990; Evans 1991).

By the 1990s, the convergence around social institutions, not least the fascination with non-governmental organizations, citizenship and human rights, provides an opportunity to design, in the 'new' context of globalized markets and the re-emergence of civil society, new configurations of state, markets and civil organizations unencumbered by outmoded or ideological notions of central planning or unhindered free markets. What is so striking in this confluence of analytics is the centrality of civil society; markets have to be socially embedded, economic dynamism demands social capital, economies are built around trust, obligation, accountability.

Civil society is also a central focus of analytical reflection in the 'alternatives to development' paradigm. Much of the practical and political impulse for anti-development is seen to reside in the so-called 'new social movements' (NSM) (Hettne 1990; Lehmann 1990; Corbridge 1991; Escobar and Alvarez 1992; Escobar, in this volume). Unlike the critique of Eurocentrism which sees the strategic choice as socialist universalism or Western capitalist barbarism (Amin 1989), the NSM presumption is of alternatives, a move from bipolarity to 'polycentrism.' It is, nonetheless, difficult to generalize about these movements and what they represent. There is a staggering heterogeneity in form and character (liberation theology, Chipko, widows associations, *barrio* networks, ethnic organizations and so on) and a tension over whether they have similarities to social movements and identity politics in the West. There are claims that NSMs in the South are principally claiming territory from developmentalist states, and are concerned more with re-source access and control rather than the quality of life (Guha 1989), yet much of the theoretical discussion of these differences draws from the work of Claus Offe, Jürgen Habermas and Ernesto Laclau (see Escobar 1992b; this volume). According to Escobar (this volume), grassroots movements do have in common the fact that they are local, concerned not so much with access to state power as the creation of 'decentred autonomous spaces.' Pluralistic without 'one particular ideology or political party' (Escobar 1992a: 422), their economic concerns are socially embedded and expressed in cultural terms which often rest upon local 'subaltern' knowledges (Wamba 1991). The NSMs are autopoietic, that is to say self-producing and self-organizing, exercising parallel (i.e. non-state) systems of power (Fals-Borda 1992).

Drawing largely from the experience of India and Latin America (Africa for example is noticeably absent), these movements unquestionably represent powerful reconfigurations of civil society, and perhaps offer the opportunity – as the institutional economists also point out incidentally – of new social contracts between state and associational life. This has led some to suggest that the market triumphalism of the 1980s, the so-called Lost Decade, spawned a 'new mode' of doing politics, new sorts of fragmented subjec-tivities, and a bottom-up horizontal vectoring which redefines political and economic democracy (Shet 1987).

Both the new development economics of the 1990s and the anti-development paradigm stake claims for alternative strategies. Furthermore, they speak in the same register in reasserting the role of civil society and in questioning the form, function and character of the developmental state. For this reason it is not surprising that appeals are made, in both camps, to the powers of 'ordinary people.' In this regard, populist sentiments and what Cohen and Arato (1992: 29) call 'the resurrection, reemergence, rebirth . . . [and] renaissance of civil society' often go hand in hand. Populism is here understood not only as a development strategy – that is to say, in terms of small-scale, efficiency, and a broad anti-urban thrust against the ravages of industrial capitalism (Kitching 1980) – but also as a particular sort of politics in which an effort is made to manufacture a collective national-popular will (Laclau 1977). Populist strategies – and the language of populist rhetoric and appeal – reside in what Ernesto Laclau (1977: 193) refers to as 'the double articulation of discourse.' Double articulation here refers to the ways in which populisms simultaneously attempt to maintain a stable relationship between 'the people' and other powerful classes in the ruling power bloc and the various discursive ways in which 'the people' are configured with the interests of specific classes.

Populisms of various sorts have provided an important counterweight to unfettered progressivist belief, and to this extent they have constituted a pronounced undertow in the larger current of developmentalism. Indeed, tracing recursive populisms and locating them with respect to the double articulation of discourse can shed much light on a central narrative theme running through the history of developmentalism. Moore's (1989) work on the relation between smallholder strategies and a Sinhalese national identity in Sri Lanka, the 'protection' of peasants from capitalism in colonial Java (Alexander and Alexander 1992), and the Fabian vision of agrarian proprietorship in West Africa (Cowen and Shenton 1991) all speak to the enduring nature of populist discourses. In this regard, the informal sector debate, the current popularity of indigenous technical knowledge, and the new concern with flexible specialization in the Third World can be situated with respect to an intellectual lineage of some depth (Hart and Watts 1992; IDS 1992). Indeed, it is unlikely that populism will ever go out of fashion precisely because in the context of the paradoxical unity of modernity it provides, as Christopher Lasch (1991: 532) recently wrote in his defence of United States populism as a credo for the twenty-first century, 'a wide ranging critique of progress, enlightenment and unlimited ambition ... [and] a distinctive tradition of moral speculation drawn from everyday experience.'

CONCLUSION: *FIN DE SIÈCLE* DEVELOPMENT

What is striking about these visions of alternative development amidst the turmoils of *fin de siècle* capitalism? There is certainly an emphasis on difference and fragmentation as though they are 'goods' in themselves; as though, as Eagleton (1990: 88) dryly notes, we have far too little variety, few social classes, that we should strive to generate 'two or three new bourgeoisies and a fresh clutch of aristocracies.' Furthermore, some of the claims made are so ambitious – Vandana Shiva (1991) for example on nature, women and traditional science or Escobar (1992a: 430) on 'the new form of the institution of the social' – that they do not look too different from the totalizing and essentialist visions of the old sort. And not least, some of these movements in structure and character strike me at least as populist in character and hence part of a long lineage within modernity itself, which raises the question, central to populist visions and utopias, of their relation to class and forces of co-optation (i.e. Peronism, Narodnism, Bonapartism and so on). At the very least, careful analyses are needed of the relations between NSMs and the hegemonic class forces of capitalism. Some post-modernist work, for example, claims not to reject class (Prakash 1992) yet it is unclear how it re-enters their post-colonial vision. Is not one of the conditions of living in an era of market idolatry and utopian capitalism – a period of creative destruction – that the new social movements, as the embodiment of an alternative to development, may also take on utopian and fantastic projections?

I am not seeking to denigrate these alternatives by virtue of their utopian characteristics but rather to situate them historically and analytically. I have tried to argue that development as one face of modernity has always contained within it what Marshall Berman calls 'the tragedy of underdevelopment.' Through a sort of historical recovery of development one can, I believe, see firstly that the origins of development are within modernity but that modernity cannot be unproblematically located within the West. The modern (and developed) require the non-modern (and undeveloped). Second, development, to paraphrase David Harvey (1993), and alternative development are dialectically organized oppositions within the history of modernity, to be seen less as mutually exclusive but as 'oppositions that contain the other' (Harvey 1993: 15).

The crisis of development in the 1990s is not, therefore, rendered distinctive by virtue of the existence of alternative visions, however apocalyptic they may be. What is distinctive it seems to me is the fact that the nineteenth-century problem of development, as Cowen and Shenton (this volume) put it, is no longer framed by the logic of the post-1945 Cold War but is attached to a growing sensitivity to the ecological consequences of unfettered growth coupled with unprecedented global inequalities. Amidst grotesque economic global inequalities and the wreckage of non-capitalist development strategies, there exists an environment in which capitalist triumphalism and post-modern alternatives have hardened into rigid, incommensurable forms. A striking irony of this incommensurability in the 1990s is how both sides of the debate – modernist and post-modernist – focus on civil society as the New Jerusalem, the touchstone for sustainable development in the next millennium. This confluence is, to return to Harvey, how modernism and its discontents contain one another at the *fin de siècle*. In the North and in the South the call for 'a New Deal in emotions' employs the powerful language of civil society. The question I would pose, therefore, is similar to that of Lasch (1991) in his analysis of late nineteenth-century critics of the new industrial and imperial order: will the misgivings be submerged – and how quickly – in a renewed celebration of progress? As in the case of development past, this question will be answered less at the level of the grand abstractions of state, capital and market than in the 'myriad of sites where the modern is produced and transformed in its encounter with the production of the non-modern' (Mitchell and Abu-Lughod 1993: 82).

61

MICHAEL WATTS

ACKNOWLEDGEMENT

An earlier version of this chapter appeared in *Progress in Human Geography* 17(2) (1993): 257–72. I am grateful to the publishers for permission to publish this version here.

3

SCENES FROM CHILDHOOD
The homesickness of development
discourses

Doug J. Porter

Concepts, like individuals, have their histories and are just as incapable
of withstanding the ravages of time as are individuals. But in and
through all this they retain a kind of homesickness for the scenes of
their childhood.

(Kierkegaard 1965: 47)

INTRODUCTION

What has discourse analysis to offer our understanding and practice of
development? One easy response is that in a field so characterized by rhetoric
and persuasion, critical awareness of ideological processes in discourse is
essential. At a minimum it ensures that aspiring development workers are
more aware of their own practices. At best it enables those eventually afflicted
by their services critically to confirm the ideological investments they are
inevitably being persuaded to make in development. In other words, discourse analysis reiterates an old responsibility of ideology critique to
challenge 'what is given, in the head as well as in reality' and to deconstruct
'these categories (which) express the forms of being, the characteristics of
existence' (Marx 1973: 106–8).

This justification for attention to the discourse of development should not
be underrated. It is ironic that there is a drift towards insularity in development just at the time that it is being challenged by social movements and
voices that previously could be kept outside the development industry.
Official discourses – from the New World Order of Global Modelling to the
'technically necessary' conditionality of Structural Adjustment agreements –
appear more and more as an expert culture, divorced from the practical moral
concerns of daily life and the major disputes in particular places. The
hierarchy of this discourse ensures that mastery of technical metaphor
paradoxically rules a debate that increasingly evokes populist images of
participation and democracy for its legitimacy.

But there is much more at contest today. Development discourse is marked by a greatly enlarged set of ambitions and obligations in the form of sustainable development and the slogans of Our Common Future. The imperatives of 'sustainability' have meant that the imperatives of development are being pushed into the furthest recesses of our lives, ideals and conduct. Given that the designs of sustainable development are so far-reaching and interventionist, we must address the question of whether discursive practices (reports, policies, projects, guidelines) are able to transform the conditions apparently leading to global collapse, or are more likely to reproduce the planetary malaise. Discourse analysis asks whether these interpretative texts and rhetorical, mobilizing devices are constitutive of the reality we seek to manipulate through development, or are part of a Foucauldian world of power with little room for manoeuvre but to reproduce the historical conditions that will hurry along global demise.

This chapter will focus on the metaphorical aspects of development discourses, in particular, metaphors which convey meanings associated with order and stability, and with constraints. It will establish that there is a historical continuity and persistence in the underlying metaphors, despite significant changes in the fashion-conscious institutional language of development since World War II. This continuity bears witness to the power of master metaphors and, more importantly, to their critical importance in maintaining and legitimating the regimes of governance and economy which prevail in development practice. Throughout the post-war period, diverse metaphors have been introduced which promote the impression of radical change without threatening the basic project of controlled and orderly manipulation of change.

Second, this historic continuity is paralleled by continuity in geographic scales of development discourse and practice. There is continuity of master metaphors from global through to local scales of practice. This will be illustrated in one integrated rural development project in the Philippines which is typical of local-level practice during the 1980s and which is currently being reasserted in rural development today. Metaphors from global development discourse are invoked to order and control the diversity of local situations. In this example, we can see both the present being disciplined by the past, and the local being integrated within a globally universal rationality of development.

Whether in the localized routines of development projects or in higher level policy rhetoric, I argue these metaphors have the same authoritarian consequences. There are two aspects of this. The centralization of authority, the first aspect, is a relatively well-versed feature of development practice. Indeed, in some earlier formulations, institutional authoritarianism was understood as a prerequisite of development. However, these tendencies in practice need not occur as a product of conscious political design but may be seen as necessarily reproduced by everyday practices.

The second aspect of authoritarianism is less widely understood since it appears to run counter to the professed intentions of recent shifts in development discourse. A theme of the past ten to fifteen years has been the importance of participation. Calls for 'people-centred' development, or community participation or the importance of indigenous knowledge, express numerous motives. But all, in some form, underlay concern with human agency in development. I argue that the metaphors of practice ultimately result in a radical de-authoring and denial of agency. Moreover, these consequences are not mollified by the resurgence of participation and democracy in development discourse. Indeed, the contemporary recasting of the original master metaphors and their buttressing by neo-populist sentiments can be read as a new attempt to reassert order, stability and continuity in the face of global biospheric and societal uncertainty and discontinuity that constitute the call-signs of the post-modern epoch.

In conclusion, this chapter reflects on the challenge of post-modernity and the implied rejection of the modernist project of development. This draws on two features of the chapter that are usefully stated at this point. It is, first, written within a post-modern epoch, acknowledging that difference is essential for human agency and democracy of diversity and divergence, in a manner similar to Deleuze and Guattari's (1977) celebration of the multiple voices now possible in the post-modern world. It also rejects, with Foucault (1979: 158), any vestiges of an 'originating subject' from which the appropriate measures of difference, divergence, and so on, can be determined a priori. However, the discussion is resolutely committed to the modernist concern with emancipation and intellectual responsibility. New forms of discourse, new voices, and new rules of the development game are emerging. Truth is contingent, but development practice cannot afford to be ambushed by relativism or immobilized by nihilism.

MASTER METAPHORS IN DEVELOPMENT

Metaphors establish authority and provide a device for making sense, creating order and certainty. They structure the way we think and the way we act, and our systems of knowledge and belief, in a pervasive and fundamental way. Tacitly metaphors constitute the starting point of 'what is' and simultaneously convey a normative sense of 'what ought to be done' (Fairclough 1992: 37ff., 194). The quest for an infallible and universal criterion of 'metaphorhood' is doomed to failure. There are no necessary and sufficient conditions for establishing what constitutes a metaphor, the focus of this analysis, since their meaning is contextual and depends on the sense made of them by the 'interpretive community' in that context. However, some key features of metaphors can be stated.

The object of analysis is linguistic texts, wherein written and spoken metaphors appear both as products and predicates of institutional processes. Metaphors figure in discourses as comparisons between two (or more) unlike entities to convey a meaning in excess of, or differing from, that which could be deduced from the literal meaning of the statement. Two kinds of metaphor will prove of interest here; simple metaphors and open-ended metaphors (Harrison and Livingstone 1982: 11–12). Open-ended metaphors cannot be paraphrased in literal expression since they convey an indefinite range of meanings which frequently move amongst different levels of meaning. For instance, 'electricity is a key link in development,' an open-ended metaphor, is polyvalent. At one level it evinces the idea that electricity is 'an indispensable component' in development, a component that 'comes after others and before yet others.' At another level, meaning nests within cluster of metaphors purporting that 'development is a linked chain of components' or yet other metaphors which evoke an essentialism of historical linearity and causal association.

Polyvalence requires that metaphors are studied historically and dynamically, in terms of shifting configurations of discourse types. They are full of 'snatches' of other texts which may merge, assimilate, contradict or ironically echo yet others – what Foucault refers to as 'tactical polyvalence' (Fairclough 1992: 84). Metaphors can adopt apparently contradictory forms and yet still prosecute the same interests and power – a point discussed later with respect to the apparent contradictions of the amalgamation of economic rationalism and neo-populism in recent development discourse.

For heuristic purposes, a distinction will be made between three different, but not discrete, types of metaphor. 'Organizing' metaphors are those which to a great extent are peculiar to the post-World War II phenomenon of development. These articulate in various ways a range, secondly, of 'master metaphors' which transcend the historical period of development and frequently provide the fount for a diverse range of discourses which may overlap with but which maintain a separate discursive identity from development. Finally, 'metaphors of practice' are features of project-level discourse which sometimes are current only in particular projects or geographic regions, but increasingly characterize the profession or interpretive community of development workers irrespective of location.

Take, for example, the metaphors in the famous Point Four of Harry S. Truman's 1949 speech:

> We must embark on a bold new program for making the benefits of our scientific advances and industrial progress available for the improvement and growth of underdeveloped areas. More than half of the people of the world are living in conditions approaching misery. Their food is inadequate, they are victims of disease. Their economic life is primitive and stagnant. Their poverty is a handicap and a threat both to them and more prosperous areas. For the first time in history, humanity possesses

the knowledge and the skill to relieve the suffering of these people . . .
our imponderable resources in the technical knowledge are constantly
growing and are inexhaustible. . . . The old imperialism – exploitation
for foreign profit – has no place in our plans. . . . Greater production is
the key to prosperity and peace. And the key to greater production is
a wider and more vigorous application of modern scientific and
technical knowledge.

(Truman 1967[1949])

Most of the metaphors were already current but their amalgamation is
insightful. Truman imbued the various connotations of 'development' with
the revolutionary fervour and sense of historical inexorability of Marxism,
but changed the prime motivator of agency from the Party and intentionality
of the proletariat to the benign role of capital, scientific technique and the
carriers of expertise. The sense of Prometheanism is almost obsessive; no
longer were international relations to be marked by domination and subordination; instead 'partners in progress' would work together under the
banner of development to take advantage of scientific and technological
progress for the global rise to prosperity (Ullrich 1992: 275). Development
became, with Truman's statement, the most violent, colonizing metaphor of
contemporary life.

Here we will discuss the metaphorical underpinnings of Point Four in
terms of two distinct historical channels. One involves a clutch of biological,
organic and evolutionary metaphors; the second draws on the 'colonial
economics' which subsequently coalesced into neo-classical economics.
Following this, links will be made with the metaphors of mid-nineteenth-
century physics, prior to drawing two implications about the master metaphors of development.

Marx's conception of development drew on a fashionable, biological
metaphor which located rationality at the level of historical structure.
Economic laws work with the 'iron necessity' of an incessant and inherent
logic on the 'structure of history' (Marx and Engels 1969, vol. 2: 87).
Development, the central theme in Marx's dialectics of historical stages
adopted Hegel's directional view of history. Hegel's teleology ('the history
of the world . . . is the process of development') goes back to Aristotle's
Politics, where development was the realization of 'actual' form from
'potential' matter. This tendency to detect large-scale patterns and laws hailed
from an early Christian sense of history as 'the great melody of some ineffable
composer.' It continued into the twentieth century in the work of writers like
Spengler and Toynbee, the former emphasizing again a history of biological
metaphors of growth and decay.

Others have linked Marx's circular view of the reproduction of capital to
the Physiocrats' biological metaphors of economic rationality drawn from
the pulmonary circulation system (Barnes 1992: 128, 130). The importance of

this is not to claim (wrongly) that Marx exclusively used biological meta-phors, but rather to point to the links between the biological metaphor and functionalism in social and economic sciences, as well as development discourses. The key point here, which features also in the master metaphors of neo-classically inspired development, is that the role of constituent parts in society is determined by their function within the whole, that is, the 'functional prerequisites' (as in Talcott Parsons) of 'order' for the system to survive and reproduce over time.

The interpretive community of neo-classical economics is different from Marxism. Their concepts of rationality differ fundamentally, basically as a result of different master metaphors; physical versus biological. But we shall not overstress this difference, for there are similarities in the practical, political consequences of utilizing either set of metaphors. In the past twenty years, there has been great interest in the conceptual connections between thermo-dynamics and neo-classical economic theory (Georgescu-Roegen 1971, 1976; Mirowski 1984a, 1984b). Drawing evidence from published works of the first neo-classicists, from biographies of the principals, and from the laws of thermodynamics, Mirowski explains that 'both the timing [of neo-classical economics] and intellectual content can be explained by parallel develop-ments in physics of the mid-ninteenth century.' Indeed, he argues that neo-classical theory redefined utility so as to be identical with energy without 'any assessment or critique of the sources of its analytical inspiration' (Mirowski 1984b: 363ff., 373).

The *homo economicus* postulate is central to neo-classical economics. The optimizing behaviour of economic man is the anchor for economics as a science concerned with the neo-classical problem of the allocation of scarce resources to optimum ends. The basis for the mathematical technique of constrained maximization is directly provided by the first law of thermo-dynamics: the 'law of the conservation of energy.' This law states that all forms of energy are able to be converted, from one form to another, at fixed ratios of transformation and that the sum total of converted energy in a closed system is a constant. By substituting utility for energy in the constrained maximization equation, neo-classical economics was able to discover the most efficient and economically rational actions of producers and consumers. This made the postulate of economic rationality equivalent to the physicists' principle of least effort (Barnes 1992: 125). The second law of thermo-dynamics implies the existence of a stable equilibrium state which does not change with time while the system is isolated. For example, a given amount of gas may assume any of a number of states, but if confined in a gas-tight container of fixed volume, it is restricted to states with volumes smaller than or equal to the volume of the container. Change can occur, and any number of equilibrium states may be assumed, but any one of these may be prohibited by 'constraints.'

The idea of 'constraints' became particularly important when the metaphors of physics passed into development discourse prior to World War II. Whereas for Marx and economists of the late nineteenth century, development referred to a historical process that 'progressed' without being consciously willed by anyone, by the 1920s, colonial economists and administrators perceived development as requiring the alleviation of constraints which fettered the transition from one stage of development to another, more advanced stage. Prior to this, government's role was that of policeman, to preserve law and order, to uphold private property, to protect the integrity of coinage but otherwise, to 'let well alone.' As Arndt (1981) points out, the sense of development say, in the line from Hegel to Marx, derives from the intransitive verb. But by the 1920s, especially for economic historians of the British Empire, development was a transitive verb (Arndt 1981: 460).

The transitive sense of development was clearest in colonial policy, particularly with respect to dominions like Canada and Australia, where the government facilitated the transfer of people and capital for resource development. Differences in rates of development were attributed, in the neo-classical schema, to the role played by government; including, in addition to the policing of order roles, developing resource-extraction infrastructure (for the removal of constraints), and provision of incentives (ultimately reducing the effect of other constraints), as in taxation or subsidization policy. The development of material resources and the social wellbeing of people was kept separate, as evident in the colonial 'dual mandate' policy applied in the British African colonies and in special development grants made to colonies by usually parsimonious metropolitan governments. However, during the 1920–30 period, the positive connotation of development was expanded. The trusteeship responsibilities of the conqueror for 'the welfare of natives' was transformed into a positive regard for minimum standards of nutrition, health and education as part and parcel of the role of government in fostering development. The 1930s depression persuaded many that although the *laissez-faire* system fostered trade and commerce, the essential engine of development, it did not necessarily guarantee overall social stability. From this point, the transitive and intransitive meanings of development became blurred to the point of obliteration.

Truman was as conscious of the expansionary needs of United States corporations and the popular sense of urgency of depriving the communists of a mass base as he was of the 1945 United Nations Charter (Article 55) to 'promote higher standards of living' (Watnick 1952: 36). But the amalgamation of the transitive and intransitive senses of development was, by that time, no longer conscious. Whereas countries were previously spatially and substantively different, with Truman time became institutionalized as the dominant parameter of difference. Differences between countries were perceived only in terms of the teleological metaphor; difference was a matter of 'productivity.' Progress was inexorable; time could be compressed by the

69

application of scientific technology and capital according to 'inexhaustible technical knowledge.' Poverty was both a handicap and threat to the orderly realization of this universal history. Accelerating history was a bold endeavour, but not a political one, since it transcended old imperialisms. It was Promethean, stealing the fire from the heavens of history.

From this time, the development metaphor in practice expressed two key meanings. One was the vending machine mentality: construct the machine-model according to ineluctable historical laws, then 'you put in the money, press the button, and get growth' (Brookfield 1975: 39). The second meaning equated development practice with the removal of constraints on order and stability and on economic productivity. Two implications of these metaphors are important for the argument developed here. First there is the universal nature of development. It is not context dependent. The historical process of development, according to the metaphor of constrained maximization/*homo economicus*, involves a rationality that applies everywhere in time and space. Indeed, if it is not applicable everywhere (which it is by metaphorical definition) and if not applied everywhere, then global maximization will not occur, thereby producing sub-optimality, irrationality, and threatening the 'functional prerequisites' of the global order.

The second implication is the denial of human agency. In the biological–physical world metaphor, particles move without consciousness, simply in conformity with the principles of least effort and optimization. The intentionality of human agency is equivalent to the desire to maximize utility. Agency is a 'function' and, as with all forms of functionalism, the character of human agency is circumscribed by the requirements of the fundamental metaphor; as a repository of externally, organically determined needs. Other values, intentions or desires become impediments and obstacles and therefore constraints to utility maximization or, at worst, threats to orderly functioning of the system. In the neo-classical scheme 'man is not an economic agent because there is no economic process. There is only the jigsaw puzzle of fitting given means to given ends, which requires a computer and not an agent' (Georgescu-Roegen cited in Barnes 1992: 125). This is what E.P. Thompson refers to as the real silence in Marxism about cultural and moral mediations on historically inexorable laws. This silence occurs also in Western capitalist discourse where agency is lost in 'the notion of the maximisation of productive growth as being the inner motor of a machine that people trail along behind' (Thompson cited in Abelove *et al.* 1983: 20–1).

Here is the power of the organic metaphors where 'all variations', according to Hepple (1992: 141), 'tend to share an emphasis on the "natural order" . . . and also on the central role of harmony and common purpose within the state, with each social group fulfilling its appropriate role.' Shorn of their moral and political dimensions, these early statements about development expressed a basic idea that persists to this day. Development practice is understood as the systematic application of a universal rationality at a societal level to achieve desired states of affairs through the control of human as well as natural processes. Development came to be seen as a 'problem' which could be broken down into a series of constraints, like savings, growth rates, or literacy, according to known causal relations between them. Once identified, these elements could be reassembled and manipulated in a controllable and predictable manner (Porter *et al.* 1991: 93–4).

Emboldened by Truman-type speeches there followed a period of wide-sweeping generalizations about underdevelopment through all-embracing theoretical constructs focusing on the removal of constraints. The shortage of some strategic input (such as the supply of savings, foreign exchange or technical skills) was the main constraint. Once these constraints were overcome, development could proceed according to fixed qualitative relationships between input and output (the capital: output ratio being the best known of these coefficients). Of course, human entrepreneurship was considered important – as in the work of Schumpeter – and in a sense here is 'agency' being promoted. However, this form of agency was operationally transformed as a matter of technical skills and education and was incorporated in manpower planning.

The usual method of incorporation was to determine a target rate of output or growth and then estimate the quantities of technical skills required on the basis of fixed coefficients between inputs of skills and output. Consequently, agency was once again defined by the system prerequisites. Overcoming the impediments to planned and controlled manipulation of the future called for authoritative intervention on an international scale. The institutional implications – the Bretton Woods international financial and development institutions, the strengthening of the nation state and, more recently, the panoply of non-governmental organizations – have been well documented elsewhere (Fowler 1992; Pottier 1992; Sachs 1992). By design, these agencies possessed the degree of leverage necessary to ensure that the traditional, particularist or politically idiosyncratic features of a situation ('obstacles and barriers') would not hold sway. Thus, as an Asian Development Bank historian remarks, they played a 'significant part in the erosion of certain unfortunate attitudes and practices which have long been a serious barrier to effective modern growth' (Huang 1975: 4, 145). Whilst projecting an 'objective and politically neutral image,' the politically authoritarian implications of development legitimated by the master metaphors are clear in the rapid concentration of power that proceeded at international levels and in the form of the local, nation state.

Affirming the master metaphors, and emboldened by the optimism of the post-war years, the classic positivist project began to unfold during the 1950s. Problems were self-evident, reducible to 'elements' and 'components,' then capable of reassembly at successive stages comprising projects, comprising sectors, comprising the National Plan, according to known causal relations, and manipulated in controlled and predictable ways.

DOUG J. PORTER

'PRIVILEGED PARTICLES' IN PRACTICE

Development discourse gave special privilege to projects and development professionals, two phenomena so essential for local-level articulation of the global task of development that they have formed the most commonly invoked conditions placed on 'aiding' development. Indeed, to aid was less a response to a call for help, than an initiative designed according to a larger pattern of needs, projects and sectors consistent with the functional pre-requisites of the master metaphor. Calculations of coefficients were supple-mented by universal standards to determine real needs – as opposed to profligate wants – and these were attributed with unwaivering benevolence to all people and places (Gronemeyer 1992: 53, 55; Porter *et al.* 1991: 91–9). For many observers, projects became synonymous with development itself. The traces of the master metaphors are evident in Albert Hirschman's oft-quoted definition from *Development Projects Observed*:

> The development project is a special kind of investment. The term connotes purposefulness, some minimum size, a specific location, the introduction of something qualitatively new, and the expectation that a sequence of further development moves will be set in motion .... Development projects ... are the privileged particles of the develop-ment process.
>
> (Hirschman 1967: 1)

Development practitioners (or technical assistance experts in the parlance of the 1950s) provided, in practice, the rational nexus of capital, technology and rationality represented by projects. UN resolutions of the early 1950s endorsed the view that 'defective knowledge and consequent inability to make rational plans' was a major constraint; technical assistance was projected as 'democracy's route for expediting' development with 'administrative integrity'; an important attraction given the sensitivities of newly independent nations to neo-colonialism (Hoselitz 1952: 11; Alexander 1966: 5). Variously, development agents were seen as, and remain, the keys to successful develop-ment and styled as linchpins of development, builders of order, catalysts and inducers of economic and social change.

It would be a taxonomic exaggeration to refer to a distinct culture of development practitioners since there are no overarching norms or dedication to common values. Queries about common definitions are most often greeted with a bemused response from practitioners of all shades: commercial consultants, NGOs, academics doing reality checks, or government officials. Yet there is a common interpretive community discernible in key notions about the special skills development practitioners believe they bring to bear in development; the attitude in which these are imparted; and the purposes of their activities. Although one cannot see the body language that helps interpret the meaning of these statements, one may hear amalgamated the three kinds of discourse that Dear (1989) talks of: (a) instrumentalism whereby professionalism is reproduced; (b) the rhetorics of persuasion; and (c) of performance whereby the interests of the individuals and their clients are advanced. Development practitioners' statements about their specialized competence (such as engineering, agronomy or finance planning) rest on a tacit agreement about their mandate. A World Bank consultant explains:

> If capital and technology were the towels and hot water needed for a smooth developmental birth, then a mid-wife was also needed. The countries concerned were inexperienced in dealing with such changes and neither could they keep up with the anticipated birth rate of projects. Besides, like traditional healers, they would probably mess it up by deviating from set procedures in preference for custom, or sorcery and family tradition.

Applied regardless of geography, the political context or the specific nature of the problem, the universal rationality both enables and sustains the claim to provide the cutting edge in situations confusing to laypersons. 'The critical constraints' of any problem can, according to one practitioner, be 'zeroed in on.' Thus, 'when we look at a development problem, what do we find? Low production levels, low living standards, erosion problems, poor health.' These, he said, 'translate to poor project management and massive inefficiency.' His job is 'to get projects finely tuned. After all, we're the ones who make it work.'[1] The managing director of a project management company suggests the cutting edge is 'a special skill (in the) ability to isolate specific projects, to help local people get a coherent view, to sell the project. . . . His motives of survival and profit make him biased, but they also make him mentally aggressive and inventive and can if properly used give a cutting edge to all phases of the project.'

If efficiency and the expeditious application of particular skills are the cutting edge of practice then discipline is the primary vehicle; most often referred to as 'arm's length advice.' The dispassionate pursuit of rationality, independence and objectivity amount to a deeply internalized set of symbols. These are essential to the professional promise and to the exclusion of the possibility of any other means whereby the challenge of development can be met. Development practitioners, according to the director of the project highlighted below, offer 'the Velvet Glove Solution, the middle path between revolution and resignation' in the face of what the former chairman of the Australian Professional Consultant's Council sees as 'the colossal dangers of social and political unrest.' This requires 'that we muster the right combination of professional skills' in order, somewhat alarmingly, to achieve 'mastery

---

1 The examples in this section derive from field diaries compiled during the 1980s. I have endeavoured to draw references only from development workers associated with the project examined later in the chapter.

over or manipulation of the natural environment, in which term I would include for this purpose all social interaction within the human species' (Faithful 1982).

The power of the professional promise is not derived from the actual results. The legitimacy of practice is derived from a kind of promissory note on which appear metaphors such as 'arm's length advice' and 'the cutting edge.' In turn, the promissory note authorizes the development worker's participation in a range of rituals. The purpose of rituals, of which cost-benefit analysis is the most infamous example, is to screen out the uncertainty and instability caused by contending voices or interpretations of development (Porter et al. 1991: 192–3). In rituals metaphors articulate otherwise distant and opaque master metaphors and thereby assist in the critical process of creating certainty, of turning arbitrariness into givenness and actuality.

AUTHORITY IN PRACTICE

All development projects reflect particular constellations of global and local political, strategic and commercial imperatives. The Zamboanga del Sur Development Project (ZDSDP) is no exception. The project was implemented in one province on the Philippine island of Mindanao (see Fig. 3.1) and involved around Australian $50 million of bilateral aid over its life from 1976 to 1985 devoted to a wide range of infrastructural (roads, irrigation, water supply), agricultural and community development activities. Development was here occurring in the conflictual circumstances typically associated with cultural invasion and migration, contest over productive land, and un-accountable, unrepresentative institutions of governance. Socio-economic indicators revealed 46 per cent of provincial families living below the generous World Bank poverty line, with 5 per cent of households enjoying incomes in excess of ten times greater than the bottom 50 per cent, around 38 per cent of families with zero or negative annual incomes, and a provincial Gini coefficient of 0.549. ZDSDP was a controversial project and as a consequence the views of project workers and critics were considerably more elaborated and publicly available than is normal practice. ZDSDP provides a remarkable record of the key debates in rural development practice over the 1970s and 1980s period. As the Executive Director remarked, 'There's ten years of development history written into this project.'

Political events sourced at local, national and global scales during the mid-1970s had great bearing on the local interpretive community constituted by this project. The project's conception shortly after the 1972 Declaration of Martial Law in the Philippines intruded greatly. 'Peace and Order in the Province of Zamboanga del Sur' featured in the 'Overall Goal' and the project bore many of the charges of counter-insurgency development popularized by Walden Bello and World Bank colleagues (Bello et al. 1982). Here was a classic integrated rural development project, based essentially on metaphorical

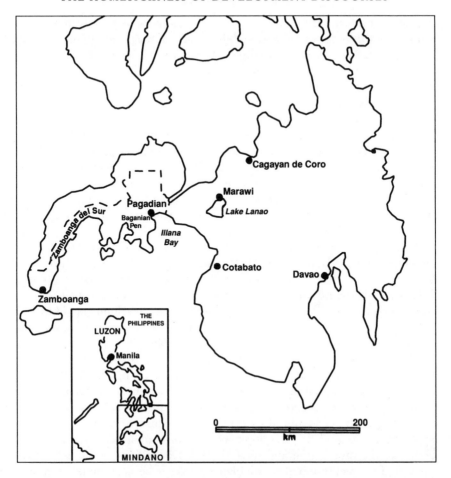

Figure 3.1 Location of ZDSDP

commitments about 'roads being the key link in alleviating constraints on rural development,' yet stumbling also to give expression to the confusing mandate coinciding with Robert McNamara's term as World Bank President.

McNamara's ascendancy brought speeches, rather like Truman's, resplendent with phrases that quickly passed into the idiom of development practice. Two were prominent, concerning the environment and the social. McNamara's way of dealing with them was hailed as a fundamental shift in development thinking, yet each was successfully factored-in to the existing development calculus. Events such as the Stockholm Conference on the Environment (1972), the OPEC oil price hikes of the early 1970s and media events like the Meadows *et al.*'s (1972) *Limits to Growth* quite rapidly penetrated to the project level. The demise of the gold standard and controlled currency devaluations in the USA forced awareness of how scarcity constituted a threat to order and stability and of the need to incorporate environmental and social externalities. 'There is a recurring tendency,' noted Apthorpe (1970: 7) 'to explain the failure of predominantly economic development plans by invoking the following reason. There must have been a troublesome knob on the development machine, marked "the human factor," which was twiddled wrongly, inadequately, or not at all, and therefore, somehow, "the non-economic variables" were left out of account.'

Development agencies hurried to recruit anthropologists 'in order to understand', according to the Australian official aid bureau's first anthropologist, 'the minimum social conditions that must exist for a project to attain its goals.' The new 'social inputs, properly tailored to project resources and needs, are not "icing on the cake" but basic ingredients' (Brouwer 1982: 135; see also Claasz 1979).

Although the vending machine mentality had been sorely tested by the first two Development Decades, for McNamara, matters were far more pressing. The concern for the social factor underwriting the era of 'Basic Human Needs' and 'Redistribution with Growth' strategies in development discourse was received with acclaim by the progressive liberal development aid lobby. However, in the context of perceived social and environmental threats to global stability, these strategies are more adequately understood as defensive modernization. Politically, the poor had become too great in number to ignore; consequently, they became explicit target groups in development discourse. New social movements were blossoming; anti-technocratic, autonomy-seeking, decentrist and rejecting the institutional practices of the modern state and economy, their threat too had to be incorporated. It was explicitly recognized that development, left to its own devices, would not eliminate the disaffection and poverty of 1,000 million people below the poverty line much vaunted by McNamara. Needs became the key requirements for system functioning and these were articulated as a 'certain specific minimum standard of living' (ILO 1977) defined in terms of the organizing metaphors of development. Un-met 'needs' were therefore new 'constraints' on order and stability that had to be 'alleviated.'

The adoption of a 'socially oriented measure of economic performance' under McNamara profoundly increased the complexity of project practice (McNamara 1981: 12). The bewilderment caused by multiple social and environmental, as well as economic, objectives is palpable in a remark from one project worker during the mid-1970s: 'I don't know what the hell the goals are, but I'm moving ahead with projects' (Moss 1978: 94). How was it still possible to 'move ahead with projects'? Three characteristics of the ZDSDP indicate the importance of metaphors of practice in this respect.

Of primary importance was the recasting of the original functionalist, organic metaphors into systems language and the application of allied techniques (PERT, Logframe, critical path analyses) derived from new systems mandarins of RAND, NASA and the Indochina theatre. The derivative metaphors of 'system' provided the context in which diverse needs could be incorporated or integrated. According to UNESCO, integrated development is 'a total, multi-relational process that includes all aspects of the life of a collectivity, or its relations with the outside world and of its own consciousness' (cited in Esteva 1992: 15). The systematic integration of diverse needs and objectives featured prominently in the Zamboanga project, as this remark of the Project Director attests:

> The rural development program contains, in addition to the infra-structural element, the initial agricultural element. Incorporated into that was what we call an engineering extension element, a community health element was built into it, a rural water supply element was also built into it. There may be a number of small elements in there that have slipped my mind.

A second underpinning was the way in which order and stability metaphors provided both an environment of certainty and boundaries within which all needs could be considered in a disciplined way. As in all projects, order and stability were essential prerequisites to the subsequent removal of constraints. Project Goals were explicitly oriented towards stability and order, and therefore were self-evident ends towards which all activities were directed. The goals were those of 'the project' and by definition bipartisan expressions of the common interest. They were recursive, at once both project effects and causal conditions. Thus, it became unreasonable to query what was simultaneously a condition of project success and an end towards which it was directed.

The project goals, as both cause and effect, and the integration of diverse elements, were combined in a key metaphor of practice – called the Agri-business System Methodology – and operationalized through a technique known as Logical Framework Analysis, or Logframe. The details of these techniques are not important here. The Agribusiness System metaphor readily provided for incorporation of all elements as well as their prioritization according to the central problem of agricultural production or, in terms of the master metaphor, system output. The Logframe, a management device designed to 'focus attention on the salient characteristics and relationships which may be blurred in the mind of man' (USAID 1973: 3) served to further disassemble each element and centre attention upon what was known as 'the farm yields problem.' This focus was quite consistent with the functionalism of the master metaphors and can be seen as the local derivative of constrained maximization in neo-classical economics. The yields problem was so taken for granted that no query was brooked. Indeed, any query of 'elements,' of the project goals, or of the focus on yields was colourfully chastised by the project workers as 'something Blind Freddy and his Dog could see.'

There was no confusion thereafter; problem analysis was 'a matter of getting your facts straight,' that is, establishing scientific research trials on isolated stations unfettered or blurred by extraneous variables. In an important sense, authoritative practice, the extent to which yield-specific facts were seen to provide the touchstones of judgement, is defined in terms of its distance from the target population's everyday lives, or the corruption or inefficiency of local administration. Trials on critical elements of the yields problem enabled the assembly of technical farming systems packages (seed varieties, fertilizers, credits, farm management practices) for farming systems, metaphors which resembled in microcosm the larger system that was to be manipulated.

Project workers reported an 'unshakeable faith' in the methodology and remonstrated that 'if you allow us to control all the factors of production then we can increase the production of the province two times, three times, four times.' However, when these packages were 'extended' to supposed real world farm conditions, other constraints were encountered. 'The essential point' an agriculturalist explained, 'is that the total system will only function correctly when all the elements of the system, inputs–farm production–processing–marketing–consumption, are functioning correctly.' All inputs and outputs must be delivered at the right place and time, with the right prices, quantities and qualities. However, discontinuity, irregularity and 'suboptimality' prevailed in practice. A project management training manual in use on the project stated the problem while also providing some inkling of how it should be dealt with. The problem for project management:

> which desire(s) to achieve overall effectiveness (is) not to have the parochial interests of one organizational element distort the overall performance . . . this means that some functional unit may not achieve its parochial objectives, for what is best for the whole is not necessarily best for each component of the system.
>
> (Cleland and King 1968: 10–11)

Observing this constraint of parochiality, the ZDSDP mid-term evaluation team predictably recommended 'it is desirable that centralization of service facilities should be encouraged. . . . It should be an objective of (project) planning to encourage such centralization.' No politics were thought to intrude here since control and order were simultaneously and recursively the project goal and the operational requirement and both were legitimated by the authority of the systems metaphor. Consequently, the authority of centralized institutions was enhanced, first at the provincial level and logically then nationally through the National Council of Integrated Area Development, then chaired by President Marcos.

The obvious authoritarian implications did not end at this point. As Hirschman's definition stipulates, projects are supposed to set in motion a 'sequence of further development moves' beyond the life of the project. In other words, 'institutionalization' of the system was essential. This was explained in Phase II of the project design as 'the process by which, through the instrument of organization, new ideas and functions are integrated and fitted into developing societies, are accepted and acquire the capacity to sustain themselves, and in turn influence the environment in which they function.' The irony is that this authoritarian logic was accomplished through a fundamentally de-authored procedure whereby disembodied system prerequisites were being met, not the apparent interests of any one authority. Reassuring to some, capricious to others, the context-free and universal nature of the underlying metaphors allow for almost infinite variation and incorporation of contenders.

The manner in which the concept of appropriate technology (AT) was incorporated into ZDSDP illustrates a further aspect of the authoritarian character of development metaphors, this time with regard to the concept of de-authoring. As initially conceived, the AT metaphor was active. Under E.F. Schumacher for instance, it focused on social justice considerations and the appropriation of productive means, highlighting organizational capacity as well as technique. Potentially, AT was a metaphor that affirmed human agency; it was about enhancing the control and authorship of marginal people over the means whereby they produced their livelihoods. Shortly after incorporation of AT in the official discourse, 'appropriate' came to designate techniques that fitted the cultural situation from a technical point of view. Little else could occur, given the master metaphor which conceived development in terms of fitting inputs to functions and removing constraints within predefined organizational patterns of control and authority. Events in ZDSDP illustrate this process, but they also allow the point to be pressed further; rather than enhancing human agency, control or authorship, AT had the reverse or de-authoring effect.

In project documents, it was always intended that 'the poorest farmers operating at near subsistence levels (were) the main target groups of the ZDSDP agricultural projects,' since here lay the constraints posed by the peace and order problem. Sensitive to NGO criticism that the project was not meeting the needs of the poor and oppressed, later stages of the project were modified and project workers declared that 'appropriate technology, as far as the technical packages are concerned, underlies all our activities now.' Just as there were broader constraints on institutionalizing the total package derived from the Agribusiness Systems metaphor, in practice similar human constraints were encountered amongst individual farmers to whom these packages were being extended. Just as inputs must be in the right place, time and quantity, so too must the farmer perform rationally, that is according to the prerequisites of the farming system metaphor. Whereas the subsistence mode of production is quite resilient to variations in the farmer's work patterns, cultural practices or idiosyncrasies, the technical packages require that life becomes rationalized to a new rhythm over which farmers have no effective control. As 'farm system managers' their performance must be de-authored, that is, it must correspond to another input bound by the strictures of all other inputs. The technician's role predominates, a role farmers dare not violate lest they overstep the margin between constrained maximization and financial catastrophe. Actual farmer performance so departed from prescribed routines that one agricultural economist despaired: 'They don't always realize that when you're using that package, you just can't use fertilizer and not do the weeding.' Given these presumed inefficiencies, it then became necessary for the project management to add further components which would 'change those aspects of the lives of these people which are likely to be impinged upon,' in much the same manner that the overall Agribusiness System was being authoritatively manipulated. Farmer participation was quite logically the next component incorporated into the project. In like manner, participation has become a key component in a development practice which by metaphorical design ensures that human agency does not become a constraint on development.

REASSERTING OUR COMMON FUTURE

Metaphors of practice are always being contested; not least in the adoption problem project workers attribute to Zamboanga's subsistence farmers. A similar contest is most evident in the development discourse of the late 1980s and early 1990s as a consequence of a new language of global biospheric uncertainty on the one hand, and a host of newly emerging social movements on the other. There are many positive aspects of these shifts and challenges to official development discourses, but just as the challenges of the early 1970s were by the mid-1980s fully incorporated (as another component/input in a persistent metaphor), some remarks about historical echoes in the present conjuncture are warranted.

From an analytic point of view, there is no similarity between the system dynamics modelling of the 1970s and the global biospheric modelling of the late 1980s (Porter 1993). Yet there is a sufficiently strong sense of *déjà vu* in new global modelling and recent World Bank publications about poverty, the environment and the human face being placed on structural adjustment to suggest that earlier remarks about the regime of truth generated during the McNamara period may be under renovation. Some of the more obvious parallels include the ritualistic nature of global modelling, their aura of relentless objectivity and their political function, not as an aid to discourse, but as stipulators of the boundaries of discourse itself. Scientism and universalism are being recalled anew. The President of the International Council of Scientific Unions recently hailed The International Geosphere-Biosphere Programme (IGBP) and the World Climate Programme as a 'bold and exciting new scientific venture' in the relentless search for knowledge of 'the manner in which the Earth system works' which uses 'state of the art technology, for high precision analysis, network computing and satellite remote sensing.' The promise remains: 'IGBP will significantly reduce uncertainties in predicting the future consequences of current trends' (IGBP

1992: foreword). There is little mention here of the extent to which the current consensus on global change has raced ahead of scientific data, or of the role of scientists and officials in the remarkable ascension of 'global change' ideology (Dahlberg 1992; Buttel *et al.* 1990).

Although the political dimensions of this new discourse are typically underplayed, the alliance of moral and technocratic imperatives prevalent since Truman's speech remain. First is the appeal to common, undifferentiated interests, nowadays globalized as Our Common Future and under the organic metaphor of One World. This is contrasted with the corrupt, parochial interests displayed by governance and economic life today. A moral imperative is presented with which it appears unreasonable to disagree. The moral appeal in turn legitimates a technocratic response and the authority of a new kind of superintending manager in order to analyse the system and to direct the changes needed. The master metaphor of the system now is greatly expanded under sustainable development and incorporates all futures (intergenerational equity), all people (intragenerational equity) and all scales from the miniature of locality to the global.

There are important historical antecedents which both prompt and give a new sense of urgency to the discourse of the early 1990s, just as was evident with McNamara, and Truman before him. These cannot be detailed here, but two are noteworthy. First, our appreciation of biospheric uncertainty has fundamentally changed as a product both of new metaphors of chaos in science, and events like rainforest destruction, ozone depletion, or Chernobyl. It is not that old metaphors of order are being ritualistically applied, for ritual is essential to organized social life. Rather, it is that the considerable scientific work on modelling environmental risk systems is so evidently insufficient to breach the limits imposed by countervailing metaphors of chaos or 'unpredictable uncertainty.' Patterns which are said to exist in global uncertainties defy capture on conventional statistical or mathematical grounds. Yet just as reviews of development practice demonstrate our puny capacity to manipulate historical change, our appetite for global prediction increases, simply because we can tolerate less uncertainty at this scale.

Uncertainty in global politics and economy point to a second set of antecedents which parallel those above. The way in which the post-war Fordist economy is being transformed by what Harvey called 'the speculative innovation in production processes' is relevant here, as are the manifold ways in which unilateral state sovereignty is being undermined (Harvey 1985: 126; Dalby 1992). The present generation of global modelling, as Donald Schon (1982) has noted, 'falsely reduces the full range of uncertainties to the more comforting illusion of controllable, probabilistic, but deterministic processes.' Indeed, contemporary biospheric changes and these political–economic uncertainties must effectively be ignored given the scale of technological and policy commitments we are being compelled to make Our Common Future.

Three features of this contemporary discourse evidently revisit the past

(Meadows *et al.* 1992; World Bank 1991a, 1992, 1993). First there is a strong tendency to eschew any real challenge to the host of multilateral, government and financial institutions. Real questioning at this level would so add to the uncertainties of analysis that their explanatory and prognostic ability, let alone the operational imperatives, would be immobilized. Meadows *et al.* (1992: 210, 211) inform us that 'to get to sustainability from here, the remaining material growth possible . . . would logically be allocated to those who need it most', yet follow this up with the assurance that existing institutions can cope with these changes and that 'no one need to engage in sacrifice or strong-arming . . . global transformation can be natural, evolutionary and peaceful.'

A renewed technological confidence marks the second feature of contemporary discourse. According to World Bank publications dealing with environment and development, for instance, 'the principles of sound environmental policy . . . are well understood (and) the technical solutions exist' (World Bank 1992: 19–20, 42). But as the history of attempts to deal with land degradation attest, these assertions are quite contrary to the high degree of uncertainty which serves to undermine any assurance that the main dimensions of environmental problems have been comprehended adequately. In this field the confidence that existed for some time about appropriate policies is being undermined by uncertainty not just about means, but about the central issue of land degradation (Blaikie and Brookfield 1987). But in the operational world of development practice, it is simply a matter of implementing taken-for-granted scientific knowledge. For the World Bank (1991a: 151), 'the main priority worldwide is to establish incentives, regulations and safeguards that lead to proper allocations of resources for environmental maintenance and energy conservation.' The echoes of the pre-war neo-classical concern with externalities and constraints are strong: taxation regimes, subsidies, market-based pricing systems, private property and, where these fail, regulation of resource use and 'special development strategies' to deal with environmental and economic cases judged to be beyond market repair (World Bank 1990a: 71–2; 1991a: 151).

The third feature of contemporary development discourse involves a peculiar, but ultimately quite consistent, application of the neo-classical metaphors of economic rationality to the public, political sphere. Essentially, this view is that politics should be privatized in the same fashion as the market is believed to mirror economically rational life. The mutually causal relationship between liberal politics and liberal economics has rapidly become enshrined in development policy, frequently as a new conditionality applied by major lending and aid organizations, such as the United States' democracy initiative or the Japanese Kaifu doctrine. Just as the nation state is seen to be a fetter on efficient resource allocation by the market (the decentralized decisions of individuals) so too must development problems be handed to 'the community.' Here we see a parallel between the market and the community, both relatively harmonious metaphors tending, if 'let alone,' toward wise equilibria.

However, there is no generalizable evidence to support the contention that widespread participation resolves resource conflicts, favours the poor, or makes for better planning or resource allocation processes. There is scope and value in community participation, but the bigger issues, such as access to shelter, land, capital or viable public entitlements cannot be dealt with adequately by unbridled incantations towards NGOs or local politics. Either participation is irrelevant to the major issues, or where it is relevant (such as in the demand that the local peculiarities are reflected in policy at larger scales) the fashionable focus on 'the community' or NGOs tends to neglect the essential role of the local state in articulating locality beyond the parochial to the generalizable interest (Kilby and Porter 1992). Indeed, the denial of politics is most evident in the silence about practical politics; would participation and market-led development be 'good' and under what circumstances?

Cases exist where poverty causes environmental degradation, and others where increased economic surpluses do lead to investments in environmental goods. But so too can one point to examples of an 'equilibria of poverty' where marginal people do exist sustainably, but where increased economic activity leads to greater exploitation and less sustainability. There is no space to consider whether this privatized form of political democracy is likely to be transformative with respect to the dominant development discourse. New social movements are said to reflect a political opposition to the features of neo-conservatism and they begin with a scathing critique of the centralizing, technocratic, anomie creating, materialist and like features of prevailing development practices. It is true that NGOs from this movement are assuming a level of power and influence that was previously thought to be the preserve of states and multinational bodies (Buttel 1991: 12ff.). Yet it is intriguing to note the ways in which conventional discourse is shaping the character of this new social movement politics in moderately reformist, if not conservative ways. Social conflicts are being overwritten by 'green imperatives.' Packages of technically necessary actions are being assembled into compelling comprehensive frameworks which are in turn being operationalized to deal with human constraints. And these, shades of Truman, are being legitimated by natural science, and wrapped in a bold new programme which it is morally imperative we adopt.

CONCLUSION

As part of the post-modern critique of development studies and practice, discourse analysis may offer a practical–moral critique of a powerful set of social actors that are presently almost unaccountable ideologically and institutionally. This chapter has not been self-consciously post-modern, but it has argued by example against a wholly modernist interpretation of development practice. It has asked what shapes the nature of development practice, the interpretive community of development agents, and the facts they emphasize. Against what is often presented as a history of learning and radical shifts it has shown that the supposed diversity of nearly fifty years of post-war development may be collapsed into one modernist project. In tracing the metaphors associated with stability and order, and with constraints, there is a compelling continuity between the global and the local scale, and between the past and the present through the three sets of master metaphors, organizing metaphors and metaphors of practice.

I have argued that the authoritarian character of development is reproduced by metaphors of practice. In Truman's partners in progress, in McNamara's social factor and Basic Needs Approaches, and in recent 'people-centred' and neo-populist forms of development, there has been a consistent trend for the overt markers of political power to be replaced by more covert and subtle ways in which established international interests have been maintained. The critical point, encapsulated by the concept of de-authoring, is not that authoritarian power is being exercised by forcefully dominating those who are subject to it. Metaphor does not translate the power of dominant to subordinate groups in society. Power is not naked, but incorporates them. The metaphors of development are successful to the extent that, as Foucault says, they hide their own mechanisms.

But there are some difficulties also with this approach. They are part of a much larger debate (Love 1989) but perhaps more pressing in respect to development. Two need to be mentioned in closing; one concerns agency, the other is about rationality. If one takes Foucault, the critical point is that discourse is not simply that which translates systems of domination, 'but it is the thing for which and by which there is struggle; discourse is the power which is to be seized' (Foucault 1979: 110). His 'capilliary' metaphor of power applies here. On the one hand, this account of modern power, infusing all life, seems to multiply greatly the potential situations in which dominant discourses, including the master metaphors as well as more immediate metaphors of practice, can be challenged and transformed. This view is implicit in the host of subaltern studies of people's resistance underway since the early 1980s.

On the other hand, Foucault has been charged with overstating the extent to which situations and people are manipulated by power (Macdonell 1986). Although forms of struggle do receive extended treatment in this work, the overriding impression one gets is of the pervasive nature of metaphor-power, in which resistance is incorporated. The space for human agency to, in Anthony Gidden's terms, 'make a difference' is analytically circumscribed. It is a short step from the capilliary metaphor to the view that all questioning of development, all searching for alternatives, is futile if not counterproductive. The simple fact of imagining other worlds inevitably contradicts and enslaves their possibility and, in this respect, post-modernism truly is 'modernism with the optimism taken out' (Hewison 1987: 132; Esteva 1992: 8).

Discursive practice constitutes development in both conventional and creative ways. It can both reproduce convention, such as by articulating predetermined policy at the local level. It can also contribute to the transformation of local situations through resistance and find expression at larger scales in the manner of 'butterfly effects on thunderstorms.' For what is characteristic of our contemporary situation is not just the playing out of powerful forces that are always beyond our control but, as Bernstein (1983: 156) remarks, 'a paradoxical situation where power creates counter-power and reveals the vulnerability of power.' The relevance of post-modern discourse theory would be enhanced if it provided propositions which enabled us to ask of specific situations 'in what is given to us as universal, necessary, obligatory, what place is occupied by whatever is singular, contingent, and the product of arbitrary constraints?'

The potential difficulty here is the problem of relativism that underpins post-modern forms of discourse analysis. My assumption at this point is modernist: that there is still an emancipatory potential in development. For implicit in the above question is some standard whereby we can judge whether specific instances of practice are likely to be transformative and enable the promotion of people's interests that would seem to be denied by conventional metaphors of a universalist or foundational nature. It is not that this form of analysis denies physical reality, but it does argue that facts are what we make of them, that is, what is made of them by the interpretive community in which they have meaning. This form of reasoning ultimately provides only temporary shelter from relativism. Here there is only radical and ceaseless critique. There is no reconstruction, only relentless deconstruction and disengagement from the world, especially the modernist one we all, by and large, live in. Bourdieu (1988: xvii) warns how, once we become involved in the intricacies of discourse, we lapse 'into indefensible forms of internal analysis' ever spiralling in on itself, recursively examining the very auspices of analytic categories (see also Clegg 1976; Sorj 1990).

Ironically, this immobilizes our critical will and represents denial of what Max Weber called the ethic of responsibility. We are left with elites with the privilege of speaking for others about power forever at work, not just behind everyone else's backs, but in every aspect of their constitution. Post-modernism entails a fundamental rejection of the institutions, the rationality as well as the 'needs' which development seeks to address. It takes little more than a glance at the graveyard of development's past mistakes to enjoy the seductiveness of this post-modernism. However, this is halted by the realization that whilst modernist development may be dead, 'needs' undeniably persist (they at least, are not an idealism).

So, one is compelled in some form to chant again, 'long live development.' To me it is unclear whether this epistemological relativism which defends the

85

heterogeneity of moral standards and criterion of truth, actually nihilistically abandons democracy, justice and more satisfying futures. This course of relativism must be preferable to the repression of difference and diversity that is implied in the New World Order of sustainable development. But how does one say NO to the totalizing of instrumental reason, yet also say YES to possibility of justice?

ACKNOWLEDGEMENTS

I wish to record the excellent support of Yvonne Byron, and to thank Bryant Allen, Robin Hide and Dean Forbes for comments on an earlier draft presented to the Geographical Sciences Graduate Program Seminar Series, Australian National University.

4

GREEN DEVELOPMENT THEORY?

Environmentalism and sustainable development

W.M. Adams

Sustainable development seems assured of a place in the library of development truisms.

(Redclift 1987: 2)

INTRODUCTION

Changing ideas in development theory are usually, after due passage of time, reflected in the language and rhetoric of development practitioners. The reverse is also true. In recent decades a series of phrases (even slogans) has colonized academic discussion of development having first been adopted widely by development practitioners. One such phrase is 'sustainable development' which suffused development discourse in the 1980s and early 1990s following the report of the World Commission on Environment and Development (Brundtland 1987) and the United Nations Conference on Environment and Development or UNCED (Redclift 1987; McCormick 1989; Adams 1990). Not only has the phrase transformed the way in which established development institutions like the World Bank talk about their task, conferring a green hue to existing and (more arguably perhaps) new policies (Holden 1987; Rich 1991), but it has provided a means through which new institutional actors in the form of non-governmental environmental organizations have sought to have an effective voice in debating the shape of development in the Third World.

In adopting the phrase, different actors have interpreted it in different ways to suit their institutional needs and the views of their sponsors or constituents. People widely speak of a 'concept' of sustainable development, but the vast and rapidly growing literature does little to analyse critically the meanings loaded onto the phrase, and related terms like 'ecodevelopment.' What do those politicians, media commentators, development practitioners and environmentalists mean by 'sustainable development'? Do they all mean the same thing (Caccia 1990)? Lélé (1991) provides a clear analysis of the weaknesses that follow from the lack of a clear theoretical and analytical framework to thinking about sustainable development. The most widely used definition is that of *Our Common Future*: sustainable development is 'development which meets the needs of the present without compromising the ability of future generations to meet their own needs' (Brundtland 1987: 43). As a definition, this is superficially attractive, but it is more a slogan than a basis for theory. Furthermore, users tend to claim a misplaced unity for sustainable development thinking. Part of its strength is the way in which it links diverse (and sometimes divergent) ideas and blends them, often uncritically, into an apparent synthesis (Redclift 1987; O'Riordan 1988; Adams 1990).

Despite the underlying confusion, mainstream thinking about sustainable development is remarkably consistent (Adams 1993). This mainstream view has been forged through a series of international publications: the *World Conservation Strategy* (WCS) (IUCN 1980), *Our Common Future* (Brundtland 1987), the 1991 follow-up to the WCS, *Caring for the Earth: A Strategy for Sustainable Living* (IUCN 1991) and Agenda 21 from the 1992 Rio Conference (UNCED). However, a remarkable feature of this powerful current of ideas about development is its lack of engagement with development theory. This is because the new discourse of sustainable development has not evolved from within development discourse, but instead has deep roots in Northern environmentalism (Adams 1990). It has embraced ideas about development rather late and rather selectively. This has important implications for the understanding of the relationship between established debates in the development field and those generated by the new 'environmental' thinking. Some of the ideas from environmentalism that 'sustainable development' carries along within it are encoded invisibly, sometimes within the simplistic problem-solving spreadsheets of sustainable development programmes themselves. However, they are no less important for being hidden. Within environmentalism there is a tension between reformism and radicalism, and between technocentrism and ecocentrism. Because of its roots in environmentalism the same tensions exist in debates about sustainable development. This marks them out from the normal discourse of development.

The flexibility of the phrase 'sustainable development,' and the rampant enthusiasm with which it has been adopted, makes it resistant to analysis. However, that analysis is important both from a theoretical perspective and also as a necessary precursor to effective dialogue between development practitioners and those environmentalists who are exerting increasing influence upon development decisions (for example through pressure on Northern donor agencies and demands for aid conditionality). One way to approach this analysis is to situate debates about sustainable development within the context of accounts of the environmentalism from which it draws. This chapter seeks to do this.

TECHNOCENTRIST ENVIRONMENTALISM AND SUSTAINABLE DEVELOPMENT

The discourse of sustainable development takes from environmentalism a strong measure of technocentrism (O'Riordan 1981; O'Riordan and Turner 1983). Technocentrist environmentalism involves technocratic management, regulation and 'rational utilization' of the environment. Such approaches would include ideas about the role of the environment in development planning (e.g. the importance of things such as Environmental Impact Assessment), and the ways in which economic development can be achieved without undue environmental costs. Technocentrist approaches are inherently reformist, working towards improved and more 'rational' planning. They draw on many of the principles of the utilitarian 'wise use' philosophies of conservation in America in the first decades of this century (Hays 1959), most notably perhaps in the work of the first US Conservator of Forests, Gifford Pinchot – although many of these ideas are found earlier in the work of George Perkins Marsh. In *Conservation and the Gospel of Efficiency*, Samuel Hays writes:

> Its essence was rational planning to promote efficient development and use of natural resources . . . . Conservationists envisaged, even if they did not realise their aims, a political system guided by the ideal of efficiency and dominated by the technicians who could best determine how to achieve it.
>
> (Hays 1959: 2)

In both *World Conservation Strategy* (IUCN 1980) and *Our Common Future* (Brundtland 1987), sustainable development is identified as a realistic means of maximizing human benefit without significant environmental costs, and without threatening economic growth. Although 'zero growth' was a major theme of environmentalism in the 1970s (Daly 1977), it has been largely ignored in debates about sustainable development, despite the importance in those debates of work by economists (see Goodland and Ledec 1984; Pearce *et al.* 1988; Turner 1988a). As Ramphal (1990: 11) comments 'as long as large-scale poverty and rapid population growth remain, "no growth" is no solution.' The Chairman of the then World Wildlife Fund, Sir Peter Scott, said that the WCS was intended to show 'how conservation can contribute to the development objectives of governments, industry and commerce, organised labour and the professions' (Allen 1980: 7). *Our Common Future* was a successor to the Brandt Reports (Brandt 1980, 1983), and picks up the same arguments about mutuality, multilateralism, and 'environmentally sustainable' growth within a Keynesian-managed world economy. This is entirely consistent with the existing economic paradigms of the industrialized North, what Lewis (1992) would call 'Promethean environmentalism.'

Sustainable development thus shares the dominant industrialist and modernist ideology of what Aseniero (1985) calls developmentalism, defined by Friberg and Hettne (1985: 231) as 'a common corporate industrial culture based on the values of competitive individualism, rationality, growth, efficiency, specialization, centralization and big scale.' As Aseniero comments 'the processes of modernization, economic growth and nation-state building, posited now as the development goals of the underdeveloped countries, are historically the constitutive processes of the modern world-system' (Aseniero 1985: 51). Within mainstream sustainable development discourse, therefore, there are no ideological conflicts with the dominant capitalist industrializing model, only debates about methods and priorities. Thus *Our Common Future* (Brundtland 1987) focuses on the potential for fairly minor reforms of the existing economic system involving new approaches (for example rational planning of land use and ecosystem exploitation, people-orientated and 'bottom-up' development planning). The focus is on better planning techniques, on more careful use of state capital, on more careful use of economic appraisal to reduce development that causes ecological disruption. The agenda is highly reformist and technocentrist.

SCIENCE AND SUSTAINABLE DEVELOPMENT

The core of technocentrist thinking in sustainable development is a utilitarian view of science, and the application of science to 'solve' human problems. This is particularly true of the World Conservation Strategy. The WCS had three main aims: to maintain essential life-support systems, to preserve genetic diversity and to promote sustainable development of species and ecosystems (IUCN 1980). This is an agenda not only expressed but also conceived in terms of ecology. Its successor – *Caring for the Earth* – presents nine 'principles for sustainable development' in a less obviously scientific way, but many of the same principles are there. They include the need to 'improve the quality of human life,' 'conserve the Earth's vitality and diversity,' 'minimise the depletion of non-renewable resources,' and 'keep within the Earth's carrying capacity' (IUCN 1991: 9–10).

Science had earlier played a central role in the articulation of colonial discourses. As J. Mackenzie (1990: 7–8) comments: 'scientific ideas were deeply embedded in imperial rule.' Particularly in Africa, there was an influential belief 'that science was capable of unlocking redemptive and regenerative forces on a vast scale' (Mackenzie 1990: 6). Science was a major element in the new engagement with rural Africa that Low and Lonsdale (1976) label the 'second colonial occupation.' Ecologists in particular seem to be easily drawn into imperial and domestic social and economic planning. The notion of 'orderly planning' in resource development stems from the central conceptual role of ecology. Tobey (1981: 207), referring to the American ecologist F.E. Clements, writes that 'the mid-1930s were years of visionary social schemes, and Clements rose to the occasion.' Clements, the 'chief theoretician of grassland ecology,' was able to move easily into the role and ideology of social planning (Tobey 1981: 28).

The situation was similar in Africa at the end of the Second World War. In his 1938 report *Science in Africa* Barton Worthington stressed the importance of scientists to colonial government. Interrelations between the sciences had 'important practical applications' (Worthington 1938: 3). In the 1940s there were many people within African colonial governments who saw in science (particularly ecology) a new basis for development planning. The journal *Nature* noted the need for 'cooperation of the economist, the agriculturist, the medical man, the educationist and administrator' (*Nature* 1948). In 1943, A.T. Culwick argued in the context of Tanganyika that development was 'primarily a scientific problem.' He suggested that 'post-war life must be planned as an oecological whole,' and that 'haphazard excursions along different lines of advance must give way to a broadly conceived scheme based on the scientific utilisation of our resources, human and material' (Culwick 1943: 5). This in turn required changes in the organization of government such that the administration, hitherto looked upon as all-important, became 'merely the mechanism for putting the big scientific plan into action' (Culwick 1943: 5). Ecological ideas were given an important position in the First Ugandan Development Plan (Worthington 1983).

In post-war Africa, ecology was seen to provide both technical insights and a model for development planning. During the 1940s most colonial powers established new scientific research organizations: the British Colonial Research Council, the French Office de Recherche Scientifique et Technique d'Outre Mer (ORSTOM), and the Belgian Institut pour la Recherche Scientifique en Afrique Central (IRSAC) (Worthington 1983). Unfortunately, of course, colonial science found African reality rather hard to handle. Timothy Weiskel (1988: 166) comments that in the Ivory Coast 'until the creation of scientific experiment stations from the 1920s onwards European contributions to innovation in African agriculture did not match those of Africans themselves.' In many cases this continued to be true. Richards (1985, 1986), for example, describes the failures of successive attempts by colonial governments to understand and 'improve' African agriculture in wetland environments in Sierra Leone. Despite perceptions in the 1930s and before of the careful adaptation of African agricultural ecology to constraints of environment or labour shortage (for example, Jones 1936; Stamp 1938); dismissive views of both agriculture and pastoralism were widely held, and colonial scientific research substantially failed to identify or address issues relevant to African agricultural production. Colonial technicians perceived serious problems of soil erosion in East and southern Africa in the 1930s, and enforced major and highly unpopular terracing campaigns on reluctant African farmers. In retrospect it is clear that many factors owing little to science influenced their perception (Anderson 1984).

Nonetheless, the central role of science in ideas about environmental

91

management and development in the Third World persisted. Their centrality in the mainstream sustainable development documents is not surprising, for these grew out of international scientific cooperation in the 1960s. The Conference on the Human Environment in Stockholm in 1972 drew institutionally and conceptually on the international scientific collaboration in the International Biological Programme (established in 1964), the Scientific Committee on Problems of the Environment (SCOPE, established 1969) and the Man and the Biosphere Programme (MAB, established 1968). UNESCO was influential in many of these initiatives, as was the IUCN (International Union for Conservation of Nature and Natural Resources). The United Nations Environment Programme (UNEP) was established as a result of the Stockholm Conference, and with IUCN and World Wildlife Fund published the World Conservation Strategy in 1980 (IUCN 1980; McCormick 1989; Adams 1990).

CONSERVATION AND SUSTAINABLE DEVELOPMENT

The influence of technocentric colonial science and 1970s environmentalism on sustainable development discourse is matched by that of preservationist ideology. Attitudes to nature in Europe and North America were imposed on the biota, landscapes, and people of the colonial periphery in the twentieth century. They have been a powerful emotive, ideological and practical source of ideas for mainstream sustainable development. J. Mackenzie (1990: 3) argues that conservation practices were 'deeply embedded in the ideologies of imperial rule, and demonstrate its follies and fallacies, limitations and liabilities, as well as its creative concerns.' Accounts of the development of ideas about nature in Europe (Thomas 1983), and about the later institutional development of nature conservation in Europe and North America (Sheail 1976; Nash 1983; Runte 1987; Worster 1985) have not tended to stress such links. However, Grove (1987, 1990a, 1990b, 1990c) has described the origins of conservation ideas in the colonial empires of Britain and France (particularly in Cape Colony, St Helena and the West Indies) in the seventeenth century. He argues that such thinking moved as much from periphery to metropole as the other way round. Accounts of the international conservation movement of the twentieth century emphasize the continuing importance of the international dimension to conservation ideologies (Boardman 1981; Mackenzie 1989; McCormick 1989; Adams 1990).

One avenue of concern for wild nature in the industrial world at the turn of the nineteenth century was the interest of the elite in sport hunting. Mackenzie (1987) describes the changing economic role of big game hunting in colonial Africa, and the projection onto Africa of English rituals of hunting, together with associated ideologies of nature and forms of control of resources and people. Put crudely, white men hunted, Africans poached. Nash (1983) links that passion for hunting African 'big game' on the part of

Europeans and Americans to the closing of the American frontier in the nineteenth century. Wealthy Europeans came to the American West to hunt in the nineteenth century, and then, as that frontier closed, they (and those from the New World, most notably Theodore Roosevelt) transferred to the new global frontier in Africa. Africa became 'the new Mecca for nature tourists like Roosevelt who were wealthy enough to import from abroad what had become scarce at home' (Nash 1983: 343).

One of the ultimate fruits of the wealth generated by industrial capitalism was the opportunity to kill animals in areas remote from the industrial heartland. In time, as depredations on big game became more obvious, hunters turned to preservation. Organizations like the Society for the Preservation of the Wild Fauna of the Empire (the 'penitent butchers,' as they became nicknamed) and the American Boone and Crocket Club were formed. It was from these Northern roots in the concerns of wealthy men of leisure, that in due course the international conservation movement grew in the period between the World Wars (Fitter 1978; Boardman 1981; McCormick 1989; Adams 1990). The leading proponents of the WCS – the International Union for the Conservation of Nature (IUCN) – first saw the light of day (under a different name) in 1948.

By the 1970s, international conservation was as much part of the growing global hegemony of western culture as capitalist economic development and industrialization were part of the project of 'developmentalism.' Dasmann (1979) rightly identified the cultural hegemony involved in the emergence of global environmental concern. He argued that the failure to achieve nature protection across the globe lay in part 'in our attempt to transplant North American ideas to parts of the world where they are not appropriate' (Dasmann 1979: 39). Concern for the environment of the Third World was ideologically and practically the product of the evolving capitalist world economy.

RADICAL/'GREEN' IDEOLOGIES IN SUSTAINABLE DEVELOPMENT

The success of mainstream sustainable development is due very largely to the compatibility of the technocratic, managerial, capitalist and modernist ideology it draws from northern environmentalism with Western economic development theory and development practice. The picture, however, is not as simple as this might suggest, since sustainable development also carries echoes of very different ideas that come from other parts of environmentalism, particularly 'radical environmentalism' (Lewis 1992). Of particular importance here are the traditions of social anarchism or anarchocommunism, and utopian socialism (Kropotkin 1972, 1974; Galois 1976; Breitbart 1981; Pepper 1984; Adams 1990).

There is a strong ecosocialist stream within radical thought, involving

93

redefinition of needs, redistribution of resources, reassessment of the industrial mode of production, replacement of private ownership in favour of social justice, and a search for new forms of social order which 'eliminate alienation, state control, and centralisation' (Pepper 1984: 197). Ideas of this kind are found in the work of Roszak and in the social ecology of Murray Bookchin (1979). Roszak (1979: 317) argues that as long as Western society remains locked into 'the orthodox urban-industrial vision of human purpose,' there is no hope that poverty, and the injustice it brings with it, can be 'more than temporarily and partially mitigated for a fortunate nation here, a privileged class there.' Clearly he has little faith in the notion of a world enjoying economically and environmentally sustainable growth. Taking a similar line, Bookchin (1979: 22) believes that there can be no solution to environmental problems without political change: 'ecological dislocation cannot be resolved within the existing social framework.' Bookchin's 'ecological anarchism' is essentially anti-industrial, anti-bureaucratic and anti-state. It demands 'revolutionary opposition' to the norms of society.

Friberg and Hettne (1985) pick up some of these themes. They explore a 'green' development paradigm that opposes 'developmentalism' and the institutionalization of the 'modern complex' – 'the bureaucracy, the industrial system, the urban system, the market system, the techno-scientific system, and the military-industrial complex' (Friberg and Hettne 1985: 207). A green strategy of 'demodernisation' would involve gradual withdrawal from the modern capitalist world-economy and the launch of a 'new, nonmodern, non-capitalist development project.' This new project would be based on 'the "progressive" elements of pre-capitalist social orders and later innovations' (Friberg and Hettne 1985: 235), the exploitative and dehumanizing element in some small-scale pre-capitalist orders notwithstanding. 'Green' principles of 'endogenous development' are (a) that the social unit of development should be a culturally defined community, whose development should be rooted in its values and institutions; (b) self-reliance, so that each community 'relies primarily on its own strength and resources'; (c) social justice; and (d) 'ecological balance,' implying an awareness of local ecosystem potential and local and global limits.

Friberg and Hettne (1985) argue that these ideas present a 'green' challenge to the modern project which aims at control, expansion, growth and efficiency and is legitimized by evolutionist thinking. The alternative 'green project' draws its strength from an alliance of three groups. The first are 'traditionalists' who wish to resist capitalist penetration in the form of state-building, commercialization and industrialization. These groups are located mostly in the global periphery and include 'non-Western civilizations and religions, old nations and tribes, local communities, kinship groups, peasants and independent producers, informal economies, feminist culture etc.' (Friberg and Hettne 1985: 235). The second group is 'marginalised people,' including the unemployed, the mentally ill, handicapped people and people in de-humanizing jobs who have lost 'a meaningful function in the mega-machine' through pressures for increased productivity, rationalization and automation (Friberg and Hettne 1985: 264). It is important to note the increasing recognition of the ways in which Third World 'counterpoint movements' of the marginalized involve protest (see Crummey 1986; Grove 1990c). The third group consists of the 'post-materialists' who dominate western environmentalism (Cotgrove 1982): 'young, well-educated and committed to post-materialist values' (Friberg and Hettne 1985: 236). The post-materialists focus on the desirability of non-hierachical and decentralized structures of decision-making, and reject Fordist industrialization and consumerism. Ideas about intra- and inter-generational equity within sustainable development discourse echo some of these radical ideas. Although they remain divergent elements within the sustainable development debate, they offer an important challenge to the conventional mainstream.

ENVIRONMENTALIST CHALLENGES TO SUSTAINABLE DEVELOPMENT

Radical critiques of political economy and the development process fit into the normal arena of debate within academic development theory. There are, however, other strands within environmentalism that offer a different and perhaps more extensive challenge to mainstream sustainable development. They certainly provide a very marked contrast to the conventional discourses of development theory. These strands are far from completely coherent, but they are significant enough to attract rebuttal (Lewis 1992). I will consider three.

The first is ecocentrism or biocentrism (O'Riordan 1981). Ecocentrism is romantic and transcendentalist in tradition, embracing ideas of bioethics and utopianism. Biocentric and ecocentric ideas within environmentalism include 'Deep Ecology' (Devall and Sessions 1985), notably the 'ecosophy' of Arne Naess. Deep Ecology stresses the transcendental elements in relations with nature. Graber (1976: 111) suggests that what she calls 'the wilderness ethic' is 'strongly religious in character.' Her 'wilderness purists' draw on Thoreau, Muir and Aldo Leopold for inspiration and group definition. Deep Ecologists also reference themselves by the writings of such people and their sense of moral order in nature, and of the continuity between humans and other organisms (and indeed inanimate nature). Both Deep Ecology and 'bio-regionalism' (Plant and Plant 1990) call for a new relation with nature that challenges established utilitarian ideas as well as reformist and managerialist 'conservation.'

Deep Ecology is but one strand of a complex shift in ideas about nature. There has also been increasing awareness of and interest in ideas about and relations with nature of 'indigenous people.' Sessions (1985: 241) argues that 'many individuals and societies throughout history have developed an intuitive mystical sense of interpenetration with the landscape and an abiding and all-pervading "sense of place".' Katz and Kirby (1991: 262) speak of 'constructs of the Native American lifeworld' – a system in which 'there exist no dualities between humans and nature, or necessarily between animate and inanimate.' This has, of course, again chiefly been a shift in consciousness among Northern environmentalists, with all that implies in terms of their education, wealth and employment (see Cotgrove 1982), but it has significance for sustainable development because of the continuing influence of Northern environmentalism on ideas about nature (and its 'development') in the Third World.

The biocentrism of Deep Ecology challenges anthropocentric thinking and, in doing so, potentially undermines the moral basis of most development action. Thus there is a new alliance between the neo-Malthusian science-based critique of population growth (with its apparently sound concepts such as 'carrying capacity') that sees famines as somehow natural, and biocentric ideologies which identify people as organisms with no special rights, and that see intervention to sustain human lives at the expense of other organisms and inanimate objects as unacceptable. This kind of 'green' thinking is perhaps what Amin (1985: 281) has in mind when he speaks of green ideas as 'a form of religious fundamentalism.'

The 'neo-Malthusian' arguments of 1970s environmentalists such as Ehrlich and Ehrlich (1970), Ehrlich (1972), and Meadows et al. (1972) have been widely criticized (Commoner 1972; Enzensberger 1974; Harvey 1974; Bookchin 1979; Pepper 1984). However, they are still present in environmentalist thinking about development, and have been given prominence in the views of some First World environmental groups such as 'Earth First!' Pepper (1984) argues that environmentalism offers an essentially determinist analysis based on the principle of unchanging limits on human action, and a pessimistic view of the potential impact of social reform. This generates, he suggests, the deeply conservative ideology and reactionary and repressive politics of 'ecofascism.' This kind of thinking Lewis (1992) describes as 'harsh deep ecology.'

Green Feminism offers a second divergent theme within environmentalism which has challenging implications for development theory. Feminist analysis focuses on the role of patriarchy in capitalist accumulation, and its dual subjugation of both women and nature (Mies 1986; Shiva 1988). Shiva describes the struggles of women for the 'protection of nature,' particularly in India, describing women's roles in the 'chipko' movement, in social forestry, and issues of the commons. She argues that a gender-based ideology of patriarchy underlies ecological destruction, and she demands the recovery of 'the feminist principle in nature,' and of the view of the earth as sustainer and provider.

A third significant challenge to mainstream sustainable development comes from shifts within its base science, ecology. Donald Worster (1990a) draws attention to the implications of changes in ecological science for what he calls 'reform environmentalism' and its critique of human impacts on nature. In the 1960s and 1970s this suggested that 'the great coming struggle would be between what was left of pristine nature, delicately balanced in Odum's beautifully rational ecosystems, and a human race bent on mindless, greedy destruction' (Worster 1990a: 7). The ecology developed in the 1960s and 1970s focused on the ecosystem; it took words from economics to talk about producers and consumers; it analysed ecological change in terms of the common currency of energetics. Worster (1990b: 16) comments that 'nature has become very explicitly an economy, and one that looks a great deal to me like a modern consumer society, with producers and consumers all organized, circulating the commodities of the shopping malls.'

New ideas, in what he calls the 'ecology of chaos,' challenge former views about the way nature works, and as a result challenge the socio-political positions based upon them. New insights on past environmental and vegetation change stress the lack of stability and equilibrium. There is a renewed focus on change, disorder and instability, on the dynamics of ecosystems. This could lead to a new and less hierarchical view of life, or it could 'increase our alienation from the world, or withdrawal into post-modernist doubt and self-consciousness.'

Cosgrove (1990: 350) suggests that the late twentieth century is characterized by 'its fragmentation and surface-like appearance, that loss of a sense of order and direction which has followed the death of Modernism: its apparent irrationality.' However, natural rhythms, such as the seasons in the New England woods (Kohak 1984), evoke responses such that 'despite the dire warnings of both the death of nature and the alienation of human spirit we remain able to embrace and create meaning in our lives out of the fragmentation and to discern the possibilities of a moral order' (Cosgrove 1990: 357). The 'wilderness purists' described by Graber (1976: 57) recognize nature 'as a rightful object of religious contemplation and ethical behaviour, with wilderness itself as a symbol of the sacred power manifest in all creation.' Matless (1991) discusses the post-modern theme of 'the innate human need' to affiliate with other forms of life – contrasting this with the modernist project of the subjection of nature. Kohak (1984: 5) writes that 'the notion of a fundamental discontinuity between humans and their natural world . . . is primordially radically counter-intuitive.' These ideas draw on conceptions of nature that are fundamentally at odds with the principles of 'foundationalism,' linear reasoning and the scientific worldview of modernism (Cosgrove 1990).

In different ways, all three sets of ideas challenge the philosophical basis of sustainable development. In a sense they are re-creating the conflict between the rational and scientific utilitarianism of the conservation movement in the USA and the transcendental and romantic ideas of nature promoted by organizations like the Sierra Club earlier this century. The construction of a dam for water supply in Hetch Hetchy valley in the Yosemite National Park focused this debate (Runte 1987). The Yosemite Act was passed in 1864, although establishment of the National Park took a lot longer. The Hetch Hetchy Valley was granted to San Francisco in October 1913. It was rational to use the valley for this purpose, but, to others, it was simply a sacrilege. This debate, about the proper use of the remaining wilderness of the United States, mirrors the tensions within contemporary views of environment and development remarkably closely. The dam was built and 'development' and 'rational use' won the day, but within environmentalism the same debate runs on. Its existence, and the strong emotions attached to it, offers an interesting challenge to the future unity of sustainable development. The philosophical bases of environmentalist social movements in the so-called 'new environmentalism' of the 1970s (Cotgrove 1982) were complex, eclectic and confused. Sachs (1990) argues that environmentalism, or the 'ecology movement' as he calls it, combines modernism and anti-modernism, a call for a better science with a critique of the rationality of science. Sustainable development is the uncertain inheritor of this confusion.

CONCLUSION

Within the breadth of thinking about sustainability, particularly within radical theories, there is clearly some common ground. At the same time, there are currents within the general theme that are strongly divergent, and reflect radically different political, and sometimes, moral and philosophical positions. D. Young (1990: x) distinguishes between environmentalism as 'a reformist philosophy which maintains an external distinction between the human species on the one hand and "nature" on the other,' and a variety of more radical ideas which perceive humanity as 'inextricably linked with the rest of the biosphere.'

Despite the enthusiasm, hype and commitment, current visions of sustainable development are messy and 'politically treacherous' (O'Riordan 1988: 30). There are very real tensions within competing ideologies of technocentric and ecocentric environmentalism (O'Riordan 1988; Turner 1988b). Turner (1988b) suggests that a coalition may be possible between 'accommodating technocentrism' (a conservationist position of sustainable growth) and 'communalist ecocentrism' (a preservationist position emphasizing decentralization and macroenvironmental constraints on growth). Such a coalition would exclude the more radically opposing worldviews of 'cornucopian technocentrism' and 'deep ecology ecocentrism.' Sustainable development contains elements of both radicalism and reformism. As a result, ships of very varied allegiance are sailing cautiously under the same flag, and the destination is rarely debated. The resulting uncertainty is of considerable political and practical importance, as well as being of great academic interest.

Attempts to implement 'sustainable development' projects are prey to the confusions inherent in the term. Those which promise some success do so on the basis of simple agrarian populism, and a high degree of dedicated leadership by expatriate staff. They may do nothing to tackle problems that exist at a larger scale, or that are important over a longer period, or that involve complex or difficult questions of wealth and power (Adams 1990; Adams and Thomas 1993). The ideology of sustainable development provides an alluring route for policy development, promising escape from the environmentally destructive record of dominant development paradigms. However, the lack of a single coherent ideology, and the lack of understanding of the political economy of the development process and the structures of the world economy, expose those seeking to implement the muddy concept of sustainable development to significant risk of failure, and those they involve in the rural Third World to yet more of the familiar risks of engagement with the development process.

At present sustainable development is widely used as 'green' packaging by international businesses and aid donors, and as a slogan with bargaining power by Northern environmental organizations. However, its rhetorical role cannot be indefinite. In time it will, like the Emperor's new clothes, cease to cover up nakedness. It may then be seen as simply one more transient label on the trickle of capital flows of aid donors from the industrialized North, and something that allowed 'business as usual' by international capital.

Alternatively, the environmentalist concerns which have brought sustainable development to prominence in the international arena through the 'greening' of domestic politics in the North, may continue to be influential. Northern concern for Southern environments may become a new stick with which ailing Third World economics can be beaten into shapes defined from the North. Environmental aid conditionality, debt swaps and the global networking of nongovernmental environmental organizations reflect the growing power of Northern environmentalism (and the ideologies behind it) as a factor in development planning in the South. The battle over the contested meaning of sustainable development could have considerable significance for development theory, for the practice of development planning and for the Third World poor.

5

SELECTIVE SILENCE

A feminist encounter with environmental discourse in colonial Africa

Fiona Mackenzie

In our land there is a wild fig tree which grafts itself on to the other trees, such as the palms ... and having grafted itself on to the higher part of the tree, it then throws the fruit down and swamps the whole tree, so it remains having completely swamped and covered up the palm tree because it got fatter and stronger. That is a parable of what Government has done to us.

(Waiganju wa Ndotana, November 1932, Kiambu, before the Kenya Land Commission: 153)

INTRODUCTION

In this chapter, I will draw on notions of 'difference' – primarily in terms of race, African *vis-à-vis* European, and gender – to examine the politicization of an environmental discourse in colonial Africa. In broad terms, the theoretical perspective is that of political economy of the environment. Following such writers as Blaikie (1989) and Redclift (1984, 1987), the environment is defined here as socially constructed, and investigation proceeds to untangle what Blaikie (1989: 23) refers to as 'the dialectic between social and environmental change.' But, in the context of colonial Africa, the analysis is both broadened in that race, the construction of what is African as 'other' (Mudimbe 1988), assumes analytical significance, and taken 'downwards' (Blaikie 1989: 26) to a discussion of the 'deeply contested terrain' (Watts 1989: 12) of the household as its members – differentiated by gender – renegotiate relations of production with a colonial political economy. This approach, of a feminist political economy of the environment, is distinguished from 'women and environments' analysis (Dankelman and Davidson 1988; Sondheimer 1991) as it locates gender, rather than women, at the analytical centre of the relationship between economy and environment. The essentialist category 'woman' is replaced by analysis which differentiates *among* women as *among* men in terms of class, race, age, marital status.

My purpose is to examine how and why an environmental discourse in eastern and southern Africa was constructed through the politicization of an ethic of conservationism and the politics of soil erosion during the early decades of the twentieth century. I will argue that the environment, as does history in Mudimbe's (1988: 188) exegesis of *The Invention of Africa*, became 'both a discourse of knowledge and a discourse of power.' In essence, in the hands of those insecure in their claim to African territory, legitimation of rights to land lay in privileging Western-based knowledge, of presenting it as 'the virtuous face of colonialism' (Beinart 1989: 159). In the process, Africans were constructed as 'unscientific exploiters' of the resource base and their 'particularist knowledges of the environment' (Richards 1983) were silenced. In Foucault's (1980b: 81–2) terms, African knowledge was 'subjugated' both by 'functionalist and systematising theory' – the epistemological ethno-centrism of Western scholarship (Mudimbe 1988: 15) – and through its disqualification as 'unscientific.'

The construction of this discourse of conservationism was, as Beinart (1984: 82) suggests, more than a momentary response to the threat of soil erosion in eastern and southern Africa in the 1930s and 1940s. Nor was it a mere justification for restructuring peasant communities. In creating an official ideology wherein state intervention and centralized planning became the means for operationalizing 'development,' it laid the ground for much more recent action. In the process, the peasant farmer (ungendered, or assumed to be male, and economically undifferentiated except where targeted as 'progressive') became the 'other' in the face of exogeneous agricultural 'expertise.'

In the next section of the chapter, I draw on Africanist scholarship of eastern and southern Africa to illustrate how a discourse of 'discriminatory environmentalism' (Grove 1989: 184) provided the means through which the massive alienation of African land was legitimated and through which African, and particularly women's (as, in many instances, custodians of the resource base), ecologically precise knowledge was silenced. In part three, a case study of colonial Kenya explores with greater specificity the nature of the intersecting arguments over land and soil, using as primary evidence oral and written submissions to the Kenya Land Commission, 1932–34 (UK 1934). In the final section, the focus is on the gender-specificity of ecological knowledge and I illustrate, briefly, with reference to an area of the case study in Kenya, how a silencing of this knowledge was linked to changed relations of power in rural Kenya, how women renegotiated their role as farmers *vis-à-vis* men and the colonial state.

THE LANGUAGE OF LEGITIMATION

Politicization of soil erosion in eastern and southern Africa on the part of colonial states and settlers occurred at a particular historical moment. Beinart

(1989: 146) argues that it had its roots in the new social Darwinist 'science' of mid-nineteenth-century Europe, and in imperial imperative, the legitimation of control of African land. Social Darwinism provided the 'righteous language' (Grove 1989: 187) by which African territory was allocated to white settlers. Grove (1989) traces the evolution of this discourse through the actions of and state response to Scots missionaries in southern Africa between 1820 and 1900.

In contrast to earlier writings which judged indigenous land use practices to be more consonant with the semi-arid conditions of the Cape than those of settler farmers, Robert Moffat, a missionary entering the colony during the drought of 1820–23, insisted that culpability for environmental failure rested with African agriculturalists (Grove 1989: 165). Moffat dismissed African ecological knowledge as irrelevant, and drawing selectively on the new paradigms of the natural sciences in Europe, linked 'heathenness' to the 'evil' of environmental catastrophe (Grove 1989: 166). In the equation between people and their environments, drought became 'the wages of environmental sin or the sins of moral disorder' (Grove 1989: 180). Moral and environmental order or disorder, to use Grove's phrase, became fused at the level of ideology. In what Mudimbe (1988: 47) sees as the non-contradictory roles of 'envoy of God' and agent of a political empire, the missionary prepared the means to legitimate European presence. A 'language of moral disapprobation' was created to condemn African land use and justify a policy of European control (Grove 1989: 183). The environmentally destructive activities of European farmers were virtually ignored.

While there was evidence of a 'counter-colonial ecology' (Richards 1983: 57) originating from within the colonial service, as the struggle for land grew, the language of conservationism became an even more insistent component of settler discourse, a 'convenient ideology' (Ranger 1989: 227) through which to legitimate claims. The timing varied among the colonies of east and southern Africa, but from the 1920s, settlers increasingly voiced their fear that policies of land alienation and racial segregation would be undercut by the inability of Africans to survive on the 'reserves' to which they were restricted (Phimister 1986; Mlia 1987; Drinkwater 1989; Grove 1989). Settler concern reached a peak during the Depression when, for example in Kenya, the state viewed African farmers, and the African taxpayer, as providing 'the only reliable ballast left on board as the colonial state lurched its way through the storm' (Anderson and Throup 1989: 15). For the administrations, concern grew with the internationalization of the 'Dust Bowl' disaster of the United States (Anderson 1984: 327; Stocking 1985: 153; Mlia 1987: 4).

For the European, political expediency in the promotion of a policy of land alienation demanded both the creation of a conceptualization of African agriculture as 'backward' and 'inefficient' (Drinkwater 1989: 293), and the privileging of environmental knowledge based on Western experience. A conservationalist ethic drawn from these roots legitimated an intensification of administrative intervention from the 1930s onwards in the Transkei (Beinart 1984, 1989), the Cape (Grove 1987, 1989), Nyasaland (Mandala 1983; Mlia 1987), Southern Rhodesia (Phimister 1986; Drinkwater 1989; Ranger 1989), Tanganyika (Berry and Townshend 1973), and Kenya (Anderson 1984; Anderson and Throup 1989). Technological solutions became the means by which Africans would be 'shoehorned' (Phimister 1986: 271) into reserves and the discursive means whereby colonial officials could, through a process of 'mental gymnastics' (Drinkwater 1989: 298), attempt to reconcile racially skewed land allocation with what was going on in the reserves.

Choice of technology was between imported or indigenous methods, between conservation (engineering) works or biological/agronomic methods, and between run-off reducing/infiltration increasing or rainsplash reducing techniques (Blaikie 1987: 10). In each case, the first option was taken. In part, such choice reflected the pervasive influence of the United States' Department of Agriculture's Soil Conservation Service, a department created to deal with the 'Dust Bowl,' under the direction of Hugh Bennett (Anderson 1984: 326; Beinart 1984: 68; Stocking 1985: 153). An engineering approach to conservation was premised on the notion that 'global problems' demanded universal solutions and necessitated state intervention. At a deeper level, a decision to favour imported technological solutions rather than locally-specific integrated approaches and use management reflected the political agenda of the administration. This choice was viewed as 'politically neutral' by policy makers as it obviated the need to reconsider whether Africans should be allowed to grow high-value crops such as coffee, tea or tobacco, and thus antagonize settler farmers (Phimister 1986: 267).

In a very real sense, it was, as Beinart (1984: 76) explains, 'social engineering.' Technological solutions were imposed without regard to ecological specificity or to African farming systems. In some cases, faults in the technology caused failure, as Showers (1989) illustrates with respect to the construction of terraces and run-off channels in Basutoland between 1936 and 1955. Data from the Transkei (Beinart 1984; McAllister 1989) and Ciskei (de Wet 1989) reinforce this explanation. More recent research conducted through the Soil and Water Conservation and Land Utilization Unit of the Southern African Development Co-ordination Conference (SADCC) demonstrates the inefficacy of divorcing mechanical conservation measures from agronomic methods (Mlia 1987; also Osuji 1989).

In other cases, failure of conservation schemes lay in the political climate created through forced labour in their implementation. Resistance was widespread. At times it was overt, as evidenced by the Sofasonke resistance movement in the 1940s in Matobo National Park, Zimbabwe (Ranger 1989: 243). At times it was gender-specific, as occurred in Murang'a District, Kenya, in 1948 when women 'revolted' against labour coercion in terrace construction (see below) (Mackenzie 1991: 248). The so-called 'Revolt of the

Women' of 1948 (Mackenzie 1991: 248), where over 2,000 women descended on the district offices 'brandishing sticks and shouting Amazonian war cries' (Kenya 1948: 2) marked the culmination of this resistance. Much of the time, however, resistance was less overt, taking what Scott (1985) refers to as an 'everyday' form: 'footdragging, dissimulation, desertion, false compliance, pilfering, feigned ignorance, slander, arson, sabotage' or a 'refusal to under-stand' (Scott 1985: xvi, 133). McAllister's (1989: 347) research in the Shixini Administrative Area in the Transkei illustrates how peasant conservatism becomes an ideology of resistance based on networks of social organization which link kinship, neighbourhood and territory. Wilson (1989) shows how Shona in southern Zimbabwe drew on an 'environmental religion' to retain local control in the face of colonial administrative intervention. It is in this context that women's hidden and overt resistance punctuates the silence of the written historical record, the dominant environmental discourse of the period, and calls into question its epistemological assumptions.

Widespread resistance grew with an intensification of administrative control, concomitant with a restructuring of relations of power and of knowledge in rural east and southern Africa. A silencing of local knowledges was an integral component in the creation of an environmental discourse to legitimate the alienation of African land for European settlements. To illustrate the gender-specificity of this silencing, I refer to the Matengo pit system of Tanganyika, unique in that within the reports of agricultural officials of the period, the gender specificity of this form of soil conservation is evident.

The Matengo pit system is made up of a grid of three square metres of hollows and ridges. Grass is first cut and laid out in the form of a grid. The centres of the squares are dug out in the shape of a basin and the contents are piled on the grass lines. Crops are planted on the ridges, run-off collecting in the pits. In preparation for the following season, plant residues are placed in the pits where silt also accumulates (Stenhouse 1944). Its effectiveness in preventing soil erosion and in promoting sustainable agricultural production was noted by agricultural officers Pike (1938) and Stenhouse (1944). Stenhouse, writes, for example, that in Matengo county, the traveller cannot help noticing:

> that even the steepest hill slopes are cultivated. The hillside culti-vations are of very striking appearance, showing an orderly layout, with straight-cut edges, and the surface of the fields is curiously pitted. The impression gathers force that this cannot be native cultivation. But it is.
> (Stenhouse 1944: 22)

Stenhouse goes on to identify the forms of crop rotation, fallowing and composting that allow continuous cropping. With respect to soil con-servation, he writes:

Hillsides, so steep that it is impossible to ascend without using one's hands, are under cultivation and there is no erosion. No dangerous accumulation of water is possible, as any overflow from one pit is trapped by the next. Even heavy downpours are fully trapped, and the water gradually sinks into the subsoil so that even the smaller streams starting high up on the hills do not dry up in the dry season.

(Stenhouse 1944: 23)

Although men might cut the grass before digging the pits, women alone were responsible for pit construction and cultivation (Pike 1938; Stenhouse 1944). According to Pike (1938: 80–1), '[e]conomic crops are being more and more planted in this area and these are generally the concern of the man alone so this is causing a decay of the system.' In a statement ignoring the broader political economy within which a gender-based struggle over land use and control takes place, Pike states that men dislike the work involved in pit construction, so the system was used only by women producing food crops. Thus, to use Dianne Rocheleau's (1991: 161) language, are 'the boundaries of gendered knowledge' altered, and renegotiated, in the context of changing relations of power.

With few exceptions, and these appeared in Tanzania rather than elsewhere in east and southern Africa, local, frequently gender-based knowledge systems, defined by their ecological specificity and integration of biological with technical measures in ensuring environmental sustainability, were ignored in attempts to resolve what the discourse created by settler and state termed an environmental crisis. In the interest of political expediency – namely justification for the allocation of agricultural land to settler farmers – an environmental discourse, which involved the introduction of universal solutions to the resolution of widely diverse problems, and silenced African knowledges, became part and parcel of a reconstruction of relations of power – of race and of gender – in colonial Africa.

THE KENYA LAND COMMISSION 1932–34: LAND AND ENVIRONMENT

The Kenya Land Commission (KLC), chaired by Morris Carter, previously Chief Justice, Uganda, was established in April 1932 only three years after the much publicized enquiry of the Committee on 'Native Land Tenure in Kikuyu Province, 1929' (Kenya 1929). In the face of growing African unrest, particularly on the part of Kikuyu, to the lack of resolution of their claims to alienated land, and of settler fears, exaggerated by the economic crisis during the Depression years, that the legitimacy of their claims to land was open to dispute, the government established the KLC to find 'a final solution' to the growing crisis. I discuss European evidence to the KLC, from settlers, administrators and missionaries, before identifying key arguments presented by African, mainly Kikuyu, witnesses.

The settler case for maintaining the racial integrity of the White Highlands rested on the creation of a discourse which both insisted on the indispensability of settler agriculture for the colonial economy and drew a picture of environmental destruction on the part of the African farmer. One of the more measured voices heard before the KLC was that of Francis Scott, chair of the Elected Members Organization. The 'Dual Policy of Development,' he argued, demanded that settler interest should not be supported at the expense of African interest (UK Evidence 1934: 2874). But, much of Kenya was, in his view, unsuited to 'native' agriculture: the White Highlands could 'only be developed to its best advantage by the most up-to-date and scientific methods of farms, and the natives . . . [had] not yet arrived at the stage where they [could] employ those methods' (UK Evidence 1934: 2852). The 'soundness of the foundations of European settlement,' he continued, were what had staved off financial collapse in the face of steep declines of commodity prices and a chronic locust problem for the colony (UK Evidence 1934: 2874). In contrast, African destruction of the resource base had been halted only by the implementation of 'ordered government' and European example. The task was, then, to change African husbandry 'so that instead of them wanting more land to ruin they should be taught how to make better use of the land which they have got' (UK Evidence 1934: 2842). Other witnesses, with specific reference to the Kikuyu, spoke of 'soil robbery' (UK Evidence 1934: 3296), particular blame being placed on women who 'will never improve their methods' (UK Evidence 1934: 763). One spoke further of the increasing carelessness of cultivation, identifying the 'haphazard method' of planting several crops (sweet potatoes, maize and beans) on one piece of land. E.S. Grogan saw Africans as 'parasites.' 'In a mad moment of emotionalism,' he charged, 'we have endowed these [African] peoples with land worth millions' and without close administration 'disaster' was certain (UK Evidence 1934: 3047).

Administrative officers were of different views. With reference to the area of the Kikuyu, for example, those with closer (i.e. more local) administrative experience tended to express greater sympathy for local needs, both with respect to land claims and to local agricultural expertise, than those at higher levels of the bureaucracy. C.O. Oates, Agricultural Officer, Fort Hall District, in 1932, considered that 'something' could be learnt from Kikuyu agriculture, noting the benefits of intercropping maize and beans (UK Evidence 1934: 1047–8). Unlike some other agricultural officers, however, he was not of the view that the Kikuyu needed more land. W.G. Leakie, who occupied the same administrative post in 1933, considered the Kikuyu to have 'a lot of very good ideas' with respect to farming, although he considered that there was room for improvement (UK Evidence 1934: 1059). Even A. Holm, Director of Agriculture, known for his pro-settler stance, noted in his testimony that 'the natives' own primitive methods of crop cultivation' depleted the soil of nitrogen and humus far less rapidly than methods practised by the settlers

106

(UK Evidence 1934: 3050). He considered intercropping 'incomparably better' in maintaining soil fertility than European systems of monoculture.

Such views contrasted with those of other administrators. C.E. Mortimer, Commissioner for Lands, insisted that there were 'limits' to any 'moral duty' to reserve land for Africans because of their 'notorious agricultural and pastoral practices' (UK Evidence 1934: 55). 'Tribes' such as the Kikuyu, he continued, were 'reckless' in their manner of cultivation and required 'showing that their just and ample requirements [could] be met by their being restricted to much more limited areas.'

With the exception of evidence presented by Father Bernhard of the Catholic Mission, Kiambu (UK Evidence 1934: 720–4), missionary support for Kikuyu in terms of land claims and land use was frequent. The most convincing arguments about responsible African custodianship of the land were made by L.S.B. Leakey, whose family was part of the Church Missionary Society. He argued emphatically that the Kikuyu had insufficient land, that Department of Agriculture recommendations regarding crops and land management practices had a negative effect on the soil and that Africans were, in fact, superior managers of their environment. Notable by its absence in his discussion is recognition of the gender of the farmer.

Concern for the environment *per se* did not enter into the discourse created by Kikuyu in defence of their interests in the land. Only in the 1932 Memorandum of the Progressive Kikuyu Party – a moderate party encouraged to counter the influence of the more radical Kikuyu Central Association (Clough 1990: 42) – is the association between loss of grazing and forest land with definition of reserve boundaries for land use within the reserves made. The Memorandum notes that 'our cultivation does not improve because our system has been overturned by the advent of the European' (UK Evidence 1934: 96). Livestock were no longer allowed to graze outside the reserves and the government's push to increase cultivation and plant wattle, they argued, exacerbated the problem of sustainable land use (UK Evidence 1934: 97). They pointed out that 'lack of progress' in agricultural production was due not to 'stupidity, as some think, but to the smallness of [our] land.'

The African case for the return of land (and I focus here on the Kikuyu) was constructed with precision and detail in terms of a discourse of customary rights to land. Claims for Kikuyu territory were reinforced by individual *mbari* or subclan claims and were differentiated by region. In the older areas of Kikuyu settlement, Fort Hall (Murang'a) and Nyeri, claims were based on the transgenerational transmission of land through the *mbari* based on patriliny. In more recently settled Kiambu, the case rested, in the first instance, on individual purchase of land from Dorobo. At times, Kikuyu strategy involved the detailed presentation of territorial 'fact.' At others, the case was presented through a transcript of metaphor (as in the quotation at the beginning of the chapter) and deference – a thinly veiled assault on the bastions of white power (UK Evidence 1934: 191). As with the earlier Report on Kikuyu Land Tenure (Kenya 1929), African witnesses defined legitimacy in customary rights to land through generational depth of the *mbari* in Fort Hall and Nyeri, and through a case built on greater individual control in Kiambu. Male transgenerational rights to land were stressed and a multiplicity of other, non-exclusive rights under the *ng'undu* system (literally land controlled by the *mbari*) were silenced. Face to face with committees of British men whose world view of property was restricted to English common law, African men selected those components of customary practice most expedient to their ends of regaining lost land. Such strategic manipulation of 'custom' prepared the ground discursively for a dominant 'customary' discourse in Kikuyu society within which gender, as well as class, interests became increasingly differentiated (Mackenzie 1991).

GENDER, LAND AND ENVIRONMENT: MURANG'A DISTRICT

In reconstructing ecological history, in unearthing the silences imposed by the written record, the task is both to identify the local context – the political economy within with such knowledge was produced – and to investigate the gender specificity of such knowledge. The search for local or 'indigenous knowledge' is not a search for what Rocheleau (1991: 157) calls 'ethnographic artifact' or 'unconscious ecological wisdom.' Nor is it a romantic attempt to recreate an idealized past. Rather, its purpose is to tease out of a history, steeped in epistemological ethnocentrism (Mudimbe 1988: 71) and andro-centrism, the process whereby ecological knowledge was constantly created and renegotiated within a context of political rights and responsibilities – within the household, the collectivity, the nation and internationally. As Rocheleau's (1991) contemporary stories from Machakos District, Kenya, illustrate, surviving the drought of 1984 depended not only on particularistic botanical and agricultural knowledge, but also on how women, individually or in groups, mobilized political skills to access privately or publicly controlled resources – from men.

From a gendered reading of colonial texts, facilitated through oral histories collected by the author (Mackenzie 1986), it is possible to construct a case to support the view that, prior to the alienation of land in the early years of the twentieth century, a contradiction had not emerged between long-term sustainability of the resource base and agriculture production. The argument rests, first, on the security of rights to land guaranteed by the *ng'undu* system. Such rights were negotiated primarily on the basis of social adhesion either through membership in the *mbari*, subclan, the basic unit of social and political organization, or through marriage. These rights were always less than absolute (Okoth-Ogendo 1989: 7), and distinctions regarding access and

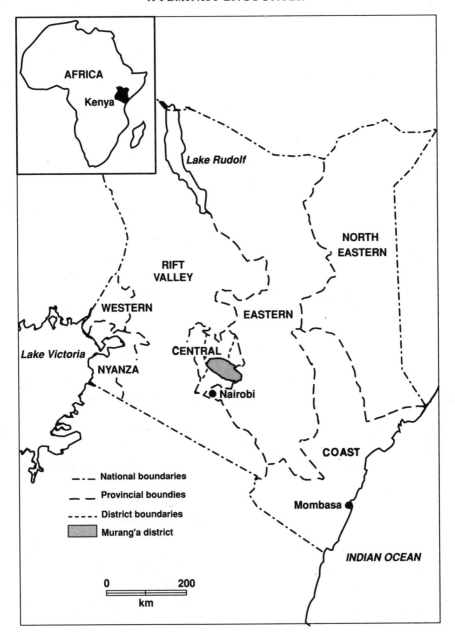

Figure 5.1 Location of Murang'a District

control – of women and men, elders and youths – led to complexity, elasticity, and interlocking interests.

Security of tenure within the *ng'undu* system was premised on the resolution of two latent tensions within Kikuyu political economy. The first concerns the tension between individual rights and corporate or *mbari* rights. The second, obscured in the historical record, centres on the relationship between women who, as wives, were producers yet non-members in the *mbari* (transgenerational rights following the male line), and men who, as husbands, were essentially non-producers but members of the *mbari*. The latent tensions provided the discursive ground for negotiating the relationships among people differentiated by age, gender and, increasingly, wealth.

Given that in Murang'a, the political economy at the turn of the century was barely touched by the demands of an incipient colonial economy, unlike Kiambu to the south (Rogers 1979: 262–3), and that the option of migration was available, it may be argued that rights to land were worked out without major social disruption. Differentiation (some individuals in some *mbari*, and some *mbari* controlled more land than others), while evident, had not reached a point where land was bought and sold. Production was primarily, but not exclusively, of use value and, with the exception of certain crops of ceremonial significance, agricultural production was in the hands of women (Kershaw 1975–76) who were guaranteed use rights in their husbands' *mbari*. In effect, a woman was 'given' land on marriage by her mother-in-law to cultivate (F. Mackenzie 1990). Women determined seasonal crop choice, the production of surplus, and within certain limits rights of disposal (Kershaw 1975–76; Mackenzie 1986). In large part, maintenance of the resource base was their responsibility. The complex division of labour between women and men implied a 'complementary, mutually dependent relationship' rather than 'hierarchical order' (Kershaw 1975–76: 179–80). As I have shown elsewhere (F. Mackenzie 1990), women were not without authority in defining 'customary' law in their interest – although the contest was undoubtedly not an equal one.

Second, the case for sustainable land use is supported by evidence of the detailed empirical knowledge of the resource base and the ecological soundness of the integrated system of land use management in place. The former is apparent in the agro-ecological specificity of language pertaining to soil and slope; the latter rests on the efficacy of primarily biological but also mechanical methods of soil conservation. Fisher's (1954) *Kikuyu Domesticity and Husbandry*, notwithstanding the problems which arise from the theoretical assumptions of her work (Kitching 1980: 121–8), provides an unparalleled source of data from which to construct an historically-grounded argument. She displays the intricacies of ecological knowledge through description of the local soil classification based on a soil's potential fertility, texture, colour and topographical location, through the diversity of crop rotations and crop choices, variations in intercropping, minimum tillage and specific soil conservation practices. For example, eleven main soil types, one of which has subcategories, range from the exceptionally fertile *iganjo* (old homestead site), *tīri mūnoru* (a 'fat' soil), a fertile soil just brought into cultivation after a period of fallow, to *tīri mūhīnjū* (a 'thin' soil), a soil either of inherently low agricultural potential or a soil deprived of nutrients through constant cropping. Crop choice was closely related to soil type, beans and bulrush and foxtail millet being grown by choice in *iganjo* or beans and bulrush millet on *tīri mūnoru*. A picture is built up, reinforced by my own research, of practices that sustained intensive agricultural production, with a system of fallowing, over a long period of time (Mackenzie 1991).

The written accounts of European travellers such as John Thomson (1885), Ludwig von Höhnel (1894) and James Macdonald (1897) attest to the productivity of the area prior to the turn of the century. This situation changed dramatically with the introduction of a colonial regime in the first years of the twentieth century. A colonial economy premised on African male migrant labour, which in Murang'a reached 41.7 per cent of the adult population by 1927 (Kitching 1980: 250), and an intensification of female labour within the Kikuyu reserve (Kitching 1980: 16), initiated growing African social differentiation in terms both of class and of gender. A growth in women's labour input into agriculture proceeded in tandem with the commoditization of production, and, as Kitching's (1980: 121–8) exegesis of Fisher's data demonstrates, the household became increasingly a site of struggle over rights to land and to the products of women's labour (Mackenzie 1991: 246). Undoubtedly, women's insecurity *vis-à-vis* the land was greatest in households which, through smallness of land area or access to land of limited agro-ecological potential, were unable to ensure household reproduction. In such circumstances, from the 1920s onwards, the sale of land was a strategy of last resort.

If Sara Berry's (1989: 44) arguments are followed with respect to a channelling of investment in 'strengthening social relations' which continue to define rights to land, rather than in directly productive activities with commercialization of agriculture, then it may be argued that insecurity in rights to land for women was far more pervasive. As I have argued elsewhere (Mackenzie 1991: 247), it is not difficult to envisage a situation where the pressure to maintain rights accorded by social adhesion would play itself out in terms of demands on women farmers to divert any 'surplus' from farm or off-farm production to meet these ends. Conflict over who controlled the surplus and its distribution is likely, at times, to have been fierce. One result is that women's labour – stretched through multiple responsibilities which increased within the colonial political economy – was less available for maintenance of the resource base.

The threat to a breakdown in ecological equilibrium rested, in part then, on changes in relations of production within the household in the context of a colonial political economy. Environmental balance was also threatened through the silencing of local agricultural knowledge in the construction of a new agriculture. New seeds – bred for uniformity and the market – and new crops (maizes rather than millets) intended to service the colonial economy (and perceived as non-threatening to settler interests), and introduced forms of (mechanical) soil conservation were an additional and integral part of a reconstruction of knowledge/power relations in Kikuyu society.

Part II

GEOGRAPHIES OF DEVELOPMENT

6

SUSTAINABLE DISASTERS?

Perspectives and powers in the discourse of calamity

Kenneth Hewitt

> The modern ideologies of prevention are overarched by a grandiose
> technocratic rationalising dream of absolute control of the accidental,
> understood as the irruption of the unpredictable. . . . It pretends to
> eradicate risk as though one were pulling up weeds.
>
> (Castel 1991)

INTRODUCTION: DISASTER AND DEVELOPMENT

My main concern in this chapter is with the concepts, practices and ethical
underpinnings of hazards research. Hazards and disaster have a place in this
volume if you accept the argument that there is a close interdependence
between risk and 'development' (Morren 1983). Moreover, if there could be
such a thing as sustainable development, disasters would represent a major
threat to it, or a sign of its failures.

A particularly convincing case has been made regarding famine and the role
of agricultural development in food security. Researchers have shown that
modern famines are not the result of failure to increase food production as
fast as mouths to be fed, nor of absolute food shortage in the economic sphere
to which the starving belong, nor primarily due to short-term crises brought
on by, say, droughts. Rather, the *already hungry* are the main casualties of
famine because of social and economic conditions that prevent them gaining
access to available food (Dreze and Sen 1990). Thus it appears as an archetypal
case of disaster prefigured by 'underdevelopment' (Susman *et al.* 1983).
Furthermore, social changes related to 'development' are integral to the
problem. In Africa, famines have been identified with 'the cost of development,' the role of droughts made far worse by '"planning," inappropriate
technology and administrative weaknesses' (Baker 1974: 170; Cuny 1983). In
India, at least since 1850, 'famine ceased to be a natural calamity and was
transformed into a social problem of poverty and dearth' (Bhatia 1991: 2).

115

Famines and related uprooting are usually exacerbated, if not triggered, by social upheavals and armed conflicts surrounding the struggles to 'modernize' (George 1984; Kent 1984; Bohle 1993).

Although we are concerned with a great many other hazards and forms of disaster, this view of famine as the 'bottom line' is important in the more general debate about hazards and development. On the one hand, hunger, famine and some of the massive uprootings of people associated with them, are the largest sources of emiseration and death. On the other, like most forms, they are disasters stereotypically identified with natural hazards or the impersonal forces of sociobiology, notably 'population.' This is a quite general feature and problem of hazards discourse.

Now, risk is present in everything and it was always so. But most risks are dealt with routinely, whether in households or by governments, and through well-established fields, practices and responsible agencies. Today that includes many hazards that were once intractable, or did not exist. The question is, therefore, why hazards research has in fact become preoccupied with those dangers and damages that lie outside, are missed by, or overwhelm routine ways of coping. As the field broadens and matures, it does so essentially by recognizing that what had seemed to apply to the more extreme natural hazards, has its counterparts in a range of technological risks and social violence. Initially, it seemed that methods developed in looking at the more extreme natural hazards had wider applicability. Now it seems that all of these problems require our attention because they are 'out-of-control' or 'out-of-place' dangers *within* modern life (Kirby 1990). They involve forms of damaging event that repeatedly overwhelm routine and centrally organized methods of coping. They are found to include not just disastrous events, but dangers in widely disseminated products and practices, due to unsatisfactory or unacceptable measures to protect the general public and the habitat. All involve undue and inequitable exposure and harm for certain people and places.

Damages and survival in disaster of all forms, have been found to relate above all to the social geography of settlement, land uses and well-being, and especially to recent changes in these – though there may be disagreements as to how and why this is so (Erikson 1976; Waddell 1977; Torry 1978; Davis 1981; Hewitt 1983, 1992; Morren 1983; Oliver-Smith 1986; Mitchell et al. 1989). Many were sufficiently impressed by the connection that the notion of 'unnatural disasters' of earthquake and flood, frost and storm came into wide use (O'Keefe et al. 1976).

Without anyone planning it, therefore, hazards research has drawn us into two related crises of modernity. The first involves the subject matter upon which the work is based – the evidence of a continuing inability to protect the greater number of citizens in all countries from repetitive forms of extreme harm. More problematic is the way the planned innovations and developments that are the special promise of modern life, go on increasing the range and severity, or the concentrations of risk (Wijkman and Timberlake 1987; UNDRO 1990; Smith 1992; Cutter 1993). The second crisis involves the way in which established disciplines and responsible institutions have responded to this challenge.

From the perspective of 'hazards discourse,' it is more the way these problems are approached and discussed as research or managerial problems that concerns us. Established disciplines and dominant institutions of government have chosen to treat these *not* as crises of modernity or the predicaments modernity creates on the ground, *nor* as failures of a research paradigm or policies and organization. Rather they have been constructed as problems due to external factors beyond their control – natural extremes, impersonal forces of demography, 'accident' and 'error,' even 'human nature.' More exactly, these hazards are placed, intellectually, socially and geographically, at the frontiers, as part of the unfinished business of modernization. The disasters are seen to arise from forces carefully situated beyond or in a highly uncertain relationship to the rest of modern life – as challenges remaining for that which is, indeed, 'modern.'

One of the the starting points for this discussion is therefore the way the preponderance of hazards and disasters studies, and action related to them, ignore, if they do not deny, an explanatory connection with economic development and, more generally, with everyday life. Their discourse accepts development and disaster as antonyms. It constructs them as opposites or adversaries. Disasters are to be dealt with as a quite separate issue or special 'application' of science – as 'weeds' to be eradicated. This recalls the literatures of conflict that treat – or rather, *exclude* – 'peace' as simply the opposite or absence of war. They only study war and even the hopes of preventing wars, as problems of conflict and war. In the field of hazards studies, the polarization is identified by calling hazards 'negative resources.' That might seem to establish a relationship with resource development. However, the intention or result has been to establish boundaries between them as distinct problem fields (Hewitt 1983). None of this work actually demonstrates that the roots of disaster or failure to prevent it lie anywhere else than in the conduct of everyday life or its transformations. It simply constructs the problem differently.

In introducing my arguments in this way, I am also assuming that 'discourse critique' does not happen in a vacuum or in terms of its own making. When directed towards problem-solving fields, it depends especially upon examining the relations or arguments running between goals and outcomes, evidence and priorities, concepts and practices. It also involves arguing the relations between the geography of phenomena and experience on the ground and, in the modern world, 'expertise,' its institutionalized forms, preoccupations, relevance and, perhaps, honesty.

'MASTER TEXTS'

The mainstream technocratic view of natural hazards, though it largely misses the point, is alive, dominant and well in the 1990s. This is most evident in the agendas of the United Nations and related national 'Decades' of natural disaster reduction (UNDRO 1990). Key proposals for such a Decade included:

> Strategies for cooperative worldwide endeavours in the collection, dissemination, and application of existing knowledge in loss-reduction measures; the identification of gaps in knowledge and the initiation of new research to fill them; the acceleration of continuing research that can yield additional insight into the physical processes of natural disasters; and transfer of technology.
>
> (Eighth World Conference on Earthquake Engineering 1984)

In a further refinement, the stated objective of the IDNHR (International Decade for Natural Hazards Reduction) was to 'reduce catastrophic life loss, property damage, and social and economic disruption from natural hazards.' This would be accomplished through 'cooperative research, demonstration projects, information dissemination, technical assistance, technology transfer, and education and training.' These activities would be 'tailored to specific hazards and locations, allowing for cultural and economic diversity' (National Research Council 1987).

The goals are entirely laudable. It is the assumptions about disaster behind the 'strategies', the reasonableness and likelihood that these priorities and activities could achieve the goals, that are questionable. They obviously rest upon some powerful assertions about what 'we' supposedly know, and can or need to do. At root, they reassert a top-down, technocratic and Western expert vision. The underlying geography of the initiative was for 'advanced' nation expertise to be deployed and passed to other governments and projects, and to train 'disaster managers' in other countries. This is taken as transparently the best or only source of improvement. It is indifferent to social and ecological contexts, to the sources of vulnerability that are largely social, and the relation with 'development.'

On the ground, however, these technocratic approaches and the high-profile, high-tech initiatives they seem to require, are often more part of the problem than the solution to the burgeoning numbers and shifting toll of disasters that seemingly have inspired the Decade. Technocratic approaches have been associated with failures in the past, even in cases like the Mississippi flood plains, where vast expenditures have been made upon them. That is evidently viewed as the result of still insufficient organization and commitment. It may be noted that IDNHR is likely to have less rather than more real resources, in an environment of enormous government debt and global economic 'turn-down' (see J. K. Mitchell 1988). However, the problems are as much with the preferred perspective, epistemology and organizational arrangements, as with what it says and fails to say about disaster.

DISTANCING AND DICHOTOMIZING: THE VIEW FROM THE CENTRE

The view upon which the main thrust of IDNHR rests treats hazards and disasters as solely problems of damaging agents, of extreme natural events and extreme experience. Disaster is thus separated off from the rest of life in a language of unscheduled, uncertain, unmanageable or counterproductive 'events.' By contrast, development is portrayed as being about system and continuity, the *planned* extension or improvement of orderly, managed and, especially, *productive* life. Pre-existing conditions may be invoked to the extent that overpopulation or underdevelopment, and other measures of 'instability' in impersonal factors, are seen as cause or accompaniment of greater risk and worse disasters.

Equally important, the most desirable if not the only response to disasters is seen to lie in advanced and expert counterforce. This combines scientific and technical understanding of natural processes such as exists in the more 'advanced' countries or institutions, with the kinds of technical and emergency measures also in place there. These are seen as offering the only chance to comprehend better, predict and, where possible, control the natural agents. This kind of knowledge is expected to provide the best ways to inform and warn those at risk, or to plan and manage their responses. Such is the technical vision of misfortune and human misery. And it may well be necessary and relatively effective in the urban-industrial heartlands of its origin. Wisely tapped into it might have wider value. As it is, the further removed in space or society are the persons and conditions of concern, the less convincing the view or its results become.

The consequence is a dichotomized 'reality' in which the risks, incidence and phenomena of disaster have been segregated and 'targeted' separately from 'normal life' and its development. But this dichotomized vision of reality, also means an internal split in the discourses of disaster. A broad range of studies in the 1970s and 1980s demonstrated that most risks and damages in disaster arise from, or depend mainly upon, ongoing material life, the human circumstances or choices within that. Natural disasters are not simply due to natural extremes, nor do damage patterns simply or mainly reflect their intensity. They also do not appear indiscriminate or due to mere chance but, as noted, relate closely to the social order and forms of development.

I have referred to the approach, in which the polarization of calamity and everyday existence is a major construct, as the 'dominant' consensus (Hewitt 1983). Like any other, it emerges as socially constructed knowledge. Without doubt, it reflects what *is* commonplace in the urban-industrial heartlands and wealthier enclaves, in the divisions of labour and responsibility that have

119

developed in them and what is widely perceived as the hallmark of modernity. To define this view as 'technocratic,' is to recognize and label that societal context.

In this case, however, 'social context' differs in meaning and form from the usual perception of 'embeddedness' in the shared world view of a community. Rather, we are dealing with the dissociation of the technical from the communal. In particular, the main technocratic actors and institutions are those at 'the Centre,' or with the greatest prestige and influence there. And they, or their more influential activities, are separated, usually in space as well as perspective, from conditions and even those with technical roles on the ground. The geography of such a situation can, perhaps, be best illustrated by a parallel example.

Late twentieth-century wars, or the thirty-five or so going on as I write, occur almost wholly in so-called Third World countries, and mainly in their remoter hinterlands. But what of the weapons in use, from rifles to jet fighters? What of the drugs and hospital facilities, telephone and satellite communications, of which armies everywhere tend to get the best and first service? They are mostly developed and manufactured in the industrial heartlands of the 'First World.' That is where the R&D, testing and training, promotion and marketing, occur or emanate from. Financial assistance, without which most of the governments involved would have few or none of these arms, also depend upon the powerful states. The same applies to training and 'technical assistance.' Meanwhile, negotiations, purchasing, intelligence gathering, plans of war, deployments of troops to remote areas, are decided in the capitals of these 'client states,' again often helped by their 'First World' sponsors.

All of this may seem obvious, and is certainly taken for granted. A good deal of the modern literature and critique of war concerns the way modern weapons and command systems require the main actors or powers to act 'at a distance'; how key decisions are radically separated in context and terms, from the places and contexts of their application. And this allows other factors or actors, other priorities, to intervene between what would seem, in terms of a strict 'fighting' and battlefield focus, the only issues. Many of the worst scandals, disaffection of fighting men and debates by military historians involve the consequences of the different horizons of 'Headquarters,' 'the control tower,' or 'the government,' and what happens and is needed in the front line.

Much the same top-heavy structure and geography applies to official and 'expert' disaster research and management as, indeed, the control of most development programmes. To be sure, if one's task is to determine the geotectonic conditions involved in earthquakes, or to analyse and model data from arrays of seismographs; to design or assess the seismic safety of high-rise buildings or dams; to plan communications for emergencies; or to draft legislation for a national approach to disaster assistance – one will be working within a more or less complex, techno-managerial organization. A detached, technical, strategic and standardizing style of work will be favoured. At least, that seems to describe how such tasks are carried out in the modern world. They are not unimportant tasks in urban-industrial societies, and perhaps this is the way to carry them out. However, it encourages a view of problems from the outside and 'from above'; one attuned to the preoccupations of those in authority. As a result, the most sophisticated and effective technical work is socially constructed around institutions and viewpoints of power.

The hierarchically organized and overdetermined form of the modern military machine actually resembles the arrangements most favoured by IDNHR – an overriding concern with 'The General Assembly,' governments, 'A Special High Level Council,' 'international scientific and engineering organizations,' 'inter-agency' committees and liaison, and so forth. It may sound convincing and is possibly a very necessary gambit in the committee and assembly rooms in Geneva, New York or Rome. But the main result is that those expected to get the job done are remote from, quite subordinate to and powerless compared to ruling politicians, committees and the advisers to high administration. The practitioners on the ground are likely to get the poorest and least adequate share of the resources.

Another illusion is that, somehow, by passing on information and training, by sending experts, existing resources will become magically less embedded in existing institutions and social practices, or more effective in terms enjoyed in places of wealth. All of this represents a view from above, epitomized by technocratic centralism, its obsession with 'objectivity,' statistics or information generally, with tools and models, a faith in the latest technologies and fashionable terms. It is strategic vision and identifies technostructures for mastering 'the opponent' from a distance, and the impact of teleculture. But these views are, in turn, underwritten by a very partial view of risk and disaster. For professionals and agencies, it allows disaster to be treated, as Castel (1991: 284) says of 'care,' generally, 'a relationship constituted among the different expert assessments.'

The Decade seems intended to extend or define further dimensions of this ruler's science, through a discourse of projects and strategy, of cooperation among governments and agencies rather than people, of monitoring, control and targets. An oxymoronic term like 'sustainable development,' in such a framework, is in danger of becoming a prescription for better-orchestrated rape; 'Disaster Management' for Orwellian interventions by professionals as an adjunct or integral part of national and international security.

SEGREGATING DISASTERS AND DEFINING TERRITORIES

Geography is quite central for the strategic thinking of the dominant view. Invariably, the map of natural agents and their relative intensity and frequencies defines the basic pattern of risk for natural disasters. It is defined in terms of 'high hazard' areas, perhaps flood plains or seismic zones, drought 'polygons' and 'tornado alleys.' Disasters themselves are treated as places or archipelagos of extreme and more or less random events. They tend to be cordoned off as areas of spatial disorganization or national security crisis. Thus, the geography of disaster is also constructed in stark contrast to the *patterns* of 'land utilization,' the space economy, urban systems and development issues.

If we are talking of natural hazards this might seem logical. Obviously an 'earthquake disaster' cannot happen without an earthquake, and most occur within the more seismically active regions of the earth. However, the spatial distributions of actual disasters, even in the more casual view, do not follow those of natural extremes closely. In detail they depart more or less remarkably from them. It is not only that the vast majority of all avalanches or landslides do no recorded damage. The same applies to most of the area defined between the limit of recorded damage in earthquakes and their epicentres, or along destructive storm paths. The patterns of damage within so-called 'disaster zones' vary in a complex way, and not primarily as a function of the intensity of the natural agent that initiated the damaging episode. This follows because the pattern of survival and loss actually reflects most closely aspects of social geography within the area where damages occur. These elementary observations of damage geography are generally ignored in maps of risk, yet risk is a relation between potentially damaging conditions and what is or may be at risk from it. Indeed, one of the strangest things about mainstream disaster views of risk is how they depart from an actuarial, or damage evidence basis, such as is used in almost every other area of risk assessment.

In trying to uncover the inner logic of this dominant view found in IDNHR, Foucault's (1984) studies of the historiography of other modern fields of practice are helpful. He showed how, in the urbanizing, centralizing state apparatuses of eighteenth-century Europe, modern scientific applications were crafted to suit emerging professional classes of 'minders.' The way some founding notions of modern science emerged from the debates about 'the poor,' the Poor Laws and pauperism in the rising industrial powers, touches closely upon our concerns here. According to one recent analysis:

> Poverty constitutes a development area for techniques designed to structure the organic social order which, whatever the *concrete localisation of the human subjects it deals* with, is able to bring under its management those zones of social life that have *hitherto remained formless*. What is involved is the *constituting of a different subject from the Productive subject*: a subject 'aware of its duties', a civil and political subject, one might say; it is not poverty as the stigma of inequality that

is combated, but pauperism *understood as a cluster of behaviours, a carrier of difference.*

(Procacci 1991: 164, emphasis added)

Try substituting 'disaster' and its victims or, say, 'women in development' for poverty. Replace 'government of poverty' with disaster management. It begins to sound like a proving ground for the present deployment of 'techniques' in the globalized arenas of development and disaster. Generally, the early nexus of relations between 'enlightened' knowledge and government laid the foundations for today's powerful, techno-systems science – rather than the inevitable march of scientific discovery. The latter emerges not as the product of itself – i.e. of reason, experiment, the uncovering of natural laws and predictive success of models – but of the discourses and practices of institutions, personalities, interests, opportunities and power-contexts.

There is, however, a particular, largely unnoticed social context that frames the field and concerns of 'hazards and disasters.' It is those areas that have been defined and treated as 'acceptable,' 'managed,' 'normal' risks by modern communities, governments, mutual societies, public and private corporations. Disaster involves a focus upon risks: (a) lying outside the generally acknowledged realms of the 'accident,' culpable or criminal harm; (b) that are not part of the (usual) responsibilities of professions and agencies dealing with everyday crisis and trauma among the general public, and; (c) that have been outside insurable risks. They have lain or been described as lying outside or at the uncertain and unmanageable bounds of the responsibilities of civil societies for the security of citizens.

'Natural disasters' have been placed, for the moment, beyond the bounds of the responsibilities of civil societies or their main institutions, for the security of citizens. They are explained by extraordinary events, to be combated by extraordinary measures. The other side of the same coin is how risks, treated as integral to the productive economy, its labour, machinery, infrastructure and pleasures, are largely excluded from the agenda of disaster in the wealthier urban-industrial nations. Some see this as simply part of the unfinished business of the risk-mitigating institutions of modernity. There are arguments for shifting a variety of natural disasters and some novel risks such as the AIDS epidemic or nuclear technology, off the list of special dangers, and into the 'normal' spectrum of treatment, prevention and insurable risks. Others are appalled by the way societies or their authorities refuse to treat some everyday sources of far greater misery and death than most or all natural disasters, as calamities requiring extraordinary action. These include traffic 'accidents,' cancers induced by environment or lifestyle, and heart disease.

Of course, the risks that have become part of the ongoing systems of care, services and compensation within modern societies had to be fought over, and the arrangements are under continual threat of disarray, downgrading, or dismantling (Defert 1991). The work of Robert Castel (1991) is among that which sees the subordination of practitioners to administrators, and subsequent disillusionment of the former, as having much the same root in centralized, technocratic 'rationalizations' of these networks of care. Yet, I would also argue that in such societies it is these everyday arrangements, their institutions and personnel, that are critical to disaster response. They are central to the much smaller loss of life and emiseration in disasters, than in countries which lack them or groups which are not served by them (Hewitt 1987). Then again, there is evidence that all societies have had their own equivalents: built-in ways to cope with crisis and personal trauma. The disasters of poorer nations tend to be focused in those communities and groups where such arrangements have been undermined or destroyed by social and political changes.

By insisting upon an integral relation between everyday life and the risks of disaster, I do not deny the uniquely terrible realities of disaster experience for its victims, nor the possibility that it calls upon or releases thoughts and activities otherwise rare or absent. Rather, this recasts the problems of interpretation and organized response. More often than not, disasters, or most of the harm done in them, are 'tragedies' in that peculiarly modern sense that A.C. Bradley (1904: 11) describes: 'The calamities . . . do not simply happen, nor are they sent, they proceed from actions, and those the actions of men.' In particular, this identifies how danger is distributed, and destruction discriminates within or is channelled by social geography; by gender, age and class; by control of land, land use and how power and projects intervene in them. It is also to recognize that survival is equally a function of pre-existing conditions, including well-thought-out risk avoiding and crisis arrangements. 'Actions' are, of course, different in meaning and content from impersonal forces of nature, human sociobiology or the vagaries of 'human error.' Again, that contrasts with and becomes engaged in critique of a view of hazards as essentially 'indiscriminate' and 'unscheduled.'

HIDDEN DAMAGES: THE VIEW FROM BELOW

Disaster is also an experience of its victims, in the places where disaster occurs, and of those who are the 'targets' of development and coercion. Their views may be articulated by others who see themselves speaking for those who are not given a voice of their own. If risk and damage are primarily shaped by social conditions, it is hard to see how we can understand how they happen and people act or are allowed to act, without 'being there.' We require the evidence given only by close association with and awareness of the circumstances and concerns of those threatened with, or suffering from severe earthquake losses, starvation or extrajudicial terror. In that regard, outsiders, however well-trained, may not achieve an adequate degree of insight without eye-witness accounts; without considering the plight and stories of distinctly more vulnerable members of society – perhaps children or the elderly, especially of disadvantaged groups (see Hewitt 1994a). We would need to pay attention wherever certain families, neighbourhoods and larger communities suffer unusual losses or are threatened by them.

This is not a view which has found widespread favour, so that it is often placed or forced into an adversarial role with respect to the prevailing view – a dichotomy of discourse within a dichotomized reality. We feel obliged to speak of missing persons or unheard voices; of 'hidden damage' and 'shadow risks' and, more severely, of 'silent' or 'quiet violence' (Watts 1983; Kent 1984; Hartman and Boyce 1983; Hewitt 1994b). We identify 'voiceless' and 'invisible' presences; conditions and people ignored or marginalized. Issues are found to be 'hidden,' 'masked,' 'obfuscated' or redescribed to suit other – also often 'hidden' – agendas. This is not an unusual situation in studies that confront prevailing or dominant paradigms that ignore their concerns and discoveries. Yet it in no sense implies minor or superficial items. On the contrary, in echoing the literatures of the 'hidden economy' or 'shadow work' I am arguing about some fundamental and pervasive matters. They are 'absent' only in the sense of disappearing or being given highly partial treatment in the most influential work. In the present case, the social conditions ignored tend to recur as basic features of most disasters; the people kept silent about are often most at risk and most severely affected by disaster.

However, those who have given attention to conditions on the ground and in disasters find that 'ordinary' persons are rarely mere passive and pathetic victims of disaster, unless incapacitated by it. Many survivors act with courage to carry out some – often most of the immediate – life-saving responses, and the longer-term humanitarian actions in their own homes and community centres. Societies classed as 'traditional,' no less than some that are 'modern,' prove to have developed rather robust and humane ways of coping with drought or flood (Waddell 1983; Oliver-Smith 1986; Torry 1986; Bode 1989).

Among other things, participant science and the defining of risks in context seem to involve and require a different *modus operandi*, methodologies and perspectives: a view not merely 'from below' but from within rather than outside communities, a capacity for sharing in their sense of crisis. One requires *in*sight rather than *over*sight; a capacity to listen to, comprehend and interpret experience and circumstances expressed in the local languages and vernacular 'discourse,' rather than technical ones. Even if we accept this alternative vision of what is required, modern conditions may, of course, oblige anyone who has come at the problem as a 'researcher,' official or expert to provide 'translations' from one to the other discourse. To the extent that it is possible and desirable, that may be the most useful thing one can do – at least until those at risk are empowered to influence, and provided with the means to be major actors in, risk assessment and responses.

In the disasters field, one is concerned particularly with the neglect of cultural, demographic, and status differences that underpin the degree and forms of risk to which people are subject. These constrain their responses and the possibilities of outside or exotic assistance. Of course, researchers or professionals in the disasters field are actually 'caught in the middle' – neither in charge nor usually speaking as victims of calamity. If we are to incorporate the view from below it will be as 'go-betweens' if not advocates. But this and the direct empowerment of those at risk to participate in enquiries and planning have received little 'expert' or official support.

As Gilbert F. White showed, there is almost invariably a wide range of potential responses to and means to prevent or reduce disaster. We would expect a society or responsible institutions to evaluate them so as to arrive at the best choices (White 1961). Of course, this in turn reflects the Western tradition which says life in a 'free' society or for its citizens should be one of open discussion, and decisions based on consensus. Unfortunately, wherever we look, choices are usually much more constrained than the range of possibilities would suggest. The vast majority of people have far fewer options, and often little or no say in major decisions affecting risks to their lives.

Again, a common assumption has been that people at risk or harmed in disaster must have made bad or uninformed choices – since they invariably depart from the ones expert opinion favours. This would only be valid if, in fact, people had the option to choose well, the information and resources to implement a range of adjustments. Almost invariably, we find the victims of disaster have little or no choice about where they live, what structures they live in, how better to protect members of their family and what they can do when disaster strikes, unless it is to embrace still greater risks. Increasingly, those harmed in disasters are second or third stage victims of economic disadvantage, uprooting, urbanization, and political change: 'In such a choice, preferences one way or another reveal little or nothing. Much more important than the choices people make are the constraints under which they choose' (Marglin 1990: 8).

What is also implied is that whether you realize it or not, any change in disaster response means social and political change. And just as is increasingly said in critiques of development, real improvements in disaster prevention and mitigation will only come about if they originate in or are in step with social and political improvements. The more vulnerable only have a chance to become less vulnerable with greater empowerment (Sen 1981; Watts and Bohle 1993). The dominant view, with its support of top–down and expert systems, not only fails to offer such social and political advancement but is almost certain to prevent it. For the essential result of such centralized and paternalistic arrangements is to create greater dependency of the vulnerable and hence greater vulnerability.

CONCLUSION: BAD GEOGRAPHIES

An integral part of the problem is that we have hardly begun to define or accept geographies of conditions and cultural contexts on the ground. We need to incorporate realistically spatial perspectives from below as well as the variety of perspectives that converge in problems of disaster (Wijkman and Timberlake 1987; Watts and Bohle 1993: 64). Instead, the prevailing development and disaster geographies have not been created by geographers and are only rarely subjected to critique by them. Our maps and problem spaces reflect other people's bad geography. We have constructed our views through such notions as Third World, North–South, Cold War, Orientalist and Africanist, not to speak of 'globalization' geographies. The one thing these have in common is that they are 'master geographies' in keeping with the master texts of dominant views. They were crafted in the hegemonic and stereotyping style of United Nations and dominant nation statistics, of atlas-gazers and intelligence-gatherers' visions of geography. They express and maintain a view from above and from The Centre; from the metropolis, the dominant states and institutions. They reflect the concerns and measures suitable for the work of the bureau, the technocrat, the strategist. Within them is the beguilement of the 'administrative' region and science for government.

Since modern geography grew up primarily as a servant of state, imperial and corporate powers, it was bound to embrace their strategic vision of the deployment of science (Godlewska and Smith 1994). However, its increasingly civil roles in education and research, its public duties in problem-solving and humanitarian domains, should have led to a critique if not a rejection of those master geographies in favour of others more in keeping with our new roles and the ostensible preoccupations with peaceful and civilized life. Perhaps this is what we are beginning to see happening (Wisner 1988; Pratt 1992).

The maps and geographies, the notions of geographical information and knowledge currently favoured are singularly inadequate and misleading in the contexts of development and disaster. These constructs, intended to reify or preserve old forms – perhaps changing the names to protect the guilty – are travesties of the human geography of all but a very few, if dominating, geopolitical concerns. But perhaps this leads to the kind of science predicated on cynical reason; one 'afflicted with the compulsion to put up with pre-established relations that it finds dubious, to accommodate itself to them, and finally even to carry out their business' (Sloterdijk 1987: 6).

Here we confront a final problem. If master geographies prevail in development and disaster, it leaves little doubt that these are subjects of compelling concern for powerful states and interests. In the realm of development most geographers draw their portraits around the image of wealthy nation philanthropy cultivated around the idea of development assistance. They do little to dispel widespread ignorance of the far greater scales of wealth flowing into them from the less wealthy countries, not least the greater payments in debt-servicing than the original loans. Likewise we have promoted a view of development as demanding state intervention by the Centre, informing, teaching or legislating for those who might be seen to need it. That most of the benefits actually flow from the groups and areas 'targeted' to the capitals, wealthier enclaves, families and their technical advisers, is not part of the map of development. So it seems very likely that the absence or obfuscation of the political economy of disaster from the dominant paradigm reflects a similar coyness about the interests of power and the geography of benefits in a technocratic, top–down vision of initiatives and capabilities.

7

THE OBJECT OF DEVELOPMENT
America's Egypt

Timothy Mitchell

The geographical and demographic characteristics of Egypt delineate its basic economic problem. Although the country contains about 386,000 square miles, . . . only a narrow strip in the Nile Valley and its Delta is usable. This area of 15,000 square miles – less than 4 percent of the land – is but an elongated oasis in the midst of desert. Without the Nile, which flows through Egypt for about a thousand miles without being joined by a single tributary, the country would be part of the Sahara. Crammed into the habitable area is 98 per cent of the population. . . . The population has been growing rapidly and is estimated to have doubled since 1947.

(Ikram 1980: 3)

INTRODUCTION

Open almost any study of Egypt produced by an American or international development agency and you are likely to find it starting with the same simple image. The question of Egypt's economic development is almost invariably introduced as a problem of geography versus demography, pictured by describing the narrow valley of the Nile River, surrounded by desert, crowded with rapidly multiplying millions of inhabitants. The 1980 World Bank report on Egypt quoted above provides a typical example.

The visual simplicity of the image, spread out like a map before the reader's eye, combines with the arithmetical certainty of population figures, surface areas and growth rates to lay down the logic of the analysis to follow: 'One of the world's oldest agricultural economies,' a report written for the United States Agency for International Development (USAID) begins:

Egypt depends upon the fruits of the narrow ribbon of cultivated land adjacent to the Nile and to that river's rich fan-shaped delta. For more than 5,000 years agriculture has sustained Egypt. During the first half of this century, however, . . . the growth of agriculture failed to keep up with the needs of a population which doubled, then nearly tripled. It is a matter of simple arithmetic.

(Johnson *et al.* 1983: 1)

The popularity of this image of space and numbers is summed up in the World Bank report: 'These two themes – the relatively fixed amount of usable land and the rapid growth of the population – will be seen as leitmotifs in the discussion of Egypt's economic problems' (Ikram 1980: 3).

Fields of analysis often develop a convention for introducing their object. Such tropes come to seem too obvious and straightforward to question. The somewhat poetic imagery favoured by writings on Egyptian development seldom lasts beyond the opening paragraph, and the text moves quickly on to the serious business of social or economic argument. Yet the visual imagery of an opening paragraph can establish the entire relationship between the textual analysis and its object. Such relationships are never simple. Objects of analysis do not occur as natural phenomena, but are partly constructed by the discourse that describes them. The more natural the object appears, the less obvious this discursive construction will be.

The description that invariably begins the study of Egypt's economic development constructs its object in two respects. In the first place, the topographic image of the river, the desert surrounding it, and the population jammed within its banks defines the object to be analysed in terms of the tangible limits of nature, physical space and human reproduction. These apparently natural boundaries shape the kinds of solutions that will follow: improved management of resources and technology to overcome their natural limits. Yet the apparent naturalness of this imagery is misleading. The assumptions and figures on which it is based can be examined and re-interpreted to reveal a very different picture. The limits of this alternative picture are not those of geography and nature but of powerlessness and social inequality. The solutions that follow are not just technological and managerial, but social and political.

In the second place, the naturalness of the topographic image, so easily pictured, sets up the object of development as just that – an object, out there, not a part of the study but external to it. The discourse of international development constitutes itself in this way as an expertise and intelligence that stands completely apart from the country and the people it describes. Much of this intelligence is generated inside organizations such as the World Bank and USAID, which play a powerful economic and political role within countries like Egypt. International development has a special need to over-look this internal involvement in the places and problems it analyses, and present itself instead as an external intelligence that stands outside the objects it describes. The geographical realism with which Egypt is so often intro-duced helps establish this deceptively simple relationship.

TOO MANY PEOPLE?

We can start with the basic image of overpopulation and land shortage. Whenever you hear the word 'overpopulation,' Susan George suggests, 'you should reach, if not for your revolver, at least for your calculator' (George 1990: 18). It is seldom clear, as she points out, to what the prefix 'over' refers. What is the norm or the comparison to which it relates? 'Egypt has the largest population . . . in the Middle East,' notes the World Bank (1989a: 129). 'Its 52 million people are crowded into the Nile delta and valley . . . with a density higher than that of Bangladesh or Indonesia.' Why Bangladesh and Indonesia? The World Bank might equally have mentioned Belgium, say, or South Korea, where population densities are respectively three and four times higher than Indonesia – but where the comparison would have a less negative implication.

It is true that Egypt's level of agricultural population per hectare of arable land is similar to that of Bangladesh, and about double that of Indonesia (FAO 1989b: Annex table 12b). But this comparison is misleading, for arable land in Egypt is vastly more productive. It is estimated that Egyptian agricultural output per hectare is more than three times that of both Bangladesh and Indonesia (FAO 1986: 36, table 5.7). So it is not clear that Egypt is overpopulated in relation to either of these countries.

Perhaps it would be more realistic to gauge Egypt's land shortage by comparing it not with poorer countries but with places that have a similar total population and per capita gross national product, combined with far greater areas of cultivated land. The Philippines and Thailand are the two closest examples in population size and GNP, and have cultivated areas respectively three times and eight times that of Egypt (FAO 1988: table 1). Yet despite having far less land to farm, Egypt's agricultural population produces more crops per person than either of these countries. Egyptian agricultural output per worker is perhaps 8 per cent higher than that of the Philippines and 73 per cent higher than that of Thailand (FAO 1986: 34, table 5.6).

Despite the visual power of the image of more than 50 million Egyptians crowded into the valley of the Nile, there is no *prima facie* evidence for the assumption that this population is too large for its cultivable area. Perhaps it might be argued in more general terms that the world's population exceeds some equilibrium in relation to its resources (George 1990: 18). In that case, however, there is no reason to single out Egyptians. On the contrary, Egyptians make very modest demands on the world's resources (measured in terms of energy consumption per capita) compared with inhabitants of Western Europe, Japan and North America. One inhabitant of the United Kingdom, for example, requires more of the world's energy per year than six Egyptians, and one American is more expensive in energy terms than a dozen Egyptians (World Bank 1989b: table 5). So it can hardly be the latter who are threatening the world's limited resources.

Perhaps it can be agreed that having more than 50 million inhabitants does not necessarily make Egypt 'overpopulated.' Development experts might insist, however, that the problem is not the size of Egypt's population but the rate at which it is growing. A United States Department of Agriculture report asserts that the country's 'exploding population is the most serious problem facing Egypt today' (USDA 1976: 48). The rapid growth in population appears to have outstripped the country's ability to feed itself, and since 1974 Egypt has been a net importer of agricultural commodities. Food today accounts for almost 30 per cent of Egypt's merchandise imports, a higher proportion than for all except one of the 100 countries for which figures are available (World Bank 1993: table 15). It would appear from these figures that the case for an imbalance between population figures and agricultural resources has been established after all. But before accepting this conclusion we should reach, once again, for the calculator.

NOT ENOUGH FOOD?

Between 1965 and 1980, according to World Bank tables, the population of Egypt grew at an annual rate of 2.1 per cent. Yet during the same period, the World Bank also shows, agricultural production grew at the even faster rate of 2.7 per cent a year. During the 1980s, when the population growth rate increased to 2.4 per cent a year, agricultural growth continued to keep ahead (World Bank 1991b: tables 2, 26). In 1991, food production per capita was 17 per cent higher than at the start of the previous decade (FAO 1993: table 9). So it is not true that the population has been growing faster than the country's ability to feed itself. If this is the case, then why has the country had to import ever increasing amounts of food? The answer is to be found by looking at the kinds of food being eaten, and at who gets to eat it. Official statistics suggest that Egyptians consume relatively large amounts of food. The World Bank ranks Egypt as a 'low income' country, yet the country's daily calorie supply per capita is estimated to be higher than all except one of the 'lower middle-income' countries, and indeed higher than a majority of the world's upper-middle and high-income countries (World Bank 1990a: table 28). The daily protein supply per capita also far exceeds the level of most middle-income countries and rivals that of many high-income countries (FAO 1989a: table 107). Despite these figures, Egyptians suffer from high rates of undernutrition. A 1988 survey found that 29 per cent of children suffered from mild undernutrition and another 31 per cent from moderate or severe under-nutrition (Studies in Family Planning 1990: 351). Between 1978 and 1986, the prevalence of acute undernutrition may have more than doubled (CAPMAS and UNICEF 1988). A study of anaemia (probably caused by the interaction of malnutrition and infection) in Cairo found the condition in 80 per cent of children under 2 years old and in 90 per cent of pregnant women (Nockrashy et al. 1987: 30–2), rates that the World Bank described as 'alarmingly high'

(World Bank 1991b: 23). Clearly the high figures for calorie and protein supply per capita do not reflect the actual distribution or consumption of food.

What the calorie supply figures probably reflect is high levels of food consumption among the better off, a shift in what they consume towards more expensive foods, especially meat, and a significant diversion of food supplies from humans to animals. A World Bank study of agricultural pricing policy in Egypt notes that there is an extremely high variation in the value of food consumed between rich and poor, which it attributes to the low per capita level of income and its unequal distribution (Dethier 1989: 20). This inequality was already increasing in the decade from 1964/65 to 1974/75, when the last comprehensive income distribution figures were published: in the countryside the share of household expenditure of the lowest 20 per cent of households decreased from 7 to less than 6 per cent in that decade, while in urban areas the share of the top 20 per cent of households increased from 47 to 51 per cent (Ibrahim 1979: table 2). During the brief oil boom from the late 1970s to the mid-1980s, the income of the poor improved and the gap between low- and middle-income families may have narrowed. But the wealthiest 5 per cent increased their income share between 1974/75 and 1981/82 from 22 per cent to 25 per cent in the case of rural and to 29 per cent in the case of urban households (Richards and Waterbury 1993: 282). From the late 1980s, as USAID and the International Monetary Fund (IMF) imposed policies that removed price subsidies, increased unemployment, and brought economic recession, the degree of inequality almost certainly increased.

Increasing wealth, together with increasing numbers of resident foreigners and tourists, led to an enormous increase in the demand for meat and other animal products, which 'are chiefly consumed by tourists and other non-Egyptians, plus middle- and upper-class urban residents' (USDA 1976: 23). A 1981/82 household survey showed that the richest 25 per cent consumed more than three times as much chicken and beef as the poorest 25 per cent (Alderman and von Braun 1984: table 12). In the subsequent oil-boom years, income growth together with extensive US and Egyptian government subsidies encouraged a broader switch from diets based on legumes and maize (corn) to less healthy diets of wheat and meat products. From 1970 to 1980, while crop production grew in real value by 17 per cent, livestock production grew almost twice as much, by 32 per cent (Dethier 1989: 19). In the following seven years, crop production grew by 10 per cent, but livestock production by almost 50 per cent (FAO 1988). To produce one kilogram of red meat requires 10 kg of cereals, so feeding these animals has required an enormous and costly diversion of staple food supplies from human to animal consumption.

FODDER FOR PEACE

It is this switch to meat consumption, rather than the increase in population, that has required the dramatic increase in food imports, particularly of grains.

Between 1966 and 1988, the population of Egypt grew by 75 per cent. In the same period, the domestic production of grains increased by 77 per cent but total grain consumption increased by 148 per cent, or almost twice the rate of population increase. Egypt began to import enormous and ever increasing quantities of grain, becoming the world's third largest importer after Japan and China. A small proportion of the increase in imports reflects an increase in per capita human consumption, which grew by 12 per cent in this 22-year period. But the bulk of the new imports was required to cover the increasing use of grains to feed animals. Grain imports grew by 5.9 million metric tons between 1966 and 1988, to cover an estimated increase in non-food consumption of grains (mostly animal feed, but also seed use and wastage) of 5.3 million tons, or 268 per cent (see Figure 7.1).

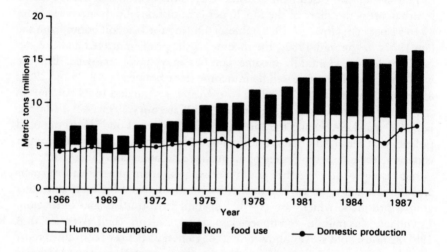

Figure 7.1 Supply and consumption of grains
Source: calculated from USAID/c 1989a

The dependence on grain imports since 1974 has been caused not by population growth, which lagged behind the growth of domestic grain production, but by a shift to the consumption of meat. This shift has been obscured, however, by the way different grains have been used. Rather than importing animal feed directly, Egypt has diverted domestic production from human to animal consumption. Human consumption of maize (corn) and other coarse grains (barley, sorghum) dropped from 53 per cent in 1966 to 6 per cent in 1988 (USAID/c 1989a: 209). Human supplies were made up with imports, largely of wheat for bread making. So it appears as though the imports were required not to feed animals supplying the increased demand for meat, but because the people needed more bread. USAID has supported the shift to meat consumption among the better off since 1975 by financing at reduced interest rates over US$3 billion worth of Egyptian grain purchases from the United States, making Egypt the world's largest importer of subsidized grains. Yet the agency claims that the purpose of these subsidies has been 'to help the poor' (USAID/c 1989b).

Subsidized American loans have financed only a part of the grain imports. The rest have required further borrowing, contributing to a total external debt that in 1989 reached US$51.5 billion, a figure surpassed that year by only Brazil, Mexico, and Argentina. Whereas the debt levels of the three Latin American countries ranged between 26 and 93 per cent of GNP, Egypt's debt as a percentage of GNP was 165 per cent (EIU 1993: 46; World Bank 1991a: tables 23, 24). Egypt began to default on the debt and required large loans just to keep up payments on its earlier loans. To meet this crisis, the United States used the pretext of Egyptian support in 1990–91 for the war against Iraq to write off Egypt's US$7 billion military debt and to arrange for a relaxation of the remaining US$28 billion of long-term bilateral debt, half of which was written off and half rescheduled (EIU 1993: 45–6). As a condition of this refinancing, the United States insisted on a further shift towards export crops, away from staple foods, to produce more hard currency to pay the debts.

The transformation in food consumption habits has affected not only agricultural imports and the balance of payments, but also domestic agriculture. It is no longer accurate to write that Egyptian capitalist agriculture 'still is to a large extent the cultivation of cotton' (Stauth 1989: 122). In terms of the commitment of land and labour, the priority is now the production of meat, poultry and dairy products. In 1989 cotton occupied only about one million of Egypt's 6 million feddans (1 feddan = 1.038 acres = 0.42 hectares). The other major industrial crop, sugar cane, occupied a little over 250,000 feddans. Of the remaining 4.75 million feddans, more than half was used to grow animal fodder – principally Egyptian clover (berseem) in the winter and maize and sorghum in the summer and autumn (USAID/c 1989a). Egypt was now growing more food for animals to consume than for humans.

The shift to the production of meat and other animal products (which has been accompanied by an increased production of other more expensive, non-staple agricultural products, particularly fruit and vegetables) has two prin-cipal causes. First, as the World Bank puts it, 'effective demand has been modified by a change in income distribution' (Ikram 1980: 175). In other words, the growing disparity in income between rich and poor enabled the better off to divert the country's resources from the production of staples to the production of luxury items. Second, the Egyptian government, supported by the large American loans already mentioned (so-called 'Food For Peace'), encouraged this diversion by subsidizing the import of staples for consumers, heavily taxing the production of staples by farmers, and subsidizing the production of meat, poultry and dairy products (see Dethier 1989: 246–65). Livestock raising is particularly concentrated on large farms, those over

10 feddans, where there are three to four times as many cattle per feddan as on farms of 1 to 10 feddans (Commander 1987: 80, table 4–13). Yet government food policy forced even the smallest farmers to shift from self-provisioning to the production of animal products and to rely increasingly on subsidized imported flour for their staple diet.

The image of a vast population packed within a limited agricultural area and increasing in size at a rate that outpaces its ability to feed itself is therefore quite misleading. The growth in agricultural production has always been ahead of population growth. Egypt's food problem is the result not of too many people occupying too little land, but of the power of a certain part of that population, supported by the prevailing domestic and international regime, to shift the country's resources from staple foods to more expensive items of consumption.

Population growth rates of over 2.5 per cent a year, some might argue, are nevertheless still very high. Surely it would be better to produce fewer children and more buffaloes, cows and chickens – as in fact a 1990 family planning initiative proposed. But this depends on one's point of view. Such a proposal would probably seem reasonable to an upper-class or middle-class family in Cairo, and indeed the birth rate among such families is already much lower. But to a rural family or among the urban poor it might seem far less reasonable.

In a social world where daughters leave their parents' family at marriage to join their husband's household and where there is virtually no system of social security to support parents when they become too old or sick to work, it can be argued that to desire a minimum of two surviving male children is not excessive. According to figures for 1980, in rural Upper Egypt, the poorest part of the country, women gave birth to an average of 7.5 children during their childbearing years. But almost one in three of their children (2.7 out of the 7.5) died in childhood (Kelley *et al.* 1982: 9). Under these circumstances, if the parents' aim was to ensure that at least two sons survived to support them in later life, then 7.5 children was not an unreasonable birth rate. Since 1980 infant deaths have been reduced, thanks largely to a simple treatment for diarrhoea, and women have begun to have smaller families (El-Rafie *et al.* 1990; Studies in Family Planning 1990). These women are unlikely to attribute their economic problems to population growth, as does the World Bank. Far more serious, perhaps, is the insecurity of their futures, their meagre share of local, national and global resources, and the political and economic power-lessness that prevents them from altering this condition. Any discussion of their situation would have to start from this question of powerlessness.

NOT ENOUGH LAND?

The sort of effect the pictorial framework has on the analysis it introduces can be seen by turning to the question that is central to the problem of rural poverty and powerlessness, that of land distribution. The image of a narrow strip of fertile land crammed with so many millions of inhabitants enables most contemporary analyses of Egyptian economic development to move very quickly past the problem of access to land. With so many people occupying so little space, the problem appears to be already explained. 'The present picture is not bright,' concludes a study for USAID, 'mainly because there is just not enough land to go around. . . . The average size of a holding is two feddans, 94 per cent of all owners have less than 5 feddans each, and only 0.2 per cent have at least 50 feddans each' (USDA 1976: 172). This picture of a countryside made up of millions of tiny parcels of land persuades us once again that if Egyptian farmers are finding things difficult, it is because there are just too many of them for the space available. As before, however, we should ignore the image and check the figures.

First of all, holdings of less than 5 feddans are not as small as they may seem. With Egypt's fertile soils, year-round sunshine and permanently available irrigation water, the country is like a vast open-air greenhouse in which high yields can be obtained from two or even three crops a year. A five-feddan holding, in other words, produces between 10 and 15 feddans of crops a year. In fact 5 feddans is reckoned to be the maximum size for a family farm – that is, the maximum area a family of five can cultivate on its own, working full time, without hired labour (Ruf 1988: 236). The minimum farm size required for such a family to feed itself, assuming an annual consumption of 250 kg of grains (or equivalent) per head and a state tax of 30 per cent of production, was estimated in 1982 to be 0.8 feddan, or just over 19 qirats (1 feddan=24 qirat) (Ruf 1988: 214, 220, 236). Given the increase in yields since then, the minimum area required by 1988 would be only 0.625 feddan, or 15 qirats.

The USAID report mentioned that 94 per cent of landholdings were of less than 5 feddans, the limit of a family farm. What it failed to mention is that the remaining 6 per cent of landholders, with holdings from 5 feddans up to the legal limit of 50 feddans per individual or 100 feddans per family with dependent children, controlled 33 per cent of the country's agricultural area (Zaytoun 1982: 277). Since the mid-1970s, moreover, these large landholdings have increased in number. By 1982 they represented 10 per cent of holdings and controlled 47.5 per cent of the country's cultivated area (Springborg 1990: 29).

These official figures, furthermore, underrepresent the concentration of landholding, for they are based on village land registers. Studies of landholding in individual villages frequently reveal a much greater concentration of ownership, with the largest farms being registered under several different names to stay within the legal limit (see Mitchell 1991). The official limits also do not apply to the large holdings of agribusiness corporations. Bechtel International Agribusiness Division, for example, manages a 10,000-feddan estate in Nubariyya owned by a Gulf investor, and the Delta Sugar Company, which is 50.3 per cent owned by the Egyptian state sugar company and 49.7

per cent by a group of Egyptian and international banks, owns a 40,000-feddan estate on irrigated land in the north-central Delta (Springborg 1990: 29; International Finance Corporation 1983).

Even if one ignores these additional forms of land holding, the official figures still represent a large concentration of land in relatively few hands. The limit of 50 to 100 feddans should be compared with the limit of around 7 feddans (3 hectares) achieved in the early 1950s by the land reform programmes of Taiwan and South Korea. In Korea, less than 20 per cent of the land in 1975 was held in farms of 2 hectares or more (approximately 5 feddans), while in Egypt almost half the land (47.5 per cent) is in holdings above this limit (Lee 1979: 510; Springborg 1990: 29). On the other hand, almost one third of landholders in Egypt (32.3 per cent) have holdings under one feddan, amounting to only 6 per cent of the agricultural area (Springborg 1990: 29). In addition, a significant but unmeasured proportion of the agricultural workforce, which totalled 4.3 million workers in 1985, still remains without any land at all (Dethier 1989: 5).

If Egypt were to carry out land reform measures comparable to those of South Korea and Taiwan, the problem of landlessness and near landlessness would be eliminated. By placing the ceiling on landholding at 3 feddans (almost five times the minimum required to feed a family), at least 2.6 million feddans of land would be available for redistribution (Springborg 1990: 29, table 1). If distributed to the landless and near landless, no agricultural household in Egypt would have less than the 15 qirats required to feed itself. Total agricultural production might also increase, as there is evidence that small farmers produce larger yields per feddan than large farmers.

The discussion of land holding usually ignores the large proportion of land held in amounts over 5 feddans, and refers to such holdings as merely 'medium' sized. Only holders of more than 50 feddans are labelled as 'large' landowners. This 50-feddan threshold, incorporated into the 1961 land reform law, was the definition of 'large' landowner formulated in 1894 by the British Consul-General in Egypt, Lord Cromer, in accordance with British political and fiscal interests (Gran 1978: 370). It takes no account of the contemporary interests of most Egyptian farmers. Nor does its continued use reflect the fact that crop yields have increased by a factor of 4.5 over the last 100 years (Ruf 1988: 214; USAID/c 1989a: 224). A 50-feddan farm today produces as much output as a 225-feddan farm of the 1890s, or perhaps a 500-feddan farm if one takes into account the spread of perennial irrigation and double and triple cropping.

The question of additional land reform is simply never raised in official studies of the obstacles to Egypt's further economic development. USAID has refused to support detailed independent proposals for land reform and instead, as we will see, supports a 'free market' programme for rural Egypt that is undoing earlier reforms and consolidating land into larger farms. Thanks to the powerful image of millions of Egyptian peasants squeezed into a narrow river valley, it seems natural to assume that land holdings are already smaller than is practicable and that other sorts of solutions are required.

HIGH-PAYOFF INPUTS

Once the problems Egypt faces are defined as natural rather than political, questions of social inequality and powerlessness disappear into the background. The analysis can then focus instead on how to overcome these 'natural' limits of geography and demography. The international development industry in Egypt proposes and funds two complementary sets of methods for the solution of Egypt's problems, the technological and the managerial. One requires the massive input of capital resources from the West, the other of expertise. 'The development problem is essentially a question of the quantity, quality and proportion of resources to be devoted to development on the one hand,' according to the World Bank (Ikram 1980: 5), 'and to economic management on the other.' The productive limits set by nature, in other words, will be overcome by the forces of technology, while existing natural resources will be made more productive by more efficient management – in particular by dismantling the bureaucracy of the Egyptian state and recasting its power in the form of 'market forces.'

The naturalized image of the River Nile and its inhabitants often introduces a certain construction of history, from which will follow the need for technological rather than political solutions. The geographical determinism of the image implies an agricultural order that remains in essential ways unchanged since antiquity. Only recently, it seems, has this ancient world discovered the West – or its synonym, 'the twentieth century.' This relationship between nature and an unchanging history was expressed in part of one of the passages already quoted from a USAID report: 'One of the world's oldest agricultural economies, Egypt depends upon the fruits of the narrow ribbon of cultivated land adjacent to the Nile and to that river's rich fan-shaped delta. For more than 5,000 years agriculture has sustained Egypt' (Johnson et al. 1983: 1). A similar theme, and similar words, are found introducing an earlier report for USAID: 'The Nile Delta and its lifeline, the Nile River Valley extending southward some 600 miles, is one of the oldest agricultural areas of the world, having been under continuous cultivation for at least 5,000 years.' With this in mind we are ready to accept a few lines further down the strange idea that, 'In many respects, Egypt entered the twentieth century after the 1952 Revolution' (USDA 1976: 1). A 1977 USAID report baldly states that 'The transformation of the Egyptian village started twenty five years ago with the agrarian reform measures' (Harik 1977: 1).

The implication of these statements and images – that until the latter half of the twentieth century life in the Nile Valley had remained essentially unchanged for centuries, if not millennia – is of course highly misleading (T. Mitchell 1990). It ignores hundreds of years of far-reaching economic and political changes, such as the growth in the Middle Ages and subsequent decline of a network of world trade passing through the Nile Valley, or the consolidation in the nineteenth century of a system of export-oriented agricultural production based on the new institution of private landowner-ship, both of which involved transformations in Egyptian villages arguably at least as important as that of 1952 (Mitchell 1988). Ignoring such develop-ments creates the impression that the poverty of the Nile Valley is the 'traditional' poverty of a peasantry that has not yet or has only recently joined the 'twentieth century' – rather than very much a product of the political and economic forces of that century.

This image of a traditional rural world implies a system of agriculture that is static, and therefore cannot change itself. If Egypt 'is to fully enter the modern world,' as a report for USAID put it (USDA 1976: 25), the impetus and the means must come from outside. These external forces must carry out not simply adjustments to the existing system but what the World Bank in 1980 called a 'qualitative transformation' of Egyptian agriculture (Ikram 1980: 5). New capital investment, new irrigation methods, improved seed varieties, mechanization, and the switch to export crops such as vegetables and cut flowers to bring in the foreign capital required to finance such technologies are the principal means to achieve this transformation.

USAID's Agricultural Mechanization Project, which ran from 1979 to 1987, used just this image of a 'traditional' agricultural system to justify technological solutions to the problems of rural Egypt. The project's stated aim was to encourage the mechanization of Egyptian farming by purchasing agricultural machinery from the United States for field trials and demon-stration programmes in Egypt, financing the construction of service centres for the machinery, and sending Egyptians to the United States and other countries for training in 'the techniques of technology transfer' (USAID/c 1989c: 60). USAID awarded the US$38 million contract for this to Louis Berger International Inc. of East Orange, New Jersey. In their final report, the contractors explained the 'underlying philosophy' of the mechanization programme: 'To ensure that the project serves the purposes of development, it is necessary to relate mechanization to development theory so that mechanization does not conflict with, but rather is supportive of, develop-ment objectives' (Louis Berger International Inc. 1985: 2–1).

To this end, they drew on the ideas of T.W. Schultz (1964), whose *Transforming Traditional Agriculture* was an early classic of economic development theory. Schultz argued that farmers in 'traditional' agriculture make efficient use of their resources within the limits of the expertise and technology available to them. Through long years of trial and error, he claimed, they have eliminated inefficiencies and wastage and reached 'a particular type of equilibrium' in which the agricultural economy is 'in-capable of growth except at high cost.' Only the large-scale introduction of new technology and capital from outside this equilibrium can enable the farmer 'to transform the traditional agriculture of his forebears' (Schultz 1964: 23, 29). 'In other words,' Louis Berger International Inc. (1985: 2–1) explain, 'the continued investment in traditional inputs will produce very little in terms of an additional income stream. Consequently, the transformation from traditional agriculture is an investment problem dependent on a flow of new high-payoff inputs: the inputs of scientific agriculture.' There has probably never been a 'traditional' agriculture resembling Schultz's description. Certainly no such system has existed in Egypt in recent historical memory, still less in the 1980s when Louis Berger International Inc. arrived there from New Jersey. What is missing most of all from Schultz's account of individual farmers making rational decisions to maximize their income, as an anthropologist's critique points out (Hill 1986: 25–6), is any concept of social and economic inequality.

Schultz tested his theory using evidence from studies of a Guatemalan village by Sol Tax (1953) and a village in north India by W.D. Hopper (1957). The Guatemalan village was involved mostly in trade rather than the production of food for local consumption, so was hardly 'typical of a large class of poor agricultural communities' as Schultz claimed (Hill 1986: 25–6). The Indian village yielded evidence that the proportion of land and other resources allocated to various crops corresponded closely to their relative market prices, so that altering the allocation of inputs would not significantly increase the farmers' income (Hill 1986: 26). But this analysis pays no attention to inequality and the difference that poverty makes. 'Severely impoverished individuals,' Hill (1986: 26) notes, 'who exist in all communities, . . . are necessarily inefficient if only because they lack the resources to set themselves to work effectively.' For example, poor farmers in Egypt usually cannot afford sufficient fertilizers for their crops and get lower yields as a result. The most 'efficient' allocation of resources in Schultz's terms, Hill points out, would allocate no land at all to the poorest farmers.

Despite the lack of firm evidence for Schultz's rather dated argument, it supplied the 'philosophy' to justify American funding for the mechanization of Egyptian agriculture. Mechanization has also been heavily funded by the World Bank and by the Japanese Agency for International Cooperation (Commander 1987: 233). These external funds required large additional contributions from the Egyptian government, which was already paying for mechanization by providing farmers with subsidized loans and fuel. Consultants hired by USAID claimed that this 'high-payoff' solution to Egypt's problems would shorten the interval between crops and increase crop yields by as much as 55 per cent (ERA 2000 Inc. 1979). This claim contradicted the evidence from other countries, which suggested that higher crop yields occur with mechanization only in exceptional cases, and certainly not under conditions of intensive land use as in Egypt (Binswager 1986: 30–2). It also contradicted existing experience in Egypt, where, as Alan Richards reported, 'there is no evidence that tractor farms have higher yields or cropping intensities than unmechanized farms' (Richards 1980: 11). A subsequent study showed that indeed no increase in yields had occurred (Winrock International 1986: 41).

The demand for mechanization had intensified among large landowners in the later 1970s, due to a supposed shortage of agricultural labour which lasted into the early 1980s. This 'shortage' took the form of a temporary rise in the wages of male agricultural labourers, particularly in regions close to large cities, caused by the higher wages available for urban construction work during the building boom of that period and by labour migration to the Gulf (Commander 1987: 162–6). Agricultural wages, having averaged only one third of the average real wage for all economic sectors during the first half of the 1970s (Ikram 1980: 211), for a while began to catch up with urban wages. Large farmers, given the artificially low prices they received for their crops, were unable or unwilling to pay the higher wages. The true cause of the labour 'shortage,' in other words, was the unequal distribution of land into large farms requiring hired labour (small farms use mostly family or cooperative labour), and the low agricultural prices imposed by the state. Rather than addressing these problems, however, the state, large farmers and international development agencies turned to the high-payoff programme of mechanization. The high payoffs did not take the form of increased yields, as we have seen, but of higher profits to the new machine owners and their foreign manufacturers. The demand for rural male labour was reduced once again, and the inequalities between agricultural labourers and landowners were kept in place. It is these inequalities that mechanization and other 'high-payoff' inputs consolidate, and that accounts of the Nile Valley and the need to transform its 'traditional' agriculture keep from view.

DECENTRALIZATION AND THE MARKET

There is a second dimension to rural inequality in Egypt, and a second aspect to the historical image of the Nile Valley that tends to naturalize it. The rural poor have suffered not only from local inequalities in distribution of land and other resources, but also from the inequality of central government policies that transfer wealth from the rural population to the state. The state has come to play a major role not just in maintaining inequality, but in producing it. This is a political question, requiring an analysis of whom the state represents and who benefits from the wealth it appropriates. International development, with its naturalized images of the Nile Valley and its limited resources, depoliticizes this issue and transforms it into a question of the proper management of resources. The solutions that follow are those that are supposed to increase efficiency: decentralize the state and reconfigure some of its powers as forces of 'the market.'

Before 1952 it was the institution of large landowning that extracted wealth from the farming population and transferred it elsewhere. The 1952 land reforms preserved significant landholding inequalities, but placed a majority of farmers directly under the control of the central government and its compulsory cropping requirements, requisitions and price policies. Even if one takes into account state investment in irrigation and the subsidizing of farm inputs, the net effect of government policies between 1960 and 1985 was to appropriate 35 per cent of agricultural GDP (Dethier 1989: 220). Small farmers, moreover, suffered more than larger landowners, as the latter had greater opportunity to invest in more profitable areas such as fruit, vegetable and dairy farming. From the mid-1970s the government began to relax the compulsory cropping and price fixing policies, and from the late 1980s began to abolish them. But the changes were carried out in a way that benefited primarily larger landowners. Smallholders continued to be disproportionately involved in cotton, rice and sugarcane production, where fixed prices and compulsory deliveries to the state were the last to be relaxed. To complete the reversal of the 1952 reforms, in 1992 the government abolished the security of tenant farmers, re-establishing the 'free market' in agricultural land and confronting more than a million small farmers with eviction (Hinnebusch 1993).

The system for appropriating wealth from the countryside needs to be examined as a political process, in which changing state policies have reflected a complex of dominant (although not always coherent) social interests – those of the state bourgeoisie, the growing state-supported private sector, and larger rural landowners. The image of the Nile valley, its population, and a 5,000-year-old agriculture makes it possible to ascribe this appropriation instead to a 'tradition' of 'strong central government' determined by the very geography of the Nile Valley and stretching back to Pharaonic times. Thus the coordinator of a USAID-funded programme at Eastern Kentucky University providing management training to Egyptian local government officials explains that:

> For centuries Egypt has been governed as a political system with a highly centralized decision making process. Although there have been a few minor exceptions, this statement is valid for the period since the unification of Upper and Lower Egypt was accomplished late in the fourth millennium B.C. – i.e. for at least the past 5 thousand years.
>
> (Singleton 1983: i)

Drawing on familiar imagery, the author goes on to explain this centralized power in geographical and demographic terms:

> Integral to the question of administrative structure of the Arab Republic of Egypt is its principal social and economic problem – over-population – and the Nile river. Although the land mass area of the ARE includes 386,000 square miles, over 96 per cent of the population resides on the 4 per cent of the land area adjacent to the Nile valley and its delta.
>
> (Singleton 1983: 2)

Depoliticized in this way, the state's role in agriculture ceases to be a question of power and control over people's resources and lives. It becomes instead a problem of management. The intervention of the state has resulted in 'disequilibrium' (Ikram 1980). In the language of neo-Ricardian economics, there is supposedly a natural balance between forces of agricultural supply and demand, a balance called 'the market.' The market is a simple image for picturing the relations between farmers, labourers, landowners, state officials, international agribusinesses, and consumers, which reduces these interrelated but very unequal concentrations of power into nominally equivalent buyers and sellers, and represents the inequality between them as the market's equilibrium. Building this imagined equilibrium, which has never existed in more than a century of Egyptian capitalist agriculture except as a sequence of dispossessions, food shortages, minor revolts, violent repressions, and urgent demands for state intervention, is the aim of the process of 'structural adjustment.'

To create such an equilibrium, alongside the supply of 'high-payoff inputs,' USAID has tried to promote in rural Egypt a gradual reorganization of the role of the state, under the slogans of 'decentralization' and 'privatization.' By increasing the role of local officials and involving the country's elected village councils, USAID has tried to encourage 'democracy and pluralism' in the provinces (USAID/c 1989c: 37). To weaken the power of the central bureaucracy might be a positive step for rural Egyptians, but the actual political outcome will depend on the distribution of resources and power at the provincial, district and village levels to which authority and funds are transferred. Local government or the private sector is not necessarily more democratic, or even more efficient, than central government. Popular village councils, if they have any role at all, are frequently controlled by powerful village landowners and local officials, largely for their own benefit. Decentralization may do no more than shift exploitation from one agency to another.

A review of decentralization projects in eight different villages found that funds had gone to improvements in infrastructure and to income-producing projects such as milk refrigeration units, animal husbandry, poultry, bee and silkworm raising, date packaging, olive canning, carpentry and furniture making, and the purchase of trucks, tractors and taxis. The report, written for USAID, notes that 'naturally, not all villagers have savings that enable them to invest' in these projects, and therefore the profits accrued to those in 'middle to upper bracket income groups more than poor folks' (Harik 1977: 22). An olive pickling and canning project in a village in Fayyum, for example, provided employment for 200 villagers but served the marketing needs of just five wealthy farmers, for only wealthy farmers can afford to grow olive trees. Likewise, only the wealthy villagers can hope to raise bees, because the economic success of such an enterprise requires raising at least twenty beehives, which is a large investment. Village officials such as agronomists often enter into partnership with such farmers and undertake such projects on their own (Harik 1977: 24).

In other words, when it transfers resources to an existing system of inequality, decentralization and privatization are liable to reinforce that inequality. The profits go to large farmers and local state officials, and the poor receive at best only certain opportunities for wage labour. The USAID report acknowledges that 'the better off, the more educated and expert officials benefit more than ordinary villagers,' but argues that this is 'developmentally advisable' (Harik 1977: 24, 26). 'It would be remiss to call such a phenomenon exploitation simply because the better-off can benefit more,' the report argues. Exploitation in rural Egypt existed only 'before 1952 where cultivators were given survival wages or shares by owners.' The relationship between rural capitalists and wage earners is termed instead 'differential advantage,' meaning 'the variable ability of individuals or groups to make better use and reap greater benefits than others from available opportunities' (Harik 1977: 24). A sure way to 'reap greater benefits' from an investment, of course, is to pay lower wages to those one employs. This 'ability' is based on a distribution of land that leaves many villagers with no resources besides their labour, on the absence of a set minimum wage, and on a system of patronage, policing and surveillance in rural Egypt that prevents 'poor folks' from protesting against or organizing to change their condition. Even when exploitation is shifted from state to local or private means and renamed 'differential advantage,' it remains a politically constructed system of inequality – which decentralization and privatization programmes will only reinforce.

The reinforcement of inequalities in the name of improved 'management' of resources and of 'removing constraints to the operation of market forces' can be seen in another major strategy for reducing the role of the state. This is what the development industry calls 'cost-recovery' in the provision of government services (USAID/w 1989: 83, Annex II). Cost recovery is a euphemism for transforming healthcare, schooling, and other public services into private, fee-based institutions as in the United States. In education, for example, USAID has been pushing for the introduction of private schooling in Egypt at the secondary and university level and, on a more modest level, for a scheme to sell advertising space on the covers of school exercise books (USAID/w 1988: 76–8). In healthcare, for which USAID budgeted only $246 million from 1975 to 1989, representing 1.6 per cent of total non-military assistance to Egypt, the sum of $95 million (almost 40 per cent of the health budget) was scheduled for privatization programmes. With technical assistance from the consulting firm of Emery Associates/Taylor Associates, USAID's aim is to 'establish a sound financial structure for the health sector emphasizing cost recovery systems.' The programmes involve pushing the Egyptian government to implement 'policy changes to allow a fee structure for curative care' and 'to convert selected hospitals and clinics to fees-for-service facilities' (USAID/c 1989c: 111). One of the advantages of selective private healthcare is its increased dependence on imported US drugs and equipment. It is worth noting that even under the existing system of public financing for healthcare and schooling, Egyptians pay large personal sums on health and education. The percentage of total household consumption expenses spent on medical care in Egypt (14 per cent) is already the second highest in the world, after Switzerland, and equal with that of the US, and the percentage spent on education (11 per cent) is the third highest in the world, after Canada and Singapore (World Bank 1989c: table 10).

Privatizing healthcare, schooling and other social services does not inherently create a 'sound financial structure.' What it does do is transfer the source of funding from government revenue, to which people contribute according to their means, to fees or insurance premiums, for which the poor must pay as much as the rich. This creates or reinforces an unequal access to healthcare and schooling. Privileged levels of education and health become, in turn, a mechanism for transferring wider social privilege from one generation to the next.

The rhetoric of management, financial soundness and market forces depoliticizes these complex issues. Programmes for decentralization and cost recovery transform questions of social inequality and powerlessness into issues of efficiency and control – in the same way as agricultural mechanization programmes transform the question of inadequate wages and landlessness into issues of technological efficiency. The underlying political issues people face can be ignored, because the naturalized imagery of the Nile and its population has reduced the topic to questions of natural resources and their more efficient control. It never need be asked at whose cost efficiencies are to be made, or in whose hands control is to be strengthened.

OBJECTS OF DEVELOPMENT

A final aspect of the geographical image of the Nile Valley is the way it removes from sight the participation of development agencies in the dynamics of Egyptian political and economic life. By portraying the country and its problems as a picture, laid out before the mind's eye like a map, the image presents Egypt itself as something natural. The particular extent of space and population denoted by the name 'Egypt' is represented as an empirical object. Development literature reproduces the convention that Egypt exists as a sort units. The workings of this unit – its economic functions, social interactions, and political processes – are understood as internal mechanisms. They constitute the unit's inside, to be distinguished from economic and political forces that may affect it from outside.

This convention of imagining countries as empirical objects is seldom recognized for what it is – a convention. The relations, forces and movements that have shaped people's lives over the last several hundred years have never in fact been confined within the limits of nation-states, or respected their borders. The value of what people produce, the cost of what they consume, and the purchasing power of their currency depend on global relationships of exchange. Movements of people and cultural commodities form international flows of tourists, television programmes, information, migrant workers, refugees, technologies and fashions. The strictly 'national' identity of a population, an economy, a language or a culture is an entity that has to be continually reinvented against the force of these transnational relations and movements. This has always been the case, for the global interrelationship of commodities, populations, languages and ideas is far older than the modern invention of nation-states.

The apparent concreteness of a modern nation state like Egypt, its appearance as a discrete object, is the result of recent methods of organizing social practice and representing it: the construction of frontiers on roads and at airports, the attempt to control the movement of people and goods across them, the producing of maps and history books for schools, the deployment of mass armies and the indoctrination of those conscripted into them, the representation of the nation-state in news broadcasts, international sports events and tourist literature, the establishing of a national currency and language, and, not least, the discourse of 'country studies' and national statistics of the American-based international development industry.

These essentially practical arrangements of language, imagery, space and movement are mostly of very recent origin (see T. Mitchell 1988, 1989). We tend to think of them as processes that merely mark out and represent the nation-state, as though the nation-state itself had some prior reality. In fact the nation-state is an effect of all these everyday forms of regulation and representation, conjured up by them in the appearance of an empirical object. The geographical imagery of the Nile and its inhabitants that introduces so many studies of Egyptian development invokes and reproduces this effect.

MODEL ANSWERS

There are two consequences of the way development economics takes for granted the nation-state as its object. The first is the illusion of the model. Portrayed as a free-standing entity, rather than a particular position within a larger arrangement of transnational economic and political forces, an individual nation-state appears to be a functional unit – something akin to a car, say, or a television set – that can be compared with and used as a model for improving other such units. This supposed comparability is emphasized by the annual volumes of statistics produced by the World Bank and other international development agencies. Economic features of one state appear to be neatly transferable to other states, ignoring their different position in larger economic and historical networks.

The example of this in Egypt's case is the way agencies like the IMF and

USAID promote the growth of exports as the solution to the country's economic problems. Egypt is to develop the export of winter vegetables and cut flowers to markets in Europe and the Gulf, along with textiles and possibly other light manufactured goods, in order to earn the hard currency to keep up interest payments on its foreign debts. The idea is that Egypt and similar countries should follow the path of the so-called economic miracles. of East Asia – Singapore, Hong Kong, Taiwan and South Korea.

The notion that solutions from East Asia provide a model for other Third World states is curious. Egypt's merchandise exports in 1987 amounted to less than one-fifth of one per cent of world trade. More than two-thirds of this merchandise consisted of oil, the supply of which will decline in coming decades. To match the per capita level of exports of Singapore, Egypt would have to expand its exports to capture 23 per cent of world trade – or significantly more than the merchandise exports of Japan and the United States combined (World Bank 1989c: table 14). Even the far more modest goal of matching South Korea, whose exports were worth $1,120 per capita in 1987, would require Egypt to capture a massive 2.35 per cent of world trade. This would involve a forty-fold increase in non-oil exports, from the present annual level of about $1.25 billion to more than $52 billion (World Bank 1989c: tables 14, 16).

There is no evidence that Europe's demand for airlifted shipments of Egyptian cut flowers and winter tomatoes might grow by even a fraction of this amount. In the absence of the kind of far-reaching land reform carried out in South Korea, there is also no evidence that such export policies would be of any benefit to the landless and near-landless majority of rural Egypt. In fact other cases of agro-export policies suggest the opposite. For example, Brazil, which is 'a stunning success as measured by investment in agrofood production and exports, is also a nightmare of evictions from the land, displacement of local food systems, hunger, and social unrest' (Friedmann 1993: 50). Finally, as Streeten (1982: 166–7) and others have noted, this export-oriented solution is supposed to occur not during a period of enormous regional and global demand, such as that generated by Japanese growth and the Vietnam war during the period when the East Asian economies began to expand, but in a period of economic retrenchment – and in a period when a dozen or more large Third World economies are adopting similar remedies and competing for the same limited market.

There is a second consequence of the way the imagery of the Nile Valley and its people – and the larger discourse of development – constitute Egypt as a self-contained object. By setting out this sort of visual image of Egypt, the country is imagined as an object that exists apart from the discourse that describes it. The geographical metaphor that introduces the reports of an organization like USAID in Cairo evokes an entity 'out there': Egypt, laid out like a map as the object of the organization's planning and knowledge. The organization itself, the metaphor suggests, is not an aspect of this object.

It stands above the map of Egypt to measure and make plans, a rational centre of expertise and policy-making that forms no part of the object observed. USAID is not marked, so to speak, on the map.

Development discourse thus practices a self-deception – what Partha Chatterjee calls 'a necessary self-deception,' for without it development could not constitute itself (Chatterjee 1993: 207). Development is a discourse of rational planning. To plan effectively, it must grasp the object of its planning in its entirety. It must represent on the plans it draws up every significant aspect of the reality with which it is dealing. A miscalculation or omission may cause the missing factor to disrupt the execution of the plan. Its calculations must even include the political forces that will affect the process of execution itself.

This calculation has a limit, however, which is where the self-deception is required. As Chatterjee (1993) points out, the political forces which rational planning must calculate affect not only the execution of plans but the planning agency itself. An organization like USAID, which must imagine itself as a rational consciousness standing outside the country, is in fact a central element in configurations of power within the country. Yet as a discourse of external rationality, symbolized as the consciousness that unfolds Egypt as a map, the literature of development can never describe its own place in this configuration of power.

Consider the case of USAID's decentralization programme, designed to reduce the role of the state and encourage 'democracy and pluralism' by channelling development funds to private initiatives at the village and district level. The report quoted earlier suggested that among the principal beneficiaries of these funds were local government officials, state agricultural engineers, and other members of the state apparatus. The other main beneficiaries, wealthy farmers, often entered into partnership with such officials (Harik 1977: 24). Far from encouraging a 'private sector' in opposition to the state, such programmes make the state an even more powerful source of funds and site of patronage. The new accumulations of wealth are never more than semi-private, for they are parasitic on this strengthened state structure.

A similar process has been described by Robert Springborg at the national level. He gives the example of one recipient of USAID funds, a man who was chair of the Foreign Relations Committee of the State Advisory Council (*Majlis al-Shura*). He was from a family long involved in Egyptian politics and business and had a personal wealth of several millions. USAID provided him with two sizeable loans to purchase American irrigation equipment for large tracts of reclaimed land he owned, parts of which he sold off immediately after the equipment was installed. Springborg (1989: 280) concludes that 'a large proportion of USAID private sector assistance has been utilized by those well connected within the state apparatus to turn quick profits' – to the extent that even USAID economists in Cairo have become disillusioned with the programme of private-sector loans.

These examples illustrate the characteristic limits of development discourse. The major goal of USAID programmes in Egypt is to develop what is termed the 'private sector.' The actual effect of these programmes, however, is to strengthen the power of the state. This is not simply some fault in the design or execution of the programmes. USAID itself is a state agency, a part of the 'public sector,' and therefore works in liaison with the public sector in Egypt. By its very presence within the Egyptian public sector it strengthens the wealth and patronage resources of the state. USAID is thus part of the problem it wishes to eradicate. Yet because the discourse of development must present itself as a rational, disinterested intelligence existing outside its object, USAID cannot diagnose itself as an integral aspect of the problem.

OPPOSED TO SUBSIDIES?

This difficulty in seeing itself as a part of the scene reflects a much larger deception. The prevailing wisdom of organizations like the World Bank, the IMF and USAID is that the problems of a country such as Egypt stem from the restrictions placed on the initiative and freedom of the private sector (Laïdi 1989). The programme of 'structural adjustment' these organizations have attempted to impose on Egypt since the late 1970s, particularly following the 1985–86 collapse of oil prices which left the country incapable of keeping up payments on its international debts, aims to dismantle the system of state subsidies and controls and enable the private sector to flourish in the unrestricted freedom of 'the market.' Prices Egyptians pay to consume, or receive for producing, food, fuel and other goods are to reflect prices in the international market.

Yet it hardly needs pointing out that world prices for many major commodities are determined not by the free interplay of 'private' market forces but by the monopolies or oligopolies organized by states and multinational corporations. Oil prices are determined not by the users of cars and electricity but by the ability of producer states to coordinate quotas and price levels. The price of raw sugar (a major Egyptian industrial crop), whose volatility is more than twice that of any other commodity monitored by the World Bank, is determined largely by US and other government price support programmes. Only about 14 per cent of world production is freely traded on the market (International Finance Corporation 1983). The international market for aluminium, one of the major heavy industries in Egypt, also operates under extensive state controls.

Perhaps the most significant example is the world grain market. One of the arguments against Egypt producing the staple foods it needs is that it cannot compete in the world market against the low grain prices of American farmers. Yet these prices are the product of subsidies and market controls.

150

American agriculture, operating under an imperative of constant growth, has come to be dominated by giant corporations that supply the inputs to farming and process and market its products. By the 1980s over three-quarters of the American farm supply industry was controlled by just four firms. Six corporations, all but one of them privately owned, controlled 95 per cent of US wheat and corn exports and 85 per cent of total world grain trade (Wessel 1983: 91–3). As Congressional investigations have shown, the monopolies these firms enjoy enable them to control the market and administer prices. Squeezed by these monopolies on both ends, inputs and marketing, American farmers have found themselves having to grow ever larger quantities of crops merely to survive, investing constantly in new technologies and getting increasingly into debt.

Since the 1930s, this accelerating treadmill has put more than two-thirds of the country's farms out of existence – and continues to ruin them today (USDA 1989: 20). To mitigate the system's effects, the state has introduced massive subsidies, starting with the price supports and crop controls of the New Deal programmes, followed by the subsidized exports of the post-war Marshall Plan, the Public Law 480 programme (which financed up to 58 per cent of US grain exports during the 1950s and 1960s), and President Nixon's 1972 New Economic Policy (which further subsidized exports, and boosted prices by paying farmers to take 62 million acres out of production, an area equal to ten times the total cultivated area of Egypt). As a result of these policies, by 1982 American grain was being sold at prices 40 per cent below estimated average production costs, and keeping farmers afloat was costing US$12 billion a year in state subsidies (Wessel 1983). Despite the low producer prices, moreover, consumer prices remained so high that 40 million Americans required government subsidies to purchase food, costing a further US$27 billion a year in Federal funds. Government export subsidies pay for middle- and upper-class consumers in non-Western countries to shift to a meat-centred diet and thus expand the market for American feed grains. As we have seen, the largest site in the world to be incorporated into this system of state-subsidized American farming has been Egypt. The arm of the state that has organized this incorporation is USAID.

The self-deception of USAID discourse is not just that it sets up an object called Egypt in which it cannot recognize its own internal role. It is that this supposed object is caught up in a much larger configuration of power, a network of monopolies and subsidies misleadingly named the world 'market,' of which USAID itself is but a subsidiary arm. An agency devoting itself to the cause of dismantling subsidies and promoting the 'private' sector, is itself an element in the most powerful system of state subsidy in the world.

USAID's role as a source of subsidies to American agriculture and industry can be seen by examining how it spent the US$15 billion budget for 'Economic Assistance' to Egypt from the start of its operations there in 1974/5 up to 1989 (see Figure 7.2). Almost every penny of this amount, it

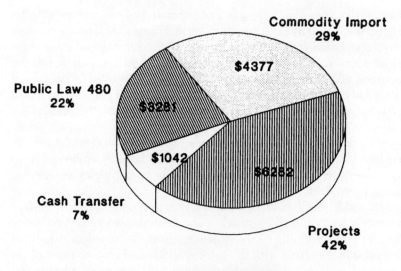

Figure 7.2 US economic assistance to Egypt ($ million)
Source: USAID/c 1989c

can be shown, was actually allocated to American corporations. Just over half the total, first of all, represents money spent by Egypt to purchase goods from the United States. The PL480 Food Aid programme and the Commodity Import Program, totalling about US$7.7 billion up to 1989, enable Egypt to purchase grain, other agricultural commodities, agricultural and industrial equipment, and other US imports. About half the commodities are paid for in dollars, with the United States providing low-interest long-term credit. The other half are paid for immediately or on short-term credit, but in Egyptian pounds.

A further US$1 billion of the total aid was also paid directly to the United States, this part by the US government itself, in the form of so-called Cash Transfers used to keep up payments on Egypt's military debt. United States law stipulates that all aid except food must be stopped to a country that falls more than a year behind in military debt repayments, as Egypt began to do in the winter of 1983–84. The US government responded to this threatened collapse of the entire system of subsidies to its own private sector by converting all subsequent military loans to grants, allocating the bulk of those grants for progress payments to itself on earlier Egyptian arms purchases (US Congress 1984: 508), and instructing USAID in the meantime to circumvent the law by setting aside about US$100 million a year from economic development funds as Cash Transfers, to be deposited in the Federal Reserve

Bank of New York and then returned to Washington as Egypt's monthly interest payments on its military debt. When this illegal diversion of economic development funds for military purposes was discovered by Congress (thanks to the leak of an AID cable to the *Washington Post*), USAID denied it was happening – but continued the practice. The law, a USAID lawyer later admitted, 'was an academic question, since actual CT [Cash Transfer] expenditures were untraceable' (US Congress 1987: 78). So a total of US$8.7 billion, or 58 per cent of all US economic assistance, was spent directly in the United States rather than on development projects in Egypt, and most of this 'American aid' in fact represents money paid by Egypt to America.

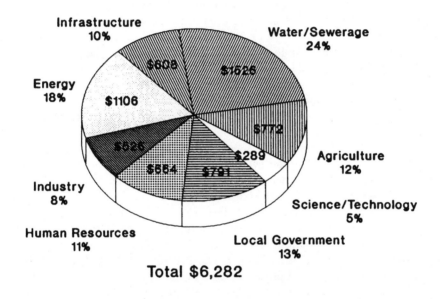

Figure 7.3 Projects by sector, FY 1975–89 ($ million)
Source: USAID/c 1989c

The remaining 42 per cent of US economic assistance funds to Egypt, totalling US$6.3 billion, were earmarked for development projects within the country (see Figure 7.3). Yet none of this money was transferred directly to Egypt. The entire amount, as far as one can tell, was spent in the United States, or on American contractors in Egypt. The major recipients of the funds were large American manufacturing, construction and consulting firms. Over US$1 billion went to corporations like General Electric, Westinghouse and Overseas Bechtel to purchase thermal power turbines and electricity distribution systems. More than US$1.5 billion went to US engineering and construction firms to build sewage networks and drinking water plants. Some US$300 million dollars went to American Telephone and Telegraph and other US communications companies to supply telephone equipment for Cairo and Alexandria. Over US$200 million went to Ferguson International of Cleveland, Ohio, and other US firms for the construction of two cement plants. American agribusiness and engineering firms received multi-million dollar contracts to supply grain silos and fats storage facilities for the country's expanded food imports. Dredging and earth-moving equipment was purchased from firms like Caterpillar, John Deere, and International Harvester. Westinghouse Health Systems received tens of millions of dollars to improve rural and urban 'health delivery.' And hundreds of millions of dollars went to American universities and research institutes to provide training in agricultural sciences, management and technology transfer (USAID/c 1989c).

Many of these projects also required local payments within Egypt in Egyptian pounds. In 1988 such local implementation costs were said to amount to about LE200 million annually (LE=Egyptian pounds), equivalent then to just over US$100 million, or about 10 per cent of annual US dollar aid for development projects (USAID/w 1988, main volume: 21). But such payments are not made from US dollar funds. Instead, local currency funds paid by the Egyptian government to purchase American imports under the Commodity Import Program, mentioned above, are used by USAID in Cairo to pay for all local costs. In other words, the local implementation expenses of development projects (and even the local operating costs of the USAID mission in Cairo) are paid by the Egyptian government, in exchange for commodities imported from the United States.

Many millions of Egyptians, needless to say, have benefited from this economic assistance, at least in the short term. The supply of power stations, sewage networks, telephone exchanges, drinking-water plants, irrigation systems and numerous other basic infrastructure projects and services has improved the deteriorated physical fabric of the Egyptian economy. At the same time, several Egyptian scholars have argued that these benefits have come at the price of a crippling dependence on imports of American food, machinery and technology (Husayn 1985; Galal 1988). In the 1980s the USA became the largest supplier of Egyptian imports, and by 1988 the country's imports from America had reached LE1.94 billion. The following year, 1989, they jumped more than 50 per cent, to LE2.93 billion.

This dependence, and the astronomical levels of debt it caused, has given the United States a powerful position of influence within the Egyptian state. USAID conducts what it terms 'cabinet-level dialogue' on macroeconomic policy with the Egyptian government. At times, USAID reports, when this 'dialogue' has not been 'completely successful' – meaning that the Egyptian government has rejected or delayed implementing American demands – 'annual releases of funds have been delayed' (USAID/c 1989c: 8). Acquiring at every level of the Egyptian bureaucracy this sort of 'policy leverage,' as it is called, has now become the principal criterion according to which USAID development projects in Egypt are evaluated (Springborg 1989: 258–9). And all this is achieved by a programme whose larger effect is to provide vast subsidies to the so-called private sector in the United States – both directly, by the purchase of billions of dollars of its products, and indirectly, by converting Egypt into a future US market.

OPPOSED TO THE STATE?

Thus USAID operates, more or less successfully, as a form of state support to the American private sector, while working in Egypt to dismantle state supports. None of this is explained in the discourse of USAID itself, which pretends to stand outside Egyptian politics, conducting merely a 'dialogue' at the rational, detached level of 'policy.' Yet there is even more that is missing from the discourse of development on Egypt. The US$15 billion of assistance between 1974 and 1989 represents only about one half of US aid to Egypt. The other half consists of economic assistance to the Egyptian military. From 1985 to 1990 total American aid to Egypt was more than US$15 billion, half of which consisted of military aid (EIU 1989: 53; USAID/w 1989, main volume: 293). The military aid was largely spent in the United States to purchase weapons, representing since 1985 alone a further US$7.5 billion of subsidies to US industry. The US excludes this aid from its figures for 'Economic Assistance,' however, and lists it separately as military aid.

So American aid, which describes its aims in Egypt as the support of the private sector and 'pluralism,' has in fact channelled half its funds (or more, if one includes the Cash Transfer payments with which America paid back to itself Egypt's military loans) directly into the most powerful sector of the state. The Egyptian military, with the support of American funds, has developed into a major presence within the country's economy. Its arms industries, which receive state subsidies but whose income goes into military rather than national accounts, is the country's largest manufacturing sector, producing exports estimated to be worth about three times the total of all other non-textile manufactures (Springborg 1989: 111). The army has also moved into civilian manufacturing, symbolized by the deal it negotiated with General Motors in 1986 to manufacture passenger cars. Under pressure from the American Embassy, USAID pledged US$200 million from its aid budget to subsidize this project (Springborg 1989: 110).

Agriculture is another sector in which the military has become a dominant presence, through the acquisition of reclaimed land and the development of food processing industries, particularly in meat, fruit and vegetables. Its Food Security Division represents by far the largest agro-industrial complex in the country, producing in 1985–86 LE488 million worth of food, or almost one-fifth of the total value of Egyptian food production (Springborg 1989: 113). The military has also played a major role in the construction of bridges, roads, telephone systems and other infrastructure projects. All these activities have provided plentiful opportunities for patronage and personal profit-making. Together with the construction of its own housing, hospitals, shops, resorts, tourist services and elite training colleges, such developments have transformed the military into what Springborg calls 'an almost entirely autonomous enclave of middle-class modernity in an increasingly impoverished and marginalized Third World economy' (Springborg 1989: 107).

This enclave of privilege has received substantial support from Egyptian public funds. Among the world's twenty or so lower-middle income countries, Egypt in 1987 came near the bottom of the list in the proportion of central government expenditure devoted to health and education (only Syria and Mexico spent proportionately less) and near the top in the proportion (20 per cent) devoted to the military (only Syria, Jordan, North Yemen and El Salvador spent proportionately more). At the same time, power and privilege on this scale would never have been possible without the multi-billion dollar contributions of United States aid.

Despite its massive presence in the Egyptian economy, the large proportion of government funds it consumes, and its even larger proportion of total American support, the military receives almost no attention in the literature of organizations like USAID and the World Bank. Given the supposed objectives of developing the private sector and pluralism, the silence of this discourse is astonishing. The silence reflects the necessary limits of the discourse of development. A systematic inquiry into the economy and power of the Egyptian military would reveal its relations to American military industries, to the system of state subsidies on which those industries depend, and thus to the larger object of American aid programmes. In the same way, as I have suggested, a proper analysis of Egyptian agriculture examining the causes of the shift to meat production and the country's resulting shortages of food and growing indebtedness would reveal the connections between these events and the crisis of American farming and the remedy of subsidized food exports. Such analyses would serve as a reminder that the discourse of development is situated within, not outside, such global relationships.

That is the reason for the silence. Development discourse wishes to present itself as a detached centre of rationality and intelligence. The relationship between West and non-West will be constructed in these terms. The West possesses the expertise, technology and management skills that the non-West is lacking. This lack is what has caused the problems of the non-West. Questions of power and inequality, whether on the global level of international grain markets, state subsidies, and the arms trade, or the more local level of landholding, food supplies and income distribution, will nowhere be discussed. To remain silent on such questions, in which its own existence is involved, development discourse needs an object that appears to stand outside itself. What more natural object could there be, for such a purpose, than the image of a narrow river valley, hemmed in by the desert, crowded with rapidly multiplying millions of inhabitants?

ACKNOWLEDGEMENTS

Earlier versions of this chapter were published in Arabic in *Misr fi al-Khitâb al-Amrîkî* (Nicosia, Cyprus: Dâr 'Iybâl, 1992) and in English in *Middle East Report* (March 1991). I am grateful to the editors of *Middle East Report* for permission to publish this version here. I am also grateful to Lila Abu-Lughod, Nick Hopkins, Karen Pfeifer, Marsha Pripstein, Yahya Sadowski, Joe Stork, and Jim Toth for comments on earlier drafts, and to Sophia Anninos for research assistance. Any shortcomings of the chapter remain my own responsibility.

8

MODERNIZING MALTHUS
The World Bank, population control and the African environment

Gavin Williams

The pressure of population is causing desertification to accelerate, because it forces people and their livestock farther into the marginal grassland. The productive capacity of land is failing because of shorter rotations, soil erosion and overgrazing. Growing population also raises the demand for fuelwood and cropland, and the resulting deforestation increases runoff and erosion, lowers ground water levels, and may further reduce rainfall in arid areas.

(World Bank 1989a: 22)

In several countries overpopulation is putting unsustainable pressure on agricultural land. In many places traditional farming land is already over-cultivated, and more fragile land is being exploited to meet the needs of the growing population. Without agricultural modernization the result is rapid desertification, deforestation, and loss of vegetation cover. With sound practices and technological innovations Africa might eventually accommodate several times its present population. But this will take time, and, meanwhile high population growth spells disaster.

(World Bank 1989a: 40–1)

Population pressure is pushing farmers onto marginal lands and causing deforestation, severe soil erosion and declining productivity. Poor families cut whatever wood they can for essential fuel. . . . Inevitably, the poor suffer most.

(World Bank 1989a: 44)

INTRODUCTION

The World Bank's (1989a) *Sub-Saharan Africa: From Crisis to Sustainable Growth* attracted extensive press comment which focused particularly on one of two recurrent themes of the text: the crisis of African governance (Beckman 1990) and the fact that 'Africa must grapple with two major trends: explosive

158

population growth and accelerating environmental degradation' (World Bank 1989a: 4). The World Bank thus gave authority to a common popular image of Africa as a continent confronted by a chronic problem of food production, which is in turn the outcome of the exposure of a fragile ecology to the claims of a rising population. These pressures appear to render African countries vulnerable to famine.

The World Bank's obsession with population growth is of long standing. During the 1960s, when the World Bank's attentions were focused on Asia, it diagnosed population growth as the major problem of agricultural development in the Third World. Its solution was to reduce the birth rate and to promote yield-enhancing technologies (the 'green revolution') to raise agricultural productivity. In the 1980s, the World Bank's attention, together with its diagnosis, has shifted to Africa. In 1986, the Bank declared in its study of *Population Growth and Policies in Sub-Saharan Africa* that it views population assistance as its highest priority in Africa (World Bank 1986: 6). The concern with population growth is linked to a fashionable revival of the central obsession of colonial agricultural policies – environmental degradation. As the Bank argues, 'population pressure has already contributed to ecological damage in parts of Africa ... making some populations increasingly vulnerable to the vagaries of climate' (World Bank 1986: 24, also 19, 26–7).

Population limitation is a constant refrain throughout *Sub-Saharan Africa* (World Bank 1989a: 22, 40–1, 44) where it is identified as the main cause of environmental degradation. The passages cited above all repeat the same message: more people – less land – lower productivity – less food for everyone. The fecklessness and ignorance of the poor are the source of their own suffering. To save themselves, they must adopt the contraceptive and agricultural technologies on offer from the international aid agencies. The association between the growth in population, the increasing scarcity of land, the degradation of the environment, and the fall in the *per capita* supply of food appears to be so obvious that it can be taken for granted; neither argument or evidence is needed to support this self-evident fact.

These presumptions form the 'principal thesis' of a subsequent draft World Bank report on *The Population, Agriculture and Environment Nexus in Sub-Saharan Africa* (World Bank 1990b, 1990c). Thus, 'rapid population growth rates, lack of agricultural development, and degradation of natural resources are mutually reinforcing' (World Bank 1990b: 1). Women are also brought into the equation, as contributors to and victims of the demographic, population and environmental crises. The authors of the draft report recognize other constraints on agricultural production (such as civil wars, inappropriate policies and failing international prices for exports) but these are 'not strictly part of the nexus between agricultural, environmental and population problems' (World Bank 1990b: 3). Nor are the impact of migration, AIDS, tropical diseases and illnesses related to poverty and squalid living conditions (World Bank 1990b). The impact of endemic and pandemic diseases on the age and gender structure of the population, and people's productive capacities is ignored.

The 1990 draft report interprets each of the three aspects of the 'nexus' – population, agriculture and environment – and the interactions among them in terms of a common model (World Bank 1990b). The equilibrium characteristic of a 'traditional' society breaks down, so that practices, which once formed part of a stable system, now cause problems and interact to exacerbate them. The task of governments and international agencies is to direct the transition to practices appropriate to the 'modern' world.

CONSTRUCTING AFRICAN DEMOGRAPHY

The World Bank's *Population Growth* draws attention to the high birth, fertility and population growth rates in all the countries of sub-Saharan Africa (World Bank 1986: 8). While the rate of population growth has been falling elsewhere since the 1960s, it has continued to rise or to remain very high in all Sub-Saharan African countries except Zimbabwe and Botswana. The Bank explains this high and rising rate across most African countries by generalizing common social, economic and cultural factors across the continent: 'In Africa, the prevailing young age at marriage for women, the frequency of polygamy, an unequal work burden between the sexes, and the low educational levels of women all combine to perpetuate the low status of women' (World Bank 1986: 39). 'Traditional cultures place a premium on high fertility,' perhaps in response to such 'age-old facts' as the importance placed on children as a source of labour and of security (World Bank 1986: 12). At the same time, at least among some groups of women, lengthy breastfeeding and sexual abstinence after a birth have been diminishing (World Bank, 1986: 11).

According to the World Bank (1989a: 41), the benefits attributed by Africans to large families are illusory. The view that 'land [is] abundant and labour [is] scarce' is simply an illusion shared by most Africans, who are 'not yet persuaded of the advantages of smaller families.' However, 'expanding opportunities for parents to invest in education' are beginning to erode 'the traditional benefits of large families.' Support from 'extended families,' and government provision of 'free or highly subsidized schooling and health care' means that parents 'do not themselves pay all the costs of childraising.' Further, in many African countries, the costs of bearing and rearing large families, including greater maternal, infant and child mortality, fall most heavily on women, but it is men who decide to have them – which brings us back to 'traditional culture' (World Bank 1986: 12, 19).

The authors of the 1990 World Bank draft report observe that 'women may be choosing to maintain high fertility rates to maximize the number of children, in part to provide more labour to farming, water fetching, and fuelwood gathering tasks. Additional labor may be the only factor of production which women can easily add to produce more' (World Bank, 1990c: 3–4; 1990a: 24, also 13). This approach echoes that in Caldwell's (1976) demographic transition in Africa. He argued that parents will continue to want large families as long as resources flow from the younger to the older generation. The spread of education and the nucleation of families will, conversely, encourage limitation of families.

The World Bank has two models in mind. In the first, 'traditional' society is partially exposed to the influences of 'modernity.' This is combined with a second model in which people make apparently rational decisions – but with perverse social consequences since they do not bear, or even observe, the costs of these decisions. As a result African societies have yet to make the 'demographic transition' to a low mortality/low fertility pattern. Their high rates of fertility are still those appropriate to a society with high mortality rates (World Bank 1990a: 10).

The logic of these models may be contradictory. The World Bank commits itself to 'encourage expansion of basic health services, female education, and other development programmes that generate demand for small families' (World Bank 1986: 6; see also World Bank 1990b: 12, 38; 1990c: 7). Provision of education and health facilities offers greater predictability and may encourage a transition to a 'modern' pattern of limiting family sizes. But if the government bears most of the costs of these services, individuals may continue to breed without regard to the true costs of their decisions. Improved health services, as the World Bank (1986: 11, 53) recognizes, may reduce the very high rates of primary and secondary infertility in many parts of Africa, and thus increase the birth rate. Better health services and education for women may contribute more to reducing child and infant mortality than to curbing fertility. It may not be surprising that, in Africa in recent decades, declining mortality has often been combined with rising fertility.

The World Bank's 'population assistance' is clearly primarily intended to restrain the growth of population rather than to allow women greater control over their own lives; the benefits which contraception may offer for the health of women and their children are secondary (World Bank 1990b: 36–9; see Hartmann 1987). The 1986 report emphasizes the need to remove restrictions on the use of injectable contraceptives and sterilization (World Bank 1986: 41–2) – methods which may be most effective in administering programmes to control population because they depend least on women taking responsibility for themselves. The report ignores the need to educate women about the possible adverse consequences of different contraceptive methods. Funding for health, education and family planned parenthood programmes will be safer if they are pursued for their own sakes rather than because they might lead women to reduce their fertility; otherwise, they might be cut back when they allow people to have more surviving children instead of encouraging them to have fewer.

HISTORY, GEOGRAPHY AND STATISTICS

The World Bank's (1986) generalized analysis of African demographic tendencies ignores the complex and varied historical processes which have shaped the rise, fall, and age and gender distributions of populations, and their patterns of settlement and migration during the slave trade, colonial conquest and colonial rule (Kuczynski 1948, 1949; Kjekshus 1977; Cordell and Gregory 1987; Inikori 1982; Gendreau *et al.* 1991). History is replaced by a stylized transition from tradition to modernity. Thus 'age-old facts' are used to explain relatively recent changes. Geography is simply ignored. There is no mention of the marked differences in patterns of fertility and rates of population growth within African countries, or analysis of contemporary differences in the rates of population growth between one country and another, or one region and another. Estimates aggregated at national, even continental, levels obscure the variations which we need to identify if we are to begin to understand the complex demographics of African societies.

The estimates which the World Bank cites in successive documents for population growth, fertility rates, crop yields and agricultural performance are of dubious statistical provenance. Often, they cannot be reconciled with other observations in the same set of tables or in different reports. They are appended to reports, but appear to have little bearing on their analyses or conclusions.

The World Bank's figures for African countries offer scant comfort for its diagnosis or policies. Kenya, Zimbabwe and Ghana, with the highest contraceptive use and most effective systems for providing contraceptives, had the lowest infant mortality rates. However, they also had the highest population growth rates, and between them had the highest crude birth rates and total fertility rates of all the countries cited (World Bank 1986: 8–9, 42, 53). These are not the societies where early marriage for women, or polygyny and the other stereotyped attributes of rural African culture are most prevalent. Indeed, polygyny seems to be associated with lower rather than higher rates of fertility.

The figures in the World Bank (1990b) report initially appear more promising. Thus, Botswana and Zimbabwe, who with Kenya get the highest 'Family Planning Program Effort Scores,' showed a marked decline in total fertility rates and in crude birth rates between 1965 and 1987. But the population growth rate in Kenya is among the highest and the total fertility rate has only declined marginally. However, the report states that the total fertility rate in Kenya declined from 7.7 to 6.7 between 1984 and 1989. Family planning efforts in Rwanda, Senegal, and Tanzania, who have the next highest score, have yet to stop fertility and population growth rates from rising.

These comparisons may not mean much, given the variations in the capacities of governments (and statisticians) to enumerate or estimate the number and fertility of their citizens. Many of the numbers may be no more than artefacts of procedures for recording official statistics. If anything, the figures suggest that the highest rates of population growth among and within countries are associated with relatively good health and educational services and/or with rising prosperity, especially in rural areas. These apparent simple relations among cross-sectional observations may well obscure complex patterns of change. They certainly raise doubts about the simple associations presumed in the World Bank models.

Figures for national population growth rates are complemented by arbitrary estimates for trends in agricultural production. The World Bank (1990c) baldly declares that declining soil fertility has led to a decline in crop yields, citing a table of doubtful value which does, however, show a rise in crop yields over 20 years for a majority of countries cited, including the most populous. The tables do show a rise in cereal imports to 1988 (World Bank 1990c: World Bank 1990b: Tables 6, 7). The report predicts a 'food gap' of 245 million tonnes (more than the current world trade in cereals) by the year 2020. The spectre of 'massive starvation and human degradation' threatens (World Bank 1990b: 7).

Rising food imports are invoked both as evidence and as a consequence of declining *per capita* food production, but not as one of the causes. Arguably, cheap imports at prices below production costs have displaced rather than supplemented local food production. Further, a third of the 1988 total, and over half of the increase in cereal imports since 1974, can be accounted for by four war-torn countries (Ethiopia, Sudan, Angola and Mozambique). More significantly, the levels and rate of growth since 1970 of cereal imports into sub-Saharan Africa are far less than for North Africa, despite its less fecund demographic regime (Raikes 1985: 15–16).

The World Bank (1986) study explicitly recognizes that 'rapid population growth does not necessarily prevent per capita income from rising. In some circumstances population growth may even contribute to development.' Societies may, for example, benefit from improved health, education and technologies, and from the discovery or recovery of natural resources. However, Africa poses special problems because Africans are poor and ignorant – 'a weak human resource base for agricultural research, extension and entrepreneurship' – as well as ill-governed, and thus unable to make effective use of the resources available to them (World Bank 1986: 21–3). Africans are presumed unable to adapt to the rising pressure of population and the need to feed an expanding urban population. There is no recognition of how, over the last century, African farmers have raised their productivity in different ways – growing higher-value (cocoa) and less labour-intensive (maize, cassava) crops, adopting new tools and methods of cultivation, working longer hours (especially women) and extending the areas under cultivation.

While some parts of Africa have long had dense concentrations of people relative to the resources available, most have been relatively sparsely populated. People have thus migrated to new places to grow crops and graze livestock. Denser populations reduce the costs of transport and marketing and of providing roads and services, and expand demand for local produce. Rising population may thus be of cumulative benefit in many places (World Bank 1986: 22), leading farmers to cultivate land in ways which conserve soil more effectively. Conservationary methods of land use are often labour-intensive and thus dependent on denser patterns of human settlement (Mortimore 1989: 209; Tiffen *et al.* 1994). However, these potential benefits of population growth are all discounted.

The World Bank rightly recognizes that demographic changes have important ecological and economic implications. They ought therefore to examine them more seriously. The extremely high rate of population growth in Africa does place severe strains on the capacity of societies and governments to meet the rising demands for health services, education and jobs. More people do make additional demands on land, forest and other natural resources. However, the nature, extent and impact of these demands depends on what, how and where they produce. The World Bank reports ignore the possibility that population growth rates have different consequences under different conditions and disregard the scope – and the limitations – of people's capacities to alter and adapt to them. Only imported contraceptive and agricultural technologies can save Africa from the Malthusian trap, in which the 'geometrical' expansion of the population far outstrips the 'arithmetical' increase in food resources.

MANAGING THE ENVIRONMENT

Successive World Bank reports (1989a: 95; 1990b: 1–2; 1990c: 2) reduce African agricultural practices to two forms – slash and burn and nomadic livestock raising. Once they were appropriate but, with rapid population growth, no more. The solution is new technologies:

> The necessary productivity gains can come only from technological change. This will involve a more intensive use of chemical and organic inputs, the integration of livestock into farming systems to use animal power and manure, the introduction of new higher-value crops, better irrigation methods, hand tools and crop storage techniques; and improved animal and crop husbandry.
>
> (World Bank 1989a: 90; see also World Bank 1986: 3, 37;
> World Bank 1989a: 60, 95–100)

There is no need to draw on the local knowledge and experience of farmers; enlightenment comes from above.

Development now also has to be 'sustainable.' The World Bank (1990c) argues that 'subsidies for farm inputs needed in introducing intensive sustainable agricultural techniques may be necessary, while shifting cultivation is taxed.' It is unclear what these 'intensive sustainable' technologies are. 'Improved crop variety/fertilizer/farm mechanization technologies will still be necessary'; so will a 'trade-off' with the damage they may cause to the environment (World Bank 1990c: 8).

Sub-Saharan Africa tells the same story of population growth causing environmental degradation across the continent. The report sketches, rather crudely and inaccurately, the variations in the population densities and agricultural potential of different regions of Africa and remarks on 'the enormous variety of ecological zones, microclimates and soil conditions – each requiring its own specialized crops, seeds, and farming techniques.' Fortunately, they can all be dealt with by the same policies (World Bank, 1989a: 90). The report ignores the numerous strategies which African farmers (and African pastoralists) have adopted over hundreds of years to cope with their environment – permanent cultivation of manured land, terracing of hillsides, control of pests, and adapting rice-cultivation to different sources of water (Netting 1968; Mortimore 1989; Richards 1985). The report notes the technical, economic, and environmental virtues of multi-cropping, but continues to recommend technologies adapted to mono-cropping (World Bank 1989a: 60, 100). Richards (1983) emphasizes that African farmers often combine different forms of land use – intensive/extensive, permanent/shifting, upland/floodplain, agriculture/stock rearing – to make the best use of their resources. Generalized descriptions of African farming systems ignore both their complexities and their variability.

The World Bank (1990b: 40) recognizes that 'African farmers already use a variety of techniques, highly adapted to local conditions, to manage soil fertility and conserve soil.' It admits that many of the more 'innovative practices' now being evaluated and refined at research stations are actually based on techniques developed by farmers themselves. Traditional farming practices also contain a number of practices designed, in part, to control erosion – mixed cropping, intercropping, relay cropping, various forms of fallowing, crop rotations, no-tillage and minimum tillage, multi-crop farming, various forms of agro-forestry, and so on. These techniques also spread the demand for labour over the year, make optimal use of land, and stabilize output in uncertain environments.

However, the practices of African farmers (and the policies of past colonial governments) are translated in the World Bank (1990b) study as the discoveries of research stations. These 'environmentally benign' technologies – including contour farming, mulching, minimum tillage, intensive fallowing, crop rotations, terracing and bunds, integration of livestock and cropping to maintain soil fertility, agro-forestry, integrated pest management, water harvesting, and small-scale irrigation – have 'enormous potential.' Behind each 'lies a considerable body of agricultural knowledge which finds little application in Africa outside of a few NGO projects.'

In fact, African farmers have always adopted and adapted new crops and agricultural technologies. Attempts by colonial governments and their successors to improve African farming methods have had few successes, and many failures (Heyer *et al.* 1981). The World Bank (1989a: 95) argues that attempts to introduce technology into Africa in the past 30 years have been 'disappointing.' The costs of inputs, and of labour, have been too high, the technologies have been inappropriate, yields have disappointed. Soil conservation programmes have required women and men to provide unpaid labour without generating commensurate increases in incomes (World Bank 1989a: 101). The World Bank typically overlooks its own involvement in these failures, and its responsibility for the debts left behind. The blame is placed on its partners, African governments. The World Bank does not ask whether the whole approach of exporting chemical and mechanical technologies might be inappropriate. Instead, it recommends more of the same.

Thus the World Bank (1990b: 35) identifies the continued application of 'proven yield increasing technology for several crops and the scope for double-cropping through small-scale irrigation' (promoted by the World Bank-funded Agricultural Development Projects) as the main source of agricultural growth in Nigeria. These projects have actually increased the debts of the federal and state governments without obviously increasing yields of staple grains or roots, other than yellow maize whose limited market depends heavily on poultry (Clough and Williams 1987; Williams 1988a).

Sub-Saharan Africa does not consider the environmental consequences of the World Bank's own strategies. Yellow maize, promoted by the Bank, is more demanding of soil nutrients and less drought resistant than sorghum. Chemical fertilizers replace lost nutrients but unlike organic fertilizers they do not build up the productivity of the soils. Mechanical land clearing for irrigation dams, roads and large-scale farming exposes soils to wind erosion. *Sub-Saharan Africa* warns of the dangers of environmentally-damaging irrigation and settlement projects, of destructive logging, and of noxious chemicals and pollutants, and the need for incentives and sanctions, and thorough project appraisal (World Bank 1989a: 90, 100–3). It makes no mention of the failure of past World Bank appraisal reports and projects to take account of likely harm to the environment.

In contrast, African women farmers are identified as a 'cause' of environmental damage. The World Bank (1990b: Annex B) summarizes research findings on the constraints on the time available to women and the resources they need to meet their multifarious obligations to provide directly for the needs of their families. These findings are used to show that women not only suffer the costs of environmental decline, but apparently contribute to it:

> Traditional attitudes have been translated into limited access by women to agricultural extension advice, land titles, agricultural credit. The traditional place of women, and her [*sic*] traditional role as food producer, fuelwood gatherer and water fetcher, may be impeding the intensification of women's farming operations. Women continue to

practice low-input traditional farming techniques, which prove to be harmful in terms of soil conservation, and ultimately in terms of agricultural productivity.

(World Bank 1990c: 3)

Women's 'traditional roles' render them unfit to participate in 'progressive' agriculture.

The World Bank (1990b: 45) does recognize the environmental costs of intensive agriculture when it argues that for 'environmental and cost reasons,' organic fertilizer use should be emphasized. Similarly, disease and pest control efforts should emphasize cultural and biological controls. Subsidies for fertilizers and pesticides should be removed. The report recommends 'a shift from the teaching of a large temperate climate, high input, mechanized agriculture; to a low input, labor intensive, environmentally sustainable agriculture' (World Bank 1990b: 52). Presumably researchers need to be taught this, even if farmers do not. However, these observations are not compatible with the 'improved crop variety/fertilizer/farm mechanization technologies' to which the World Bank is committed in order to raise agricultural productivity. 'Capital-intensive' agriculture will continue to take precedence over 'sustainable' agriculture.

Sub-Saharan Africa draws no distinction between pastoralists and cultivators, nor between capitalist farms and peasant smallholdings. The ecological consequences of mechanized farming and chemical fertilizers, and of dams, tubewells and irrigation projects, are overlooked. Little attention is paid to the past and present appropriation of land, fuel, and other indigenous and imported resources by large-scale farmers, mining companies, timber exporters, industrial firms, state officials and wealthy consumers, and urban populations.

Certain economic activities (such as beer-brewing and tobacco-curing) make more intensive demands on fuelwood than others; some require the cutting back of forest, others need its protection. People will be forced to cultivate marginal land and forests or denude their locations of trees where they are excluded from other lands or lack access to alternative sources of fuel. The various complex conditions which lead to environmental degradation and food scarcity are obscured by the reiteration of a simple, linear progression (Blaikie 1985).

World Bank projects consistently privilege large-scale farmers since they are politically influential, easier to reach than smallholders, and responsive to advice from project officials (Williams 1981: 40–1). The World Bank (1989a) study does not consider the consequences of land appropriation (for large-scale farms, irrigation schemes plantations and ranches) for either food production or the environment. Irrigation schemes displace farmers and pastoralists and may deny them access to downstream floodplains. The claims of large-scale farmers are not limited to land and water. They typically require a supply of cheap labour, and privileged, and usually subsidized, access to imported inputs, public services and product markets (Leo 1984; Williams 1988b). The rich and powerful may spend their money and use their influence to acquire the bulk of a country's limited resources. The opportunity costs of the use of resources by the rich falls on poorer people.

PROTECTING PROPERTY RIGHTS

The World Bank's interpretation of changes in land tenure follows the same logic as its account of fertility behaviour and agricultural practices. The 'traditional' system was once adapted to its environment, but has begun to break down, leading in many cases to 'open access' to land and a 'tragedy of the commons.' A transition to a 'modern' system of individual title will thus be necessary: 'Where property rights are well defined and the market for land is functioning well, private landowners or public managers are likely to resist degradation of their property to protect its long-term value. But where resources are held in common and traditional rules have broken down . . . individual users of the resource have very little, if any, incentive to conserve it' (World Bank 1986: 25).

Individual property rights are no guarantee against soil mining. Capitalists treat land as a source of profit. They may invest in maintaining the fertility of the soil; they may also exhaust an area of land to maximize short-term profits, which can be reinvested in acquiring land elsewhere, or outside agriculture. Only farmers who continue to depend on access to their own land to cultivate or graze stock need to concern themselves with maintaining its productivity. They need to be able to secure their rights to use the land, singly or jointly with others; this does not have to depend on individual ownership or legal title.

The virtues of exclusive rights to property and the vices of communal access are used to justify the enclosure of rangelands for commercial ranching, or creating game reserves, and damming river waters for irrigation schemes. Pastoralists are confined to grazing blocks, which cut them off from seasonal water and pastures, forced to reduce their herds or made to settle to a fixed abode and cultivation of land. However, transhumant pastoralism is better adapted than capitalist ranches to make continued use of the fluctuating resources of arid areas with uncertain rainfall. Farmers settled on irrigation projects make more sustained demands on limited water and fuelwood than pastoralists they displace.

The World Bank (1989a) argues that 'agricultural modernization combined with population pressure will make land titling necessary.' The Bank recognizes the need for caution in carrying through the 'transition to full land titling,' pending which 'traditional tenure systems need to be codified' (as if this had not been a continuing concern of colonial 'native administration').

Further, systems have to be found for adjudicating between 'legislated' and 'customary' rights in land (World Bank 1989a: 104).

This last comment reveals the real issue. Whose claims to land are to be recognized? Whose rights are to be enforced? Registration of title to land may not only extinguish the rights of women in respect of land, as the World Bank (1989a: 103) recognizes, but it also allows those with access to state power and legal forms to impose their claims to land – whether derived from inheritance, purchase or allocation by a community's authorities – at the expense of others. Indigenous forms of tenure do not necessarily preclude people from cultivating, renting, buying and selling land. They may protect people's access to land, albeit imperfectly, from individual aggrandisement, whether by outsiders or by members of local communities. Communal land may be appropriated by local chiefs and politicians or allocated to their cronies. Land titling entails considerable administrative costs and social disruption. Land titling defines and protects some people's rights. It may render some people's claims to land insecure, and exclude others from access to the resources they have hitherto enjoyed.

The World Bank often has to re-discover what has long been known. The 1990 draft report recognizes that individual land titling may lead to 'land grabbing, concentration of ownership and concomitant landlessness' (World Bank 1990b: 18–19). It cites the findings of a recent study that 'traditional land tenure systems provide sufficient security to stimulate investment in land' and that they can recognize people's rights to transfer land by sale and inheritances. Rights to land are complex: husbands and wives or sons and daughters, pastoralists and cultivators, owners and tenants may each have co-existing claims on the same resources, or their products. The report concludes that 'tenurial reform would deprive all other users of their rights at the expense of [sic!] one user.'

However, the logic of the model reasserts itself over such empirical findings. The conclusions and policy recommendations echo those of the earlier 1986 report (World Bank 1986):

> Traditional tenure systems provided sufficient security to farmers to invest when population density was low and where traditional values and customary law was respected. It breaks down with increased competition for scarce land, migration of outsiders on to the land, government claims of ownership or management, and the arrival of European land titling systems. . . . As land tenure security declines, so does the incentive for individuals to improve their land. Soil fertility maintenance requires investment of capital and labor. Such investment will not be made on collectively owned land. Traditional pastureland is becoming an open access resource, open to any user, resulting in a similar absence of incentives to conserve it.
>
> (World Bank 1990b: 20)

The report adds, quite rightly, that state ownership may reduce security of land tenure for farmers and, oddly, that it, too, leads to 'open access' (World Bank 1990c: 4–5). But the solution is still to move towards individual land title:

> Over time, population pressure and agricultural intensification will make formal land titling necessary. The machinery for land titling needs to be established to permit those seeking titles to obtain them. . . . In the interim, respect for traditional tenure systems needs to be codified.
> (World Bank 1990b: 50)

The World Bank focuses on forms of ownership and not on the distribution of assets. Exclusion from access to land and other resources denies people one of the means of providing for themselves, and may force them to cultivate or graze their cattle on marginal soils and cut down trees for fuel. They are then blamed for causing soil erosion.

A serious study of the relations of population growth, food production and the environment needs to examine the distribution of land and other assets. Registration of land title is on the World Bank's agenda; redistribution of land clearly is not:

> The focus here is on access to assets and poverty alleviation, not on the distribution of wealth. Creating wealth, in contrast to rent-seeking, is seen as essential to growth. By giving the poor access to assets and promoting their productivity, a higher level of growth can be ensured.
> (World Bank 1989a: 38)

It is not clear who is going to give assets to the poor and where they are to come from. It is clear that the rich and powerful can rest secure in the enjoyment of the proceeds of successful 'rent-seeking.'

THE ELEMENTARY FORMS OF DEVELOPMENT DISCOURSE

The publications and draft reports reviewed here are compiled by different groups of authors inside and outside the World Bank itself. Their institutional authorship emerges in the contradictions in their arguments and the divergences between the discussion of particular issues and the summary conclusions and policy recommendations. This chapter shows how evidence and arguments which question World Bank thinking – for example on land title – are either discounted or assimilated into the World Bank's underlying paradigms. Consequently, despite the multiplicity of authors and the inconsistencies, the basic diagnoses and policy recommendations of all these documents are remarkably consistent with the policies and practices which the World Bank has espoused since its foundations (Williams 1981).

The World Bank's analyses of demographic trends, environmental change and agricultural production in Africa ignore historical evidence and geographical variation. The World Bank appends to its reports copious statistical tables. They present with apparent confidence (but no identifiable sources) enumerations of population growth, crop yields or agricultural production which, in most cases, are completely unknown. However, these figures are generally disregarded in the reports; they are cited occasionally and selectively to illustrate a point, but not allowed to disturb the argument. The World Bank commissions research reports, in-house and from academics and consultants; some of it good, some of it very bad. They, too, are selectively appropriated in ways which do not question the assumptions underlying the World Bank's policies.

Similarly, the World Bank is incapable of learning from its own and others' experience. Occasionally it has recognized that its policies have not succeeded in rural Africa:

> The introduction of modern agricultural technology in the form of high yield seeds, fertilizer and farm mechanization has met stiff resistance by farmers. Population control programs based on the supply of family planning services and the distribution of contraceptives have not been very successful except in three or four African countries. Soil conservation and forestry protection projects have not had much success. New approaches are needed.
>
> (World Bank 1990c: 6)

The new approaches turn out to be the same as the old ones, repeated by successive World Bank projects, with roots in the earlier policies of colonial administrations.

The World Bank (1990c: 7) argues that 'farm productivity per unit area is to be increased.' It is confused about how this is to occur, whether it means increased inputs of labour, or of chemicals, per unit area. It clearly involves directing farmers, and pastoralists, and protecting soils from farmers:

> Agricultural intensification on a wide scale therefore requires not only better research and extension, but policies which induce farmers to sedentarize, and to intensify production. The first such policy must be the protection of forest and pasture areas from cultivation. The creation of parks, reserves and community owned pasture land will help.
>
> (World Bank 1990c: 8)

To the creation of forest reserves, the settling of pastoralists and the enclosure of pasture, the Bank adds the colonial policy of encouraging 'mixed' (livestock and arable) farming by better-off households, the planting of communal woodlots and planned settlement schemes (World Bank 1990b: 43, 48, 55). It insists that 'governments should hand over government owned farm, pasture and forest land to traditional owners' – whoever they may be.

171

Land tenure security is to be assured by codifying traditional tenure systems and facilitating a transition to individual land titling.

Population growth is to be reduced by 'stimulating a demand for smaller families.' Demand is to be increased by 'improving women's education, providing information regarding health and nutrition, providing health services and providing fiscal incentives [including the limitation of maternity benefits] for small families.' The benefits of modernization are to be extended to women, by giving them easier access to fuel and water supplies, access to land and credit, better tools and 'improved crop husbandry' (World Bank 1990c: 7, 9). How these resources are to be provided within the constrained budgets of governments implementing structural adjustment programmes is not discussed.

These policies are to be realized, in the military language of 'devspeak,' by identifying for each country an 'environment action plan' with a series of measured targets for population growth, agricultural production, reforestation, areas of land under crops and left as wilderness (World Bank 1990b: 29–35; World Bank 1990c: 10). How these plans are to be implemented is unclear. But they do justify extensive government interference in people's lives: indeed 'government must be more aggressive in developing and implementing environmental policies' (World Bank 1990b: 58). Development discourse continues to display a surprising confidence in the capacities of governments to plan people's activities and get them to follow official directives.

There are many precedents for each of the policies recommended in recent World Bank reports on Africa. Few have ever been successful. Why is the World Bank incapable of learning from past experience? Why does it not take seriously academic research which exposes the inaccuracies of its accounts of African societies and their histories? Ferguson (1990: 27–8), discussing the World Bank's (1975b) country study of Lesotho, argues that 'what is being done here is not some sort of staggeringly bad scholarship, but something else entirely,' an example of 'development' discourse quite distinct from academic discourse.

Development discourse requires whole countries be represented in standardized forms as national entities engaged in the process of development, the incorporation of 'traditional' society into the modern world. In the mid-1970s, the World Bank (1975a: 3) defined 'rural development' as 'concerned with the monetisation of rural society, and with its transition from its traditional isolation to integration with the world economy.' It was the task of international agencies, working through national governments, to promote rural development by providing the rural poor with access to markets, services and technologies. Now, twenty years later, the paradigm has developed. The institutions which were appropriate to traditional societies are unable to meet the needs of the modern world. They must now be modernized so that the institutional forms are brought into equilibrium with the changing environment. In this more developed form, the paradigm can accommodate contradictory elements, as evidence of a partial and unstable transition, and it is able to assimilate environmental and gender without bringing into question its underlying logic and the overall direction of change.

Central to the World Bank's account of the 'crisis' in African agriculture is the diagnosis that a rising ratio of population to land brings about environmental degradation, a decline in per capita food production and a rising dependence on food imports. Statistical projections represent African countries as needing US and EEC grains to feed their people (Raikes 1985). This approach ignores historical experience and variations in social structures and demographic patterns. The growth of population must have important consequences for people's access to land, incomes and public resources. Their impact will depend on what is produced, how it is produced, and how it is distributed. The implications of population growth for different people will vary according to their class situation and their gender. These considerations raise empirical questions which can only be answered by careful research in diverse contexts and not resolved by overarching generalizations. This is the task of academic enquiry. The associations between more people/less land/ lower productivity/less food for everyone appear obvious. They are validated by their apparent common sense. But they are better suited to the discourse of development agencies.

The appeal of the caricatures by which the World Bank represents Africa and Africans lies in their simplicity and generality. Development strategies rest on a common diagnosis of the problems of quite different countries, and the transfer of standardized policies and second-hand technologies from one country to another. Complexity and variation, the stuff of history, geo- graphy, sociology, or anthropology, cannot be managed within the practice, and thus the discourse of development.

Development discourse is similarly unable to accommodate indigenous or peasant knowledge, except as 'traditional' practices appropriate only to the past. As we have seen, the World Bank (1990c) appropriates research on the knowledge and practices of African farmers as 'experimental' techniques developed in research stations. If African farmers, and pastoralists, are able to respond to changing, and even extreme circumstances, they cannot be represented as in need of aid and development (see de Waal 1989). Despite an ideological distaste for government intervention, development is always brought to people from outside by the state.

The World Bank represents the problems of Africa as deriving from a generic process engendered by the reproductive and economic activities of numerous people, activities whose adverse consequences they neither intend nor recognize. Consequently, they can only be dealt with from outside, by experts who combine technical knowledge with an ability to solve the problems of 'development,' and who make up 'the development community,' whose task it is to identify and promote the common good in the form of 'development' (the thing which we are all in favour of). The process of 'development' transcends the concerns of particular places or even countries and the partisan policies of specific parties or governments. The objectives, and the agencies of development are therefore presented as apolitical. As Ferguson (1990) argues, the discourse of development acts to depoliticize issues, to take the politics out of policies.

World Bank reports do not add to our knowledge or interpretation of the problems of population growth, environmental change, or agricultural production. They do offer an authoritative view of what those problems are and how they are to be solved. They set the policy agenda for national governments and development agencies. They seek to co-opt the skills and flexible capacities of non-governmental organizations to their agenda. They delimit the bounds of practical policies for the 'development community.'

The claim to scientific knowledge is necessary to the World Bank's insistence that its actions and policies are founded on objective, economic criteria rather than political considerations. Thus 'development discourse' must emulate the forms of academic discourse. Academics are often contracted to contribute to or prepare reports for the World Bank. They are required, or their work is made, to conform to the paradigms and policy concerns of the 'development community.' The main purpose of their contributions is to give World Bank policy statements the credibility of science. Hence the apparent indifference of the World Bank to the poor quality of much of the research they undertake or commission, and their inability to appreciate the good research which some academics do provide them.

Similarly, projects must be appraised to certify that they will generate a positive net discounted rate of return on their investments. Environmentalists are called in to advise on the environmental impact of policies and projects; women are, usually, brought in to advise on gender dimensions. Here, too, the point is that the projects should be appraised. Studies of the environmental impact and gender dimensions of projects may prove more significant in justifying projects which are likely to harm the environment and exacerbate gender inequalities than in preventing fools from rushing in to spread development ever more widely.

Ashforth (1990a: 4) argues that state commissions of inquiry 'produce a discourse celebrating a marriage of truth and power in the modern State through rational identification of a purportedly objective Common Good.' Reports published, or initiated, by the World Bank lack the forensic method or judicial majesty of commissions of inquiry, but they do present themselves as offering an objective view of 'development' issues, untainted by political considerations. This view draws for its authority on the strategic position of the World Bank, not merely as creditor to governments but, in liaison with the International Monetary Fund, as able to define for national governments and commercial institutions the creditworthiness of indebted governments. 'Power' gives authority to 'truth' which offers legitimacy to 'power.' Ashforth

(1990: 15–16) further identifies the significance of 'notions of the common good' as 'a way not only of justifying state action but making normatively directed collective action on the part of those who act in the name of a State possible.' Policy reports do not decide what the World Bank and other development agencies do. They do provide an overarching framework which makes sense of current policies as a means of addressing development problems. They spell out the common sense of the development community.

9

CHANGING DISCOURSES OF DEVELOPMENT IN SOUTH AFRICA

Chris Tapscott

We come back to the same way of playing with words and phrases, where the white man in South Africa finds he has come to a cul de sac as far as certain expressions are concerned and so he resorts to another. You take for instance, the many words that have been used to refer to the black man. He was called a kaffir, then a native, then a non-white, then a bantu, then a plural and a co-operator.

(George Matanzima, in Republic of the Transkei 1982: 267)

INTRODUCTION

Development discourse occupies an ambiguous position in the South African context. It is pre-eminently an academic discipline (similar in standing to public administration, personnel management and so on). This might suggest that its contribution to ideological mobilization would be limited since, as Therborn has pointed out, the rarest and most limited form of interpellation is 'an elaborate written text speaking directly to a solitary reader' (Therborn 1982: 77). Yet its sphere of influence is more extensive and its impact on South African society has been more far-reaching than that of any conventional academic discipline.

There is also a strong materiality to the concept; academic 'development theory' (expressed as 'development thinking') is frequently translated into state policy or alternatively is deployed to translate and rationalize existing policies to a wider audience. Conceptualized as a policy discourse, it constitutes more than a mere language of legitimation. During the 1980s, 'development,' in many respects, encompassed the sum total of state policy towards the bantustans and, to a lesser extent, the African population in general. As a shorthand description of this policy, it was presented variously as 'development aid,' 'urban development,' 'regional development' and 'multi-national development'.

By the late 1980s, 'development' had become the preoccupation of a broad

176

spectrum of agencies within the state and civil society, extending from business sponsored bodies to non-profit charitable organizations. This emergent 'development' discourse formed part of a broader initiative to restructure the form of apartheid and to redirect the ideological discourse of the ruling white population. To that extent, the concept has performed an important legitimating function. Not only has it endeavoured to transform the explicitly racist and supremacist perceptions of the white population – *inter alia* by portraying Africans as underdeveloped rather than racially inferior – but it has also aimed to provide a salient ideological justification for those collaborating blacks which it has sought to incorporate, by attempting to depoliticize and technocratize matters pertaining to their socioeconomic advancement.

Construed as a policy discourse, rather than as a purely theoretical construct, however, 'development' has had substance beyond its ideological legitimating functions, and has served to inform (and in turn has been informed by) state policy *vis-à-vis* the African population. That is to say, 'development' in South Africa (as elsewhere) was never advanced purely as a means of concealing the true objectives of state policy, nor solely as a mechanism for co-opting segments of the black population. By shaping the pattern of regional development, and by determining the type of economic programmes which have been funded in the bantustans, for instance, 'development discourse' has had (and continues to have) a material impact on the lives of millions of South Africans.

In discussing the reinvention of development in South Africa in the 1980s, the intention is not to suggest that the country is particularly unique. As Ferguson maintains, 'Throughout Africa – indeed, throughout the Third World – one seems to find closely analogous or even identical "development" institutions, and along with them often a common discourse and the same way of defining "problems," a common pool of "experts," and a common stock of expertise' (Ferguson 1990: 8). What is distinctive about South Africa, however, is the process through which 'development' emerged and the particular inflection given to it by apartheid policies.

SEPARATE DEVELOPMENTS

In discussing the centrality of the concept of 'development' in the ideological armoury of a reformist discourse in South Africa in the 1980s, it is useful to chart its intellectual genesis. The Verwoerdian model of 'separate development,' as expounded from the 1950s through to the early 1970s, was premised on a belief in the national distinctiveness of the various ethnic groupings within the African population. 'The Southern African Bantu,' an official publication maintained in 1974, 'are not and have never been a homogenous nation' (RSA 1974: 10). The white population, in contrast, constituted 'a new African nation which evolved in the same way as the new nations of the

177

U.S.A., Canada, Australia and New Zealand' (RSA 1974: 15). In the period of apartheid's ascendancy, as a consequence, the development of the African population was conceived in terms of ethnic separatism, and formulated as a 'blueprint' for the bantustans. 'The structure laid down in the approved blue print,' affirmed a leading apartheid theorist, 'in so far as it affects the Bantu peoples of Southern Africa, calls for the concentration of the social, cultural and economic basis of each people in its own territory' (Lombard 1971: 24).

The socio-economic development of Africans was therefore inseparable from their political development, and both were to take place within the bantustans, 'their promised land' (Boshoff 1970: 31). 'The settlement of the Black peoples in their respective homelands' argued the Bureau for Economic Research re Bantu Development (BENBO 1976: vi):

> Embraces both social and economic processes for which a prime prerequisite is the provision of employment opportunities in order to make the settlement effective in the long term. The wielding of political power by a nation has less meaning if the vast majority of its members cannot participate because they cannot reside and work in their own country.
>
> (BENBO 1976: vi)

In the logic of apartheid, 'the unique nature of the population composition in SA implies that issues and challenges exist within its borders that are analogous to those differences presently experienced worldwide between the countries of the so-called First and Third Worlds' (BENBO 1976: vi). This dualistic global model – in which certain parts of the world are the object of 'development' – was incorporated into the apartheid vision of 'separate development' for the bantustans. Apartheid apologists made no attempt to disguise the political agenda underlying separate development, openly admitting its role in the perpetuation of white domination. Official publications explicitly emphasized the political dimensions of separate development:

> Development strategy in South Africa was never conceived in purely economic terms and it is an illusion to believe that in a multi-ethnic society such as South Africa economic growth will, *ipso facto*, solve non-economic developmental problems and that political and sociocultural factors can therefore be ignored. Far from it. Political and socio-cultural conditions and objectives can never be disregarded.
>
> (RSA 1974: 41)

The political props of apartheid, such as influx control and migrant labour, were essential for the survival of the white community, since their removal 'would in all probability mean the end of its political reign' (BENBO 1976: 39).

Economic underdevelopment and poverty was attributed to the backwardness of the Africans themselves and their resistance to modernizing forces.

Africans were blamed for the slow pace of economic development in the bantustans:

> In this respect the RSA endorses what every real authority on the development problems of the Third World knows: the success of the development schemes in economically retarded communities is to a large extent dependent upon the capacity of these peoples to learn to help themselves ... [T]here is no room for any complacent or false belief in economic short cuts to higher living standards. Nor is there any room for any baseless optimism as to how long it will take to raise the average standard of living of the Black peoples to the current level of the White population.
>
> (Bantu Investment Corporation 1975: 9)

The practice of blaming the victims of the apartheid system, in effect, obscured the relations of control in South Africa and served to legitimate the perpetuation of white domination. It did so by appealing to the supposed authority of the international expert on Third World 'problems.' The imaginary of development was used selectively to justify the development path then being mapped out for the African population; a path which viewed the African population of the country as divisible into a series of 'developing nations.' The fundamental objective was 'not so much the Westernisation of individuals as it is the development of different peoples of whom many are still only at the beginning of the long road towards democracy and economic and technological self-management' (Bantu Investment Corporation 1975: 11).

Anxious to legitimate the system of separate development in the eyes of a sceptical world, the apartheid state adopted the language of that world. In the period leading up to the 'independence' of the Transkei bantustan, for example, the Bureau for Economic Research re Bantu Development produced a series of publications designed to draw favourable comparisons 'between the homelands as developing countries and other countries in Africa' (BENBO 1976: 194). Citing various selective statistics (many of which were highly contentious), these reports purported to demonstrate that 'the development level of the homelands compares quite favourably with (and is in many cases higher than) that of independent countries in Africa' (BENBO 1976: 194).

Despite the intellectual certainty until then afforded by the apartheid 'blue print' and the analogy with Third World development, the political and economic upheavals of the second half of the 1970s profoundly shook the convictions of a number of prominent separate development theorists. During the closing stages of the decade a small number of intellectuals from Afrikaans universities and from certain state bodies, such as the Bureau for Economic Research, Co-operation and Development (known by its Afrikaans acronym BENSO) and the Human Sciences Research Council (HSRC), began to question openly the viability of existing policies (Lombard 1978;

179

Olivier 1978). Ultimately, however, it was the manifest failure of the bantustans to achieve anything approaching 'development' that concerned many theorists:

> In South Africa it has now become evident that the early expectations for the national states with regard to their political independence, economic development and especially their employment creating capacities did not (and could not) come true. The political and economic policies of the 1950s, 1960s and 1970s were inadequate to deal with the economic realities of South Africa, both inside and outside the national states.
>
> (Van der Berg and Van der Kooy 1980: 517)

'[I]t is evident', these two leading BENSO economists concluded, somewhat wistfully, 'that the Verwoerdian vision was but a dream – the development necessary to make separate development economically viable did not take place in the national states' (Van der Berg and Van der Kooy 1980: 519).

As an integral component of the overtly racist policies of apartheid, the idea of 'development' not unexpectedly attracted some opprobrium. During this period the concept was largely shunned or heavily criticized by liberal academics, the English-language press and the international community at large. As a consequence, theorizing about development in South Africa tended to be the exclusive preserve of state ideologues, Afrikaner academics and the Afrikaans press, all of whom were broadly supportive of National Party policy.

In a number of Afrikaans universities, 'Development Studies' or 'Development Administration' was, in the 1970s and 1980s, largely an upgraded version of 'Bantu Studies', preoccupied with formalistic–legalistic issues of administration and heavily oriented to the descriptive ethnography of the *volkekunde* school of ethnology. The Potchefstroom University for Christian Higher Education, for example, under the heading of 'Development Studies' throughout the 1980s offered a diploma in 'Bantuistics' which promised to 'lead the student to the living world and living conditions of the Black population group' (Geggus 1986: 77). The logic (and imperatives) of *volkekunde*, which assigned overwhelming explanatory power to the phenomenon of ethnicity, provided a salient justification for much apartheid policy and was reflected in the framework of ethnic nationalism which underpinned the bantustan system. Its intellectual legacy was also evident in the writing of a number of developmentalists who, under the guise of pseudo-scientific theory, continued to affirm the inherent fatalism, communalism and traditionality of African society.

A number of inferences were drawn from this perspective by these ethno-nationalists. Because the existence of two different societies in South Africa was held to be self-evident, it followed that the two spheres should be treated differently. 'Development,' according to Van Niekerk (1986: 64), should be seen, 'not as the process of converting one into the other, but as the search for a new structure in which there is a fairly harmonious, at least not mutually destructive, process of interaction.' At the same time, in view of the cultural backwardness of one sector of society, it was logical, according to this argument, that the process of modernization should be initiated and controlled by the more advanced component: 'development assumes a conglomeration of differentiation and specialization in a culture, and as a rule is achieved when members of a developed population actively control the process of cultural change in an underdeveloped population' (Swart 1985: 156).

Ethno-nationalist discourse has generally been reviled by more recent developmentalists, as an unpleasant reminder of their immediate intellectual past. However, the essential 'otherness' of the African population, explicit in *volkekunde* theory, has also been implicit in the bulk of subsequent 'development' writing in South Africa. The widely espoused concept of economic dualism, in particular, lent itself to the updating of racial prejudices and reinforced the distinction between white and black South Africans. While the manifestations of dualism are common to many Third World countries, in South Africa the divide, very distinctly, was seen to follow racial lines: 'the black areas in SA undoubtedly belong to the Third World and suffer from the same problems as the rest of Africa. The problems of development encountered in the black areas of SA are in all respects comparable to the problems experienced in the rest of Africa' (Van der Merwe 1980: 37; see also Lithgelm and Coetzee 1984: 6).

THE SEARCH FOR A 'NEW DEVELOPMENT PARADIGM'

With the rise of P.W. Botha to the premiership of South Africa in 1978, a concerted effort was made by state ideologues to formulate a new policy towards the African population and the bantustan system as a whole. A growing number of disaffected Afrikaner intellectuals and businessmen within civil society, as well as selected state officials, began to pose serious questions about the consequences of perpetuating apartheid policy in its existing form. To these individuals, the prevailing political dispensation, and the bantustan system in particular, threatened both the security of the state and the future stability of the economic order.

Calls for greater political participation of blacks meshed with appeals for a more incorporative ideology capable of encompassing the aspirations and winning the allegiance of segments of the black population. This was a need most strongly expressed by adherents of the Total Strategy who perceived the 'total onslaught' against South Africa to be largely ideologically motivated. 'The total onslaught,' according to General Malan the Minister of Defence, 'is an ideologically motivated struggle and the aim is the implacable and unconditional imposition of the aggressor's will on the target state' (Malan

1980: 79). A counteractive ideological offensive was essential. As the influential Afrikaner academic (and subsequent Minister of Education and Development Aid) Professor G.C. Olivier (1978: 78) asserted, 'the only effective way to counteract and perhaps neutralize such an eventuality is to find a political theme with enough symbolism, substance and common ground that will make the protection of South Africa's security a worth-while effort for all ethnic groups in this country.' Calling for a 'restructuring of the operative value hierarchy in South Africa,' he stressed the importance of rendering 'some degree of moral legitimacy to our national policies' (Olivier 1978: 79).

The National Party had two objectives. On the one hand, it set out to incorporate leading sectors of the black population by portraying its reformist policies as non-racial, impartial and an abandonment of old-style apartheid. On the other hand, it tried to convince wavering white voters that their privileges could be protected only through timely concessions. The attempt to incorporate leading strata of the black population was not without its problems. Botha's reformist rhetoric, together with the liberalization of labour laws, produced a powerful backlash among working-class, rural and petty bourgeois Afrikaners (Giliomee 1982). Attempting to stem the tide of desertions arising from the split in its ranks (and a general decline in support from the white electorate as a whole), the National Party tried to convince its followers that 'healthy power-sharing' was not only compatible with 'white self-determination' (historically a synonym for 'white superiority'), but that such reforms were imperative for white survival.

Announced as the search for a 'new economic development paradigm', the state attempted to uncouple the concept of 'development' from the now economically questionable and politically problematic notion of 'separate development.' The lexicon of 'separate development' was progressively phased out of official terminology. 'Development', from then on, was portrayed primarily as an economic activity devoid of political substance. The search for a new development paradigm (as part of a broader programme of reform) was seen as a matter of immediate concern (Van der Kooy 1979: 4; Du Pisani et al. 1980: 40; Van der Merwe 1983: 147) and quickly gained momentum among state officials allied to Botha's reformism. Commenting on this process in 1979, a BENSO official observed:

> That a search for a new paradigm has started is shown for example by the many commissions established by the RSA Government to evaluate the old paradigm and to submit suggestions which could lead to a new one. Although concerned with SA in general, this search intimately involves the Black states, directly or indirectly. . . . Another indication is the flood of publications in search of new ideas, solutions [and] paradigms.
>
> (Van der Kooy 1979: 25)

The urgency of the quest can largely be ascribed to the military, who equated the need for socio-economic development among blacks with the very survival of the white population (Lloyd 1979).

The most significant feature of the 1980s was thus the quest for a comprehensive empirical or normative theory of development. The dissolution of 'grand apartheid' with its all embracing design for social and political development left a theoretical vacuum which many academics (and others) found difficult to fill. The search for a new 'development paradigm' evolved into a quest for an 'appropriate development strategy' or ADS. The theorists strove for a comprehensive theory, 'an overarching paradigm, a stimulating ideology, a set of coherent interdependent and linked ideas to achieve development' (Van der Kooy 1988: 6). In 1984, for example, Lithgelm and Coetzee (1984: 24) argued for 'an appropriate development strategy that should take cognisance of the evolvement of thinking and the greater clarity achieved as regards the concept of development.' In the same year, the new journal *Development Southern Africa* invited its readers to participate in a Delphi study to formulate a development paradigm for Southern Africa which would lead to new development policies, strategies and programmes.

While the military was instrumental in creating a socio-political milieu in which issues of national 'development' assumed importance, it was not itself a key actor either in the elaboration of a new discourse or in the implementation of new strategies. This task was left to others, including the politicians. The military was principally concerned with the ideological impact of 'development' rather than the formulation of appropriate theory – this was left to other sectors of society to pursue, including private business.

Business people frequently exhorted their peers to sponsor community development projects in their own long-term interest. 'It matters precious little whether the motivation is pure social responsibility or naked self interest,' observed the influential business journal *Financial Mail* (1979), 'it is an option business fails to take at its peril.' Although business was not the prime mover in the generation of a new discourse of development, it managed to keep development issues at the forefront of the national political agenda. The English-language press (its major mouthpiece) frequently highlighted the plight of disadvantaged black communities, and demanded more progressive state policies in housing and employment. Big business was also represented on the boards of new development agencies and sponsored 'development' conferences, research and scholarships.

In the climate of political uncertainty (and spirit of reform) that accompanied P.W. Botha's early years in office, the 'development' movement gained momentum. As a project of 'organic intellectuals' (Gramsci 1971: 12) within both state and civil society, moreover, it drew support from a broad spectrum of white South Africans, and included within its ranks academics, business people, consultants, journalists, and clergy as well as state officials. The success of this undertaking can in large part be attributed to the fact that the idea of 'development' was advanced by its protagonists as an open debate.

Launching a new development journal in 1983, for example, the chief executive of the Development Bank stressed that it was intended to 'stimulate participation in thinking and debating about issues covering the wide field called "development"' (Brand 1984: 2).

Enthusing about the precipitate growth of the concept, the coordinator of the Programme for Development Research (PRODDER) proudly announced in 1988 that 'an enormous amount of development effort is taking place in southern Africa, possibly unsurpassed anywhere in the world. The number of development agencies is to be counted in the hundreds, available funds in the thousands of millions and development projects in the thousands' (Van der Kooy 1988: 1). In confirmation of this trend, the PRODDER annual for 1988 listed 175 'development agencies' operating in South Africa, of which at least 60 per cent had been established during the previous ten years. The Development Society too boasted that its membership had quadrupled in the four years since its establishment in 1983. In 1976, a Human Sciences Research Council report on higher education showed that no degree courses were offered on 'development' at any university in South Africa, and only one department and one institute incorporated the title 'development' (HSRC 1976). A second report, published in 1986, found that courses on 'development' were being offered in over twelve universities and that nearly all universities had some form of 'development' institute (Geggus 1986).

In the past, the application of ideas about development to the African population had been largely the preserve of state bureaucrats and National Party supporters. The new developmentalists invited participation from all academic disciplines and (ostensibly) all political persuasions. By promoting 'development' as an ongoing and unsettled debate, rather than the fixed and irreversible 'blue print' of separate development, the new development discourse effectively opened the way for a greater inclusion of dissenting parties. This process of incorporation was facilitated by the depoliticization and techocratization of all that came to be known as the 'development domain.'

BENSO was the first significant 'development agency'. It was established in 1969 as a statutory body falling under the Department of Bantu Administration and Development (Bantu Investment Corporation 1975: 201) and was initially the Bureau for Economic Research re Bantu Development – BENBO. BENSO's original mandate was to provide basic research to the para-statal Bantu Investment Corporation, the primary implementing agency in the bantustans at the time. A review of the output of BENSO in its first eight years of existence reveals that it closely followed the logic and ideology of separate development. In the second half of the 1970s (coincident with the political upheavals of the time), however, there was a discernible shift in the theoretical orientation of BENSO.

In 1978 the Bureau launched *Development Studies on Southern Africa* (DSSA), the first official journal on 'development' in South Africa. From its inception, DSSA was a platform for the dissemination of new ideas about the economic development of the bantustans, and it was through this channel that calls were first made for a 'new development paradigm.' In 1983, BENSO was disbanded and its functions and most of its staff were taken over by the newly established Development Bank of Southern Africa (DBSA). The DBSA rapidly became the most influential development institution in South Africa and played a major advocacy role in the propagation of 'development' thinking. Established by the government in 1983, the Bank was ostensibly autonomous, although it derived the majority of its share capital from the state and was governed by a body which included a number of current and former cabinet ministers. The Bank, which became the primary funding agency for economic development projects in South Africa, also served as a focal point for 'development' practitioners. Not only did it recruit large numbers of professionals from tertiary education and the private sector, but it commissioned applied research amongst academics and consultants. It also played an instrumental role in the establishment of the Development Society of Southern Africa in 1983.

The Development Society became a leading advocate of 'development' in South Africa during the 1980s. The Society drew it members widely from the private sector, the DBSA, state departments, para-statal organizations, non-governmental organizations, bantustan administrations, from Afrikaans and English-speaking universities as well as bantustan universities. At its peak in 1988 it had 650 members in six branches distributed throughout the country (Beukes 1990: 3). Through conferences, workshops and seminars the Society attempted to convert the public at large to the 'need' for 'development,' and in this role it frequently displayed the fervour of proselytism.

Other important agencies in propagating the language of development were the parastatal Human Sciences Research Council (HSRC) and the Programme for Development Research (PRODDER), established in 1987 under the auspices of the HSRC. PRODDER was set up primarily to co-ordinate development research, to 'provide support services and a network, and act as a broker between those in need of such research, the research community, development agencies and financiers' (PRODDER 1988: 8). As with other 'development' bodies at the time, PRODDER drew its support from elements within the state, the tertiary education sector and the private sector. Of the twenty-one members of the founding committee, ten were from universities, four from the HSRC, three from state departments, two from the private sector and two from 'development' agencies (Van der Kooy 1988: 166).

The development discourse of the 1980s was articulated and propagated by agencies whose personnel moved easily between the state, higher education and (to a lesser extent) the private sector. This extensive interlinkage was one of the primary reasons why 'development' thinking, once adopted, could spread so quickly. But it was also indicative of the elitism of the concept, which in its earlier phases had a limited social base among highly educated whites within the state hierarchy, the universities and big business.

'DEVELOPMENT' AS IDEOLOGICAL TRANSFORMATION

The new ideological and discursive dimensions of 'development' were not peculiar to South Africa. The use of 'development' for political ends is a feature of most Third World societies (Williams 1982). Development is frequently the vehicle through which the state consolidates its control over society: 'the "development" apparatus . . . is not a machine for eliminating poverty that is incidentally involved with the state bureaucracy; it is a machine for reinforcing and expanding the exercise of bureaucratic state power, which incidentally takes "poverty" as its point of entry' (Ferguson, 1990: 255). Development is also a malleable concept, open to contestation by different power groups in society. What is of particular interest in the South African case therefore is the way in which 'development' evolved out of 'separate development' and the discursive shifts which were necessary to legitimate this process.

Giddens (1979: 193–6) suggests that there are three mechanisms by which political domination is concealed: the representation of sectional interests as universal ones, the denial or transmutation of contradictions including the separation of the 'political' from the 'economic' domains, and the natural-ization of the present. By presenting social relations as having the fixed and immutable character of natural laws, dominant groups attempt to naturalize the present. This denies past complexity and mutability and reifies the present as historically legitimated and as a wholly normal state of affairs. These mechanisms are particularly evident in times of crisis when elites frequently engage in complex discussions to resolve contradictions and evolve new strategies. These are real debates which serve as a prism through which elites attempt to interpret their world. But they also serve the purpose of eliciting compliance from the mass of the population.

In South Africa, the 'new development paradigm' was presented as a simple evolution of past policies: 'many of the new ideas are based on the old paradigm for which new strategies, reformulations, adjustments, streamlining of existing practices are proposed' (Van der Kooy 1979: 25). The stress on continuity was particularly evident on political platforms, where the National Party was accused by its right wing of forsaking its heritage and of 'selling out to the blacks.' Developmentalists refuted these accusations, even suggest-ing in one instance that 'questions of development . . . were already being raised by numerous commissions and committees of enquiry at the time of the establishment of the Union of South Africa' (Fourie and de Vos 1986: 1).

Spatially, the new developmentalism was justified in terms of concurrent in South Africa is not unlike a similar search worldwide concerning the Third World' (Van der Kooy, 1979: 4). The necessity for a new paradigm was ascribed not to political pressure from the black majority, but to the inadequacies of the economic model then being followed. Van der Kooy (1979), for example, concluded that separate development was based on Keynesian economic principles and hence suffered from the general short-comings of this model: 'South Africa is using the same classical and Keynesian instruments and economic objectives as the West. . . . If the paradigm is the same, the same inadequacies can be expected' (Van der Kooy 1979: 21).

To claim that the pursuit of Western economic models was responsible for the state of the bantustans was obviously a gross distortion. Yet this interpretation was still being advanced well into the 1980s. According to Lithgelm and Van Wyk (1985: 327), 'the development effort in the less developed areas of Southern Africa is modelled on the traditional western capitalistic development strategy in which economic growth, rather than the structure of economic development, is regarded as the decisive factor.' By aligning 'development' strategy in South Africa in this way, critics would be compelled to query the logic and internal consistency of a particular general theory rather than the reality of political domination and underdevelopment in the bantustans.

Greenberg (1987: 391) has pointed out that the South African state's attempt to legitimate the illegitimate involved 'negating the racial character of the state, diminishing the direct and visible role of the state in the labour market and workplace and shifting the locus of prestige to the private sector.' The new developmentalism rested heavily on discursive efforts to depoliticize the social order, to transmute the racial character of the state and to argue that social life should be governed by the market.

The fiscal crisis of the mid-1980s, the continuing flight of foreign capital and the long-term effects of international sanctions, contributed to soaring unemployment among the black population and exacerbated existing political tensions. Faced with growing fiscal pressures, and anxious to restore a semblance of political stability, the state increasingly sought the aid of the private sector in providing the material concessions deemed necessary to win the allegiance of at least segments of the subordinate population. 'It is no coincidence,' Innes (1987: 566) maintained, 'that the campaign for privatization arises at precisely the moment that black people are intensifying their demands on the state.' Speaking at a meeting with the leaders of the 'independent' bantustans in November 1988, P.W. Botha confirmed the government's desire to draw support from business for its development activities. 'Private sector investment is essential,' he asserted, 'our state coffers simply cannot fund all that has to be done' (British Broadcasting Corporation, Monitoring Services, 23 November 1988).

In 1983, Dr S. S. Brand, an economic adviser to the prime minister and a leading 'developmentalist,' had observed:

As soon as the government takes responsibility for supplying houses it becomes a political matter and then people who are not in the political process, who haven't got representation in parliament for example, say it is because they don't have representation in Parliament that they have to stand at the back of the queue for houses. That is why they definitely want one-man-one-vote for this country. But if you take these things out of the political sphere as much as possible then Parliament and the Government play a less important role and it becomes a less important matter for people to necessarily have a certain form of representation in the central Parliament.

(cited in Zille 1983: 67)

In this context, the pivotal role of the private sector in the field of 'development' was axiomatic (Van der Kooy 1985).

Pursuant to this policy, the state and big business embarked on a number of new initiatives. By relaxing legislation barring the entry of Africans into the commercial and financial world, the state intended to facilitate the growth of an African entrepreneurial class outside of the bantustans. The promotion of small business ventures and the 'informal sector' was seen as a medium through which the black population could identify with the existing social and political order. The informal sector was now a 'breeding ground for an entrepreneurs' corp' (Viljoen 1984: 230). The 'informal sector' emerged in the new development discourse with the air of fresh discovery. This was in stark contrast to earlier apartheid policy. Spelling out the official position in the early 1970s, Professor Boshoff, for example, had asserted:

We can afford to offer a reasonably large number of employment opportunities for Bantu in the White areas, but we cannot afford to house a Bantu middle-class and a professional business class in addition to the labourers, because this would cause the breakdown of our entire system of civil and political rights.

(Boshoff 1970: 31)

The unfettering and acclamation of the informal sector was an attempt to make a virtue out of necessity. Economic recession and a steady decline in industrial employment forced increasing numbers of people into informal work. In 1989, up to 150,000 of the estimated 350,000 who entered the labour market each year went into informal sector activity. The words of many developmentalists, though framed in theoretical jargon, were little more than a rationalization of crisis management by the state. In openly endorsing small and informal business activity, and in stressing the opportunities (not to mention wealth) ensuing from hard work and sound financial practices, supporters of this policy also subtly shifted responsibility for success or failure over to the individual operator and, in so doing, attempted to exonerate the state from future blame (Van der Waal and Sharp 1988: 145).

DEVELOPMENT AS SCIENCE

Edwards has observed that:

> Development research is full of spurious objectivity: this is a natural consequence of divorcing subject from object in the process of education. Any hint of 'subjectivity' is seized upon immediately as 'unscientific' and therefore not worthy of inclusion in 'serious' studies of development.
>
> (Edwards 1989: 121)

This perspective is evident in South Africa, where there have been concerted attempts to portray 'development' as a science, guided by the impartiality and objectivity of scientific method.

A basic feature of South African development discourse in the 1980s was its persistent attempt to foster a technocratic justification of the social and political order. The rise of technocratic rationality has to be seen as part of a conscious effort to depoliticize politically contentious issues in the programme to restructure apartheid. As Posel (1987: 421) has shown, 'large areas of state control are depoliticized by being depicted in technical terms which disclaim their political contestability. The legitimation of such policies devolves upon 'proving' their effectiveness, rather than demonstrating their 'democratic' basis.' Such an approach, she maintains, attempts to pre-empt opposition, by presenting state policies as the objectively inescapable, expert solutions to technical problems, and hence as rationally incontestable.

A preoccupation with scientism pervaded the writing of developmentalists, who perceived underdevelopment in South Africa to be almost exclusively a function of inadequate methodology and the poor application of economic principles: 'Meaningful action in respect of rural development will only result if a hard-nosed and down to earth approach is followed in collecting data, analysing results and drawing up a cost/benefit balance sheet based on infield testing for economic viability and social acceptability' (Erskine 1985: 382). Appeals to science as the best medium for the resolution of the country's problems were most clearly evident in the 'systems approach' to 'development.'

In a survey of agricultural development in the Transkei, for example, Spies (1983: 124) identified 'three interacting forces which in combination determine the type of farming system being practised in a given agro-ecological area.' These were: physical forces and the environment, the 'human factor' and the institutional framework. While elements of this analysis are unquestionably correct, and accurately identify shortcomings in the agricultural sector of the Transkei, the overall effect of the 'systems approach' is to mystify the causes of poverty in the bantustans. As the title of the model suggests, the problems of agricultural retardation are viewed as endogenous to the bantustan and determined by the African inhabitants themselves.

In promoting a new technocratic order, the developmentalists assiduously avoided all reference to political issues. 'Development' was what has elsewhere been described as an 'anti-politics machine', an instrument designed to suspend politics from even the most sensitive of political operations (Ferguson 1990: 250). Thus, according to Louw:

The debate on development strategy in Southern Africa would be enhanced if it were separated as far as possible from the political biases of those who engage in it. Many critics of apartheid and the homeland policy seem to be unable to distinguish between the existence of the homelands *per se* and the development policies they adopt.

(Louw 1986: 18)

Developmentalists appealed for a 'rational,' 'unbiased' and 'non-political' approach. Whereas a number of developmentalists recognized the major inequalities between blacks and whites (described as the 'large discrepancies between the "haves" and the "have-nots" in SA' (Van der Kooy 1988: 469)), they seemed unwilling or incapable of making a political connection. 'Non-economic factors,' when identified, were cultural or ecological but never political. The neutrality afforded by the ostensible apolitical and technocratic nature of 'development' undoubtedly allowed a wider audience to partake of the discourse.

As in the past (Ashforth 1990b), a feature of efforts to restructure the form of white domination in South Africa was the constant referral of problematic social issues to 'experts' and 'expert commissions.' Since 'development' was viewed as a largely scientific undertaking, its advocates presumed that it should logically remain the preserve of 'experts' and 'specialists.' 'It is obvious', the Development Society of Southern Africa (1985: 3) maintained, 'that the development problem in South Africa needs much more attention by many more specialists, more research and far more interest than has been the case so far.' The rapid proliferation of development 'experts' during the course of the 1980s was a notable feature of the 'development movement' in South Africa, replicating tendencies elsewhere (Edwards 1989: 118).

The new language of legitimation, with its own depoliticized and technocratic vocabulary, was instrumental in a programme to reconstitute the subjectivity of segments of South African society. By relabelling and neutralizing politically laden terminology, the approach aimed to widen support for state reforms by appealing for 'pragmatic realism' in dealing with the country's problems. Apart from the emergence of a new lexicon (of basic needs, integrated rural development, appropriate technology, the informal sector and so on), the developmentalists attempted to eliminate offensive connotations from official nomenclature and provide a new garb for existing policies. The most explicit of these exercises was the attempt to find a politically neutral form of appellation for the African population.

While official terminology renamed 'Bantu' as 'Blacks,' developmentalists tended to shun use of such an explicitly racial term, speaking instead of the 'underdeveloped population,' 'the rural poor,' 'the inhabitants of the National States' and so forth. 'Development' would thus escape the charges of racism which had dogged the advocates of separate development. The averred neutralism of 'development' terminology was also intended to facilitate the incorporation of segments of the African population into the new ideological discourse. In support of this venture, the new development discourse was articulated largely in English, as opposed to Afrikaans which had dominated apartheid discourse. Thus, for example, the official language of the Development Bank is English despite the fact that in the 1980s the majority of professional personnel and virtually the entire executive were Afrikaners.

CONCLUSION

In South Africa, the pressing need to find new ways of managing the political demands of the black population spurred the search for a new and more incorporative development discourse capable of diffusing mass resistance without alienating the support of the white minority. In assessing the impact of this programme of hegemonic restoration, the evidence suggests a process that was both partial and uneven. Among many whites, 'development' gained wide currency during the 1980s. The 'development' of the African population, was broadly accepted as a prerequisite for a new political order and for transition to a 'new South Africa.' Development discourse thus formed an integral part of a wider debate within the white population about the political future of the country.

The openness and apparent apoliticism of the language of development unquestionably facilitated the incorporation of segments of society that might otherwise have shunned participation in such a process. During the 1980s, the logic of 'development thinking' was embraced by at least segments of the black population. But the process was limited and uneven. While the nascent African middle class appeared to accept the principles of development, many had aligned themselves with the ANC, the PAC and other black opposition forces.

The irony in the 1990s is the way in which traditional opposition forces in South Africa are themselves appropriating the language and idioms of 'development' for their own ends. Far from shunning issues of 'development' as the province of apartheid apologists, the notion is assuming increasing importance in the rhetoric of the ANC and other organizations. Paradoxically, given the origins of the concept in South Africa, 'development' is becoming a central theme in the discourse of traditional anti-apartheid forces. In a post-apartheid South Africa a 'new' version of 'development' is emerging.

10

EUROCENTRISM AND GEOGRAPHY
Reflections on Asian urbanization

T.G. McGee

From this vantage point, a question somewhat different from that of most historical sociologists arises. Almost all Western writing has been both Eurocentric and written from the hindsight of the 19th and 20th centuries. It has posed the question, crudely paraphrased: What was so special about the West that made it, rather than some other region, the 'master' of the world system? Put this way, the question has a self-congratulatory ring. It looks into the special qualities of the West to account for its success. By contrast, then, it judges the other contenders as deficient insofar as they lacked the characteristics of the West.

(Abu-Lughod 1978–79: 6)

INTRODUCTION

Two events in the last few years have brought home to me how much we in North America are in the grip of Eurocentrism. The first was the 'heroic' intervention of the West into the Middle East in the MGM-acclaimed special known as 'Desert Storm.' For months the world was bombarded with a carefully controlled version of the events which created the image of a new crusade against a cunning and insane non-Western foe. The sum total of deaths on the Allied side was small – a proportion self-inflicted – with over 100,000 on the part of the Iraqis. This was despite the use of 'clean bombs,' one of the most horrifying of all terms devised by the military complex. In contrast a devastating typhoon in Bangladesh with a figure of 150,000 deaths managed to make our headlines for only two days: just another disaster from the Third World.

The first of these events was said to be a necessary part of maintaining (perhaps creating is more accurate) a new world order in which the 'end' of the Cold War permits the emergence of an era of peace and stability. But if we are to have a new world order it will be a world order created by Eurocentric images of the world; a new order established essentially on

Eurocentric terms in which the remaining two-thirds of the world's humanity will have to adhere to these terms or suffer the consequences. I have little optimism concerning the role of disciplines like geography in adopting a critical view of this process or in training researchers who are capable of studying the process with a cultural sensitivity and knowledge that can interpret this process of incorporation from the Third World perspective. Western geography has developed limited expertise on the Third World, let alone the Asian component of it. This situation has occured for a number of reasons but mostly because of the prevailing Eurocentric mindset of the discipline and its practitioners.

In this chapter, I offer a personal challenge to the views that geographers from the European-American heartland hold – I use the term 'heartland' deliberately since it carries all the connotations of power and control that Mackinder intended. The chapter is divided into three sections. The first section analyses the relationship between Eurocentrism and Geography. The second traces my own experience of developing my awareness of Eurocentrism in geography. Finally, the chapter illustrates how an escape from Eurocentrism can release a new set of ideas for the study of urbanization in parts of Asia.

EUROCENTRISM AND GEOGRAPHY

Let me at the outset make it clear that when I am speaking about geography I am considering it as part of the broad field of social science. For many years I have regarded the splitting-up of Geography into a series of increasing specializations as quite disastrous for the intellectual core of the subject. The increasing growth of specialization represents an empirical vindication of Marx's insights on the increasing division of labour that accompanies the growth of 'capitalism' and the triumph of the 'economic mentality' dedicated to the proposition that increasing specialization is both more efficient and is closer to the truth. This raises the questions of efficiency for whom and how much closer to the truth.

The sub-theme that I wish to develop, since it is intrinsically linked to my arguments, is the need to distinguish carefully between the attempts of geographers to develop a theory which justifies their discipline and the role of geography as part of the discourse of social theory. For instance, when I was an undergraduate we spent an immense amount of time delineating the boundaries of the discipline. It was a logical step from these explanations to the 'quantitative revolution' of the 1960s to the proliferating 'isms' of the 1970s and 1980s. In a sense these developments followed the temporal pattern of Koestler's (1969) ideas on the changes that occur in artistic creation. Koestler suggests that artistic creation gets bogged down in long periods of *normality* which produces two types of reactions. The first is a tendency towards pointed emphasis characterized by a certain involution and concentration upon mannerisms, techniques and jargon. This was certainly true of Hartshorne. The second is a trend towards *economy* or *implicitness* in which the blanks must be filled in by the reader. In general the trend to normality has dominated geography but there is now a trend to implicitness which is involving geographers, making use of theories of society rather than theoretical explanations of their own subject.

What is equally important is that other social scientists are also seeking to use geography as a discourse central to many of the theoretical issues with which they are grappling (Braudel 1972; Giddens 1984, 1990; Chaudhuri 1990). The idea of geography as part of the central intellectual discourse of social sciences is important because it raises broader issues concerning 'intellectual domains' that have begun to be central to discussions of social theory in this contemporary period which have been broadly labeled post-modern. I might suggest that at present there are three major domains: ethnocentrism; feminism; and inequality. This last term may not be entirely appropriate; 'class' is often used instead. But to me it encompasses a great deal of the domains of world system analysis, the concern with the 'other voices' and the issues of relationship between international, national and local levels which Gregory (1990) labels the 'local–global dialectic.' Cutting through all these domains is the development of what Harvey has labelled 'space–time compression' which is reducing spatial barriers, creating a world market 'which not only allows a generalized access to diversified products of different regions and climes, but also puts us in direct contact with all peoples of the earth' (Harvey 1989: 109). This has created the opportunity both to escape from and reinforce the Eurocentric mindset.

But we still have to recognize that Euro-North American Geography approaches the non-Western world from a particular perspective, a vision which may be labelled 'Eurocentrism.' This is an assumption that has dominated the interpretation of the non-Western world by the Western world (consisting of Europe, North America, Australia and New Zealand) from the nineteenth century onwards. The cavalier linking of North America, Europe, Australia and New Zealand under the rubric of Eurocentrism may seem unjustified. But in the sense of the assumption of European supremacy, New Zealand, Australian and American attitudes towards their Pacific neighbours are no different from the French or the British towards Indo-China or the Middle East during the colonial period. The concept of Eurocentrism has many meanings. For some writers it is the adoption of a universal assumption that the 'imitation of the Western model by all peoples is the only solution to the challenges of our time' (Amin 1989: vii). For others Eurocentrism is essentially the ideological rationale of the expansion of capitalism (Blaut 1970, 1973, 1975; Amin 1974; Slater 1975). Finally some writers, notably Fanon (1968) and Mannoni (1956), portray Eurocentrism essentially in terms of the expression of racial superiority of whites over others. In this last sense Eurocentrism fits very clearly into the domain of 'ethnicity.' These assertions concerning Eurocentrism are very discomfiting to the liberal centralist stream of thought which dominates Western Geography (practitioners of the subject generally resident in Anglo-America, Western Europe, Australia and New Zealand, and the work they do in research, teaching and writing) for they challenge assumptions which are so deeply embedded that they are rarely questioned.

I would like to explore this issue further with reference to Geography's relationship with that part of the non-Western world which we will label Asia. Few geographers have attempted this task, even more broadly in the Third World (although see Buchanan (1973) and Slater (1976)). By Asia I refer to the geographical region stretching from North Africa, through the Middle East, to the Asian continent and Southeast Asia. This has two dimensions: Geography's involvement in the Eurocentric creation of Asia; and Geography's manipulation of Asia to fulfill its disciplinary goals.

The geographical profession played a central role in the Eurocentric creation of Asia which Edward Said (1978) has labelled 'Orientalism.' By this he means 'a style of thought based upon an ontological and epistomological distinction between the "Orient" and (most of the time) "Occident" (Said 1978: 2). Thus many novelists, philosophers, poets, geographers and administrators have 'accepted the basic distinction between East and West as the starting point for elaborate theories, epic novels, social descriptions and political accounts concerning the Orient, its people, customs, mind, destiny and so on' (Said 1978: 3). Said (1978: 1) is careful to suggest that there are different approaches to the Orient on the part of Americans which reflect their different colonial experiences. However, in the American universities it may be argued the 'Eurocentric' approach was taught by the largely 'European' academics who were imported to teach the discipline of Oriental Studies (see Anderson 1990). Said further argues that not only is Orientalism a style of thought but it also must be seen as a discourse in an institutional sense as a corporate institution for dealing with the Orient which was an 'enormously systematic discipline by which European culture was able to manage and even produce the Orient politically, sociologically, militarily, ideologically, scientifically and imaginatively during the post-Enlightenment period' (Said 1978: 3). Central to this process was the manipulation of time and space. This was based on an assumption of power, of superiority – a right to exert intellectual power (hegemony), to draw boundaries and to name places. As Gregory comments:

> For the *fin de siècle* coincided with the institutionalization of geography and its admission to the academy, often against considerable opposition, was secured in part through the recognition of its polical salience. Nation and empire, territory and frontier were as much the common currency of the discipline as the resource inventories of its self-styled 'commercial geographies'.
>
> (Gregory 1990: 59)

The term Asia (or the Orient) is Western. As Chaudhuri (1990: 22) has argued, 'the term is recognized as essentially Western. There is not an equivalent word in any Asian language, no such concept in the domain of geographical knowledge.' Thus the geographical space of Asia is given a particular cartographic identity, essentially distinguished by its non-Europeanness, so that a distinctly Western image of Asia as a distinct unit of space is developed (Chaudhuri 1990: 23). The same process occurs with respect to time. Again Asia in the nineteenth century is portrayed as a backward world of declining civilizations ignoring the fact that in the thirteenth century the world system was essentially centred on the series of trading systems in which the geographical nexus was the Muslim heartland, and in which large cities such as Baghdad, Hangchow (Hangzhou), and Cairo played a pivotal role and much of Western Europe (with the exception of Venice and Genoa) was on the periphery (see Abu-Lughod 1978–79). The historical condition changed by the ninteenth century from a situation of 'Asia and a people without history' to 'Europe and the people without history.'

How did geography come to be a central part of the system of thought known as Eurocentrism? Here one can identify two issues: how geographical space came to reflect Asia; and the manner in which Geography as a discipline helped create Asia. The latter process falls into three phases. First there is the period of geographical 'discovery' and exploitation which has its roots historically much deeper in the European 're-discoveries' of China and India. However this process was accelerated in the nineteenth century with European political and territorial expansion into the Orient. Geographers who called themselves explorers and explorers who called themselves geographers charted the dimensions of the Asian empires under the sponsorship of such institutions as the Royal Geographical Society. As colonial control expanded so geographers increasingly acted as a part of the colonial system, as evidenced in the work of Goureau (1953), Robequain (1954), and Spate (1954). After World War II geographers entered a second phase and became more concerned with interpreting Asian countries for a Western audience. The majority of geographic writing was published by Western presses for that audience. The chosen route to a chair was by establishing your geographical reputation as an expert on Asia or a part of it (see, for example, Dobby (1950); Goureau (1953); Robequain (1954); Spate (1954); Fisher, C.A. (1964); Fisher, W.B. (1971)). In the USA, Spencer (1954) and Ginsburg (1958) provide examples of this style. A feature of these works was the attempt to synthesize and legitimize the idea of Asia, to seek universals in time and space. As such, they represented in many cases superb efforts at synthesizing masses of other material within a primarily descriptive framework.

Another element of this phase of geographical relationship was the implantation of Western Geography into Asia. From the early twentieth century onwards Geography was grounded in the newly-established network of colonial universities. Of course, it did not have the prestige and importance of History but in general it tended to be the geography of the Western world taught by Western geographers to Asian students primarily at undergraduate level. A smattering of geographers were sent to colonial metropoles to obtain post-graduate degrees and in the 1960s and 1970s they became the Asian geographical elite. This process of implantation tended to be very paternalistic. There was little that Asia had to offer but students to be moulded in the Eurocentric image. In numerical terms this implantation process has been quite successful. There are almost certainly more geographers in Asia than the 'heartland' of Geography. But generally their published research is very imitative of the Eurocentric heartland. The profession has been very successful in creating a class of what Tarzie Vittachi once called 'brown sahibs.'

From the 1960s the relationship between Western Geography and Asia entered a second phase involving a more multi-stranded approach to Asia. To some extent this reflected greater uncertainty about the 'underdeveloped world' that was beginning to emerge in social science. But I must emphasize that the emergence of many different geographical approaches to Asia did not undermine the prevailing intellectual assumptions of Eurocentrism. There were three main strands in this phase. First, there was a continuation of certain elements of the first period but with a shift to more problem-oriented approaches. Representative of this strand were Ginsburg's (1960) 'Essays on geography and economic development,' many of which focused on Asia, and the efforts of the 'modernization' geographers to quantify the spatial impacts of Western indices on indigenous society (Leinbach 1972). Much of this work relied at least in part on the geographical ideas of non-geographers such as Skinner (1964–65), Hirschman (1958), and Friedmann and Alonso (1964). In the regional context Fryer's (1970) *Emerging Southeast Asia* was an aggressive defence of the Western model of development and a sharp attack on countries that want to tinker with it.

The second strand advocated and carried out increased fieldwork in Asian countries (the 'dirty boots begets wisdom' school). It was more concerned to recognize some of the indigenous validity of these societies. Precursors of this type of approach were Weurleusse's (1946) fine study of peasants in the Middle East, and Dupuis' (1960) marvellous account of Madras and its region. This research focused upon the role traditional aspects of Asia played in reaction to Western influence. As a result the reaction of Asian societies to Western impact was seen much less as a one-way process and more as an adaptative one. Most of these studies attempted a 'bottom-up' approach without imposing an ostensibly Western framework (McGee 1974b; Hugo 1975). There was also an attempt to emphasize indigenous solutions to development (Buchanan 1970; Missen 1972; Brookfield 1975).

The third strand was a more critical 'radical geography' which contributed to the dialogue concerning Asia in a less direct manner. Most of these critiques developed out of the Latin American and African contexts and, with the exception of Buchanan (1970) and Forbes (1984), never generated a sustained body of Asian geographical research (see Corbridge (1986, 1989) for an explanation).

The third, and most recent phase, of Western Geography's relationship with Asia has seen an emerging challenge from Asia itself. There are now almost certainly far more practising geographers in Asia than in the 'heartland.' A challenge to the European 'heartland' is growing, characterized by two broad trends. First there is the Asian criticism of Western academic geographers for practising academic colonialism. Part of this process is a growing realization on the part of Asian geographers that they need to develop a more indigenous geography. Second, there is the critique that Western geographers are becoming further compromised by their growing involvement in applied and policy studies as part of the aid programmes of their countries.

The charge of academic colonialism represents a logical reaction to Eurocentrism and forms a part of a much wider critique of Western social sciences, the 'takeover of Asia' as exemplified in the arguments of Said (1978). There are two grounds for complaint. First, Western social scientists have mined Asia for data which have primarily benefited themselves and the metropolitan powers. The second criticism is that they have paid insufficient attention to the voices of those they studied. The silenced voices became part of the 'colonial silence' of the Third World. But meanwhile the indigenous inhabitants of the Third World carried on their dialogues in a world of rich ethnic, cultural and political diversity. As Lasaga writes:

> The feeling associated with colonial status . . . is a feeling of minuteness in the face of what appears to be excessive affluence and power. . . . Associated with this feeling is a measure of pride in the knowledge that the expatriate cannot really grasp the inner workings and nuances of indigenous societies.
>
> (Lasaga 1973: 309)

Lasaga's arguments are important for they reinforce the attempt by Asian geographers to develop their own indigenous geographies.

This process is well exemplified by the Maghreb which is that part of Asia called 'North Africa,' in which the period of colonization had left a well-established core of indigenous geographers who worked together with European geographers. But increasingly there is a reaction to Western influences in Maghreb universities which reflects the Islamic fundamentalist movement which involves increasing cultural closure with an understandable emphasis upon Arabic literature. While this may seem very challenging to the European geographer, it is an entirely predictable response to two centuries of Eurocentrism (see Sutton and Lawless 1987). This attempt to draw upon indigenous cultural understanding represents a major element in the defeat of Eurocentrism.

On the second aspect of Western geography's relationship with Asia, I feel much more ambivalent. There is increasing activity of Western geographers in the applied programmes of the delivery of aid from their countries. On the one hand I am pleased because I feel that the range of skills that geographers (if they have the right kind of cultural sensitivity) can utilize in a wide range of areas including environmental, demographic and social analysis is beneficial (and certainly when compared to the narrow neo-classical paradigms of the economists.) On the other hand, most of this work is carried on within such a narrowly conceived set of parameters that geographers involved in it are unable to investigate the problems in an unfettered manner. They are often unable to criticize the political elites of Third World countries who are causing the problems they are investigating.

EXPERIENCING ASIA: A PERSONAL REACTION

In speaking about geography's relationship with Eurocentrism and the creation of the idea of Asia, I may seem to be absolving myself from involvement in these events. This is far from the case. Along with other social scientists who are beginning to have second thoughts on their involvement in this Eurocentric tradition (see Geertz 1988b; R. Young 1990), I should reveal my sense of guilt.

I was perhaps more fortunate than most in that much of my undergraduate and graduate work was carried out in the Department of Geography at the University of Wellington where a peculiar amalgam of Buchanan, Watters and Franklin taught a stimulating mixture of 'isms' that, while they did not completely destroy my Eurocentric assumptions, at least excited me to a concern with the 'other.' This was certainly helped by an MA thesis on the problems of East Indian assimilation into the urban comunity of Wellington in the 1950s. My involvement with the East Indian community (the majority were Gujuratis from an area north of Bombay) gave me a deep appreciation of the sense of injustice the community felt at the ill-concealed racism of white Wellington. This was a time when racial equality was a much-vaunted part of the ideology of the New Zealand state. I also developed a great taste for Indian curries consumed in long evenings of discussion with Indian families and community elders. I also learned about Indian films, for the Sunday afternoon film showing was a 'community event' followed by the family dinners. Indian films are very useful for raising questions about Eurocentrism and the Western impact.

After this experience it was perhaps hardly surprising that I chose to concentrate my PhD research and career in Southeast Asia and Malaysia in particular. I went with all the liberal-reformism of a small-town missionary quite convinced that I could help those poor, starving Malaysians and of course advance my own career by doing a good piece of research. The sense of guilt was important. Yeats' lines 'Come fix upon me that accusing eye, I thirst for accusation' was never truer of anyone setting off for a new land. I spent a good part of the 1960s and 1970s in Malaysia and Hong Kong, interspersed by a period back in Wellington. My relationship with these societies was at least in part colonial in that I worked in an 'implanted institution' conveying the ideas of Western geography. Despite the fact that the Wellington background left me with a 'critical sense' of what was happening in Southeast Asia I still went with a belief that the problems of development would be solved by imitation of the Western experience, by the application of Western models and by liberal doses of Western aid and investment. The end product of those fifteen years was complete disillusionment. First there were growing doubts about the meaning of development; in this I was simply echoing the currents which were emerging in the early 1970s. Good doses of Fromm (1966), Bernstein (1971), Goulet (1971), Freire (1972) and Illich (1973a, 1973b) were very helpful. But perhaps more important was the unquestioned persistence of poverty in less-developed countries. Even in Malaysia, a country and its people with whom I had fallen deeply in love, where ostensibly economic growth was occurring rapidly, poverty while being abated still continued.

This realization was further reinforced by involvement with my research in Southeast Asia. Initially this research grew out of a concern with the need to develop new theories to explain the process of urbanization as it was occurring in Southeast Asia. During my first few years in Southeast Asia I focused this work on the process of rural–urban migration of Malays to Kuala Lumpur City (McGee 1969) and broader studies of the Southeast Asian cities (McGee 1967) which led me to increasing dissatisfaction with the models derived from the experience of the developed Western countries (McGee, 1971). In their place I expanded the dualistic models of city–countryside division utilized by Boeke (1953) and others, and following Geertz (1963) transferred them to the urban context. This was a precursor for a whole range of dualistic models of the economic structure of Third World cities, including Hart's (1973) informal–formal sector division and Santos' (1975, 1976) lower–upper circuits model.

When I returned to Asia in 1968 to Hong Kong and a 'true' colonial experience, I came back with the idea of carrying out much more detailed work on the features of the 'bazaar' sector, as I then called it, which resulted in the study of hawkers in Hong Kong (McGee 1974a). In the initial stages, however, I was not particularly interested in the policy aspects of these theoretical models but rather in utilizing them to explain the persistence of poverty and operation of the bazaar sector in Asian cities. But as I carried out the studies of street vendors in Hong Kong I became more and more interested in the policy application of this research. Initially I was optimistic that the conclusions of such research might direct governments towards policies which would alleviate the conditions of the poor. This led me to undertake a much more challenging cross-country study of vendors in

Southeast Asian cities in conjunction with a group of Southeast Asian social scientists (McGee and Yeung 1977). It was during this period that my hair began to fall out and at sometime in the future when I am calmer I will almost certainly attempt to record the problems of that venture in a paper to be called 'The Trauma of Cross-Country Research.'

Increasingly, however, I have come to think that these studies of street vendors that have been carried out in such diverse locales as Hong Kong, Jakarta and the sleepy backwater of Santo in the New Hebrides all suggest that public policy has been developed towards street vendors on the basis of misinformation, or utilizing ideas from the Western experience which are not applicable. Even when new information became available that pointed to the 'madness' of these policies there was no guarantee that this would lead to revision. Planners and those in political power in these societies have their own ideology, their own vision, which they find difficult to change.

Also I have become increasingly sceptical of the dualistic models I was using. They held out hope that encouragement of economic activity in the 'informal sector' would lead to a betterment of conditions for the poor, when it was becoming increasingly obvious that it was beneficial to the 'modern capitalist sector' to allow the persistence of the 'informal sector' because of the advantages of savings on public welfare, cheap labour, and so on. A reading of the considerable body of dependency literature led me to place greater emphasis upon the linkages between the developed and under-developed world, and to realize that there was a distinct form of peripheral capitalism in the less-developed countries of the Third World which was a necessary part of the international capitalist system. I could dwell at length upon the mechanism of this dependency and the agents of this dependency such as the multi-nationals. I could also delineate some of the debates that have occurred among the dependency school which go a long way to explaining the growth of 'export platforms' such as Hong Kong and Singapore. But I believe that the general assumptions about the operation of international capitalism and its relationships with local capital provide a very useful set of explanatory ideas. They moved me into an increasing concern with 'local–global dialectic.'

There are, however, grave problems in the application of these models. It is particularly difficult to trace linkages between the less-developed countries and developed countries. Patterns of foreign investment, flows of capital to the metropole, for the operation of covert agencies such as the CIA are not easy to study; indeed this exploration of linkages between the macro and micro processes is perhaps the most difficult task before social scientists who are convinced of the general thrust of this set of ideas. However it is not insuperable. Buchanan's (1973) study, *The Geography of Empire*, shows how imaginative use of data from a wide variety of sources can map the macro-aspects of linkages. Armstrong and McGee (1985) have attempted to illustrate these processes with respect to the growth of urbanization in Latin America and Asia. In this book, described by one reviewer as the work of 'two pros, a little jaundiced but ever hopeful, working hard at the merger of big concepts and small street scenes,' we unwittingly fell upon a theme of post-modernism or at least that view of it as expressed by Jameson, when we chose to talk of convergence of consumption practices on a base of divergent production reflecting the division of the international division of labour. Folch-Serra makes a very similar point:

> Post-modern theory is, at one and the same time, an outcome and a reaction to modernism. Modernism seeks the integration of Third World societies into Western structures of thought, custom, production and consumption, regardless of whether or not they are suited for these societies.

(Folch-Serra 1989: 69)

Theatres was centrally concerned with impingement of these global forces upon Third World cities. As Jameson (1983: 124–5) says:

> Non-Marxists and Marxists alike have come around to the general feeling that at some point following WWII a new kind of society began to emerge. . . . New types of consumption; planned obsolescence; an ever more rapid rhythm of fashion and styling changes; the penetration of advertising, television and the media generally, to a hitherto unparalleled degree throughout society; the replacement of old tension between city and countryside, center and province, by the suburb and by universal standardization; the growth of great networks of super-highways and the arrival of the automobile culture.

(Jameson 1983: 124–5)

In some strange manner *Theatres* achieved a flawed post-modern vision of what was happening in Third World cities. At the same time its emphasis upon the local–global dialectic began to undermine my Eurocentric assumptions.

At the end of some thirty years of research one thing is now clear to me: at a very personal level I am now aware of my Eurocentric assumptions. This does not mean that I can discard them but Eurocentrism is now like feminism and the local–global dialectic: a set of intellectual ideas which must be used to raise a warning flag for every piece of research attempted. In the final section of this chapter I want to try and show how this awareness of Eurocentrism has influenced my recent research in Asia.

ESCAPE FROM EUROCENTRISM – NEW THEORIES OF ASIAN URBANIZATION?

Over the last twenty years one recurrent theme of my research has been the questioning of what may be labelled the Western bias of urban theory (McGee

1967, 1971, 1978, 1991). Let me stress 'Western bias' is simply regarded as a component of Eurocentrism. Along with many other theories of social change, the urbanization process in Asia has been portrayed as essentially repetitive of the Western experience and therefore subject to the same theoretical postulates.

For many years I have argued that this position is largely untenable for a majority of Third World countries. These ideas must be placed in the framework of the overall patterns of urbanization at a global and regional level, which are predicting a continuing increase in the proportion of the world's urban population. Thus by the year 2020, the UN Centre for Human Settlements (1987) predicts that more than 57 per cent of the world's population will be living in urban places. Of course this population will be most unevenly urbanized, with levels of urbanization at almost 77 per cent in developed countries and 53 per cent in the Third World. Within the Third World, this contrast will be even greater, with Latin America at 83 per cent urbanized, and Africa and Asia close to 50 per cent. However, Asia's urban population will account for a very large proportion of the Third World. Bangladesh, India, China, Indonesia and Pakistan alone will contain 34 per cent of the Third World's urban population.

These UN predictions are largely based on assumptions concerning the growth of population in places defined as urban. The predictions are calculated on growth rates reflecting performance in previous decades which, when projected forward, appear to suggest a successful shift to urbanized societies and a repetition of patterns of the more developed countries. As Ginsburg (1990: 21) has commented about urbanization in the United States: 'This condition reflects the progression of the . . . space-economy to a state of what one might consider "maturity," that is, to a condition whereby areas possessed of substantial comparative advantage . . . would be drawn effectively, through improved transportation networks, into the national geographic structure.' The implication for the urban systems of the largest countries is that a continued growth will create cities of immense size, between 16 and 30 million. However, this may not be the only possible outcome for Asian urbanization.

Thus I have attempted to challenge this Eurocentric view of the urban transition. The position I have taken essentially argues that in the Asian context the conventional view of the urban transition, which assumes that the widely-accepted distinction between rural and urban will persist as the urbanization process advances, needs to be re-evaluated. Distinctive areas of agricultural/non-agricultural activity are emerging, adjacent to and between urban cores, which are a direct response to pre-existing conditions, time/space compression, economic change, technological developments, and labour force change occurring in a different manner and mix from the operation of these factors in the Western industrialized countries in the nineteenth and early twentieth centuries.

To elaborate further, the conventional Eurocentric view of the urban transition is inadequate in four respects. First it is too narrow in its view that the widely-accepted spatial separation of rural and urban activities will persist as urbanization continues. Second it is inadequate in its assumption that the urbanization transition will be inevitable because of the operation of 'agglomeration economies' and comparative advantage, which are said to facilitate the concentration of the population in linked urban places. The emergence of such a system was described by Jean Gottmann (1961) as a Megalopolis which, when applied to the north-eastern United States, included a population largely concentrated in the urban and suburban areas, but interspersed with areas of low population density used for intense agriculture and as leisure spaces by the population of the megapolitan areas. In many parts of Asia, the spatial juxtaposition of many of the larger city cores within heavily-populated regions of intensive, mostly wet-rice agriculture based on a mixture of 'skill oriented' and 'mechanical' technological inputs has created densities of population which are frequently much higher than the suburban areas of the West (Bray 1986). This permits demographic densities similar to urban areas over extended zones of intensely cultivated rural areas located adjacent to urban cores. The considerable advances in transportation technology, particularly in relatively cheap intermediate transportation technology such as two-stroke motorbikes, greatly facilitate the circulation of commodities, people, and capital in such regions, creating in turn large mega-urban regions.

Third, the Western paradigm of the urban transition, which draws its rationale from the historical experience of urbanization as it has occurred in Western Europe and North America in the nineteenth and twentieth centuries, is clearly not neatly transferable to the Asian urbanization process. The uneven incorporation of these Asian countries into a world economic system from the fifteenth century onward created divergent patterns of urbanization, which reflect the different interactions between Asian countries and the world system. For example, the British, French and Dutch also developed the productivity of wet-rice agriculture in Southeast Asia. In a similar manner, Japanese rule in Korea and Taiwan further accentuated the mono-crop rice characteristics of parts of these countries as sources of supply for Japan's pre-war empire. Geo-political events meant that both these countries emerged into fragile independence with high rural densities and low levels of urbanization. On the other hand, British intervention in Malaysia created an urban system oriented to the production of export products on the west coast away from the heavily-populated rice bowls of Kedah and Kelantan, limiting the possibilities of an emergent mega-urban region.

A fourth element in this urban transition which is often ignored in these Eurocentric assumptions is the emergence of new global forces emanating from new cores of capitalist growth such as Japan. We have now moved the early models of surplus extracted from the satellite to the metropole which formed part of dependency and world-systems theory to a view of the global system which sees much more flexibility in production and the movement of capital (Storper 1991). This is facilitated by the development of more malleable communication and transportation which permit the development of a new international division of labour often located in large extended urban agglomerations such as Hong Kong, Guangzhou or São Paulo (Storper 1991).

Because of these inadequacies, it is suggested that the concept of the 'urban transition' needs to be positioned within a broader paradigm of the transition in the space-economy of countries. This would include: (1) a heightened sensitivity to the historical elements of the urban and agrarian transition within specific countries; (2) an appreciation of the ecological, demographic, and economic foundations of the urban and agrarian transition; (3) an investigation of the institutional components, particularly the role of the state in the development process; (4) a careful evaluation of the transactional components within given countries including transport, commodity and population flows; and (5) a broad understanding of the structural shifts in the labour force reflecting economic change. Essentially such an approach is attempting to investigate the manner in which particular sets of conditions in one place interact with broader processual change. Such a viewpoint is much less concerned with the contrast between rural and urban as the space-economy changes, but focuses instead on the interactions within the space-economy as they affect the emergence of particular regions of economic activity.

In the context of the Asian spatial transition it is the emergence of regions of highly-mixed rural and non-rural activity surrounding the large urban cores of many Asian countries that has most interested me. These are huge regions that are often located along fast transportation corridors between two major urban centres. Examples are the Hong Kong–Guangzhou region with more than 25 million people, and the Jabotabek region of Greater Jakarta with a population approaching 16 million. These are significant foci of industrialization and rapid economic growth. These I have labelled *desakota* from the Indonesian words *desa* for village and *kota* for town. This was done in part because of my belief that there was a need to look for terms and concepts in the languages of the countries one was studying. Reliance solely upon language and concepts of Western social science, which has dominated the analysis of non-Western societies, is a form of 'knowledge imperialism' which is a central part of Eurocentrism.

The presentation of an analysis of Asian urbanization in this manner enables one to escape from the 'vice of Eurocentricism' by emphasizing the way in which 'local' conditions interact with national and international processes in both space and time. It enables the particular role of the state as a central institutional element in the process of social change and thus urbanization to be analysed, and it permits a more 'organic' Asian-centric approach to be developed.

CONCLUSION

Clearly the emergence of geography as part of the intellectual discourse of social sciences 'uncovers' the dominance of Eurocentrism in geography but where to go from here? Post-modernists want to uncover, to read the hidden meanings of the text, but most of these theoreticians do not want to engage in any commitment to action as a consequence of their discovery. The direction I would like to see is that of increasing critical analysis which flows from concerns about issues such as gender inequality, social injustice, racism, and poverty, and which I find post-modernist theory unwilling to engage at other than the textual level.

I suspect that such an assertion will be treated with derision and a citation overkill but it is this unwillingness to 'engage' that I find most personally unacceptable in post-modern thought. In arguing for a geography of engagement I realize I am entering a plea for the impossible. Geography and other social sciences are subjects whose main task is to produce knowledge and train graduates who support the status quo. The limited development of an increasingly critical geography will be accomplished by ideological short-circuiting. First within the West there will be frequent calls for geographers to justify their subject in terms of its usefulness to society, both in training and in information gathering. Second there will be an increasing demand for geographers to play a larger role as data collectors and consultants for private and public institutions. While in some ways one may commend such developments for taking geographers out of the 'ivory tower,' these trends will certainly subvert any critical trend in Geography. What is needed now is an injection of new consciousness and concern with 'problems' of all human societies that one sees in certain aspects of the environmental movement, or feminism or Marxist humanism. I conclude with the words of the Indonesian poet, Sabagio Sastowardjo (Aveling 1975), who captures the inevitability of the second temptation for the peoples of Asia; the emergence of the Eurocentric-dominated world order:

> DI ANTARA GEDUNG PENCAKAR
> Di antara gedung pencakar
>   Tak ada cerita
> Hanya jantung berdebar menanti kehangusan
> Jerit bayi terlemper
>   pada dinding-dinding kaca
> Mukamu yang letih, ah,
>   kuburkan dalam semua peristiwa dan lupakan
>   hari
> Di sini terjadi kelahiran lagi:
> Adam terbentuk dari semen dan besi
> dan garis-garis kejang
>   memburu dengus pagi

Tubuh Hawa masih hangat
   belum terjamah tangan laki
Kandungan mandul.
      Ular naga
yang membujuk dekat puncak menara
termasuk jenis paling liar.
Dan bulan, bulanku, betapa mengerikan

AMONG SKYSCRAPERS

Among skyscrapers
   fantasy is absent
Only the heart's beat awaiting anguish
The scream of a baby thrown
   against glass walls
Your tired face
   entombed in all that happens
      obliterating day
Born again here:
Adam shaped from iron and cement
and convulsed verticals
   hunting the anger of morning
The body of Eve still warm
   untouched
Her womb barren.
      Tempted
by a dangerous snake
near the top of the tower.
The moon, my moon, how awful.

ACKNOWLEDGEMENT

An earlier version of this chapter was published in *Canadian Geographer* 35 (1991). I am grateful for permission to publish this version.

Part III

OTHER DEVELOPMENTS

11

IMAGINING A
POST-DEVELOPMENT ERA

Arturo Escobar

If I knew for a certainty that a man was coming to my house with the conscious design of doing me good, I should run for my life . . . for fear that I should get some of his good done to me.

(Thoreau 1977: 328)

INTRODUCTION

For some time now, it has been difficult to talk about development, protest or revolution with the same confidence and encompassing scope with which intellectuals and activists once spoke about these vital matters. It is as if the elegant discourses of the 1960s – the high decade of both Development and Revolution – have been suspended, caught in mid-air as they strove toward their zenith, and, like fragile bubbles, exploded, leaving a scrambled trace of their glorious path behind. Hesitantly perhaps, but with a persistence that has to be taken seriously, a new discourse has set in: that of the 'crisis of development,' on the one hand, and of 'new social actors' and 'new social movements,' on the other. Many scholars even propose a radical reinterpretation of social and political reality based on a new set of categories such as 'alternative development,' new identities, radical pluralism, historicity and hegemony.

Until recently, the relation between truth and reality that characterized political discourse was relatively clear and direct. Development was chiefly a matter of capital, technology, and education and the appropriate policy and planning mechanisms to combine these elements successfully. Resistance, on the other hand, was primarily a class issue and a question of imperialism. Nowadays this distinction has been muddled, and even imperialism and class are thought to be the object of innumerable mediations. But while innovative research into the nature of resistance and political practice is growing, the same cannot be said for development. The theory of social movements, in particular, has become one of the key arenas for social science and critical thought over the last decade or so (Touraine 1981; Laclau and Mouffe 1985; Slater 1985; Kothari 1987; Shet 1987; Melucci 1989; Shiva 1989; Calderón et al. 1992; Escobar and Alvarez 1992).

The same vitality does not characterize the second key arena with which this chapter is concerned, that of 'development.' While many consider development dead, or that it has failed miserably, few viable alternative conceptualizations and designs for social change are offered in its place. Thus the imaginary of development continues to hold sway. In social movement theory new social orders are clearly imaginable, but in the arena of develop-. ment the picture is blurred, adumbrating a future 'developed' society where only 'basic needs' are met. But to arrive at this society (assuming that it were possible) would entail that all the fuss about plurality, difference and autonomy – notions central to social movement discourse – would have been in vain.

The seeming inability to imagine a new domain which finally leaves behind the imaginary of development, and transcends development's dependence on Western modernity and historicity raises a number of questions: why has development been so resistant to radical critique? What kinds of critical thought and social practice might lead to thinking about Third World reality differently? Can the hegemonic discourses of development – inscribed in multiple forms of knowledge, political technologies and social relations – be significantly modified? The emergence of a powerful alternative social movement discourse raises further questions: how do popular actions become objects of knowledge in social movement discourse? If new discourses and practices are appearing that contribute to shaping the reality to which they refer (Foucault 1985), what is the domain that this discourse makes visible? Who can 'know,' according to what rules, and what are the pertinent objects? What criteria of politics does it put into effect, with what consequences for popular actors? Finally, what is the relationship between the demise of development and the emergence of social movements?

This chapter aims to bridge these two areas of enquiry. The argument can be summarized in three propositions. First, most critiques of development have reached an impasse. The present impasse does not call for a 'better' way of doing development, nor even for 'another development.' A critique of the discourse and practice of development can help to clear the ground for a more radical imagining of alternative futures. Second, development is not simply an instrument of economic control over the physical and social reality of Asia, Latin America and Africa. It is also an invention and strategy produced by the 'First World' about the 'underdevelopment' of the 'Third World.' Development has been the primary mechanism through which the Third World has been imagined and imagined itself, thus marginalizing or precluding other ways of seeing and doing. Third, to think about 'alternatives to development' requires a theoretical and practical transformation in existing notions of development, modernity and the economy. This can best be achieved by building upon the practices of the social movements, especially those in the Third World. These movements are essential to the creation of alternative visions of democracy, economy and society.

THE HEGEMONY OF DEVELOPMENT

The making of the Third World through development discourses and practices has to be seen in relation to the larger history of Western modernity, of which development seems to be one of the last and most insidious chapters (Escobar 1995; Watts, in this volume; Cowen and Shenton, in this volume). From this perspective, development can best be described as an apparatus that links forms of knowledge about the Third World with the deployment of forms of power and intervention, resulting in the mapping and production of Third World societies. Development constructs the contemporary Third World, silently, without our noticing it. By means of this discourse, individuals, governments and communities are seen as 'underdeveloped' and treated as such.

Needless to say, the peoples of Asia, Africa and Latin America did not always see themselves in terms of 'development.' This unifying vision goes back only as far as the post-war period, when the apparatuses of Western knowledge production and intervention (such as the World Bank, the United Nations, and bilateral development agencies) were globalized and established their new political economy of truth (see Sachs 1992; Escobar 1984, 1988). To examine development as discourse requires an analysis of why they came to see themselves as underdeveloped, how the achievement of 'development' came to be seen as a fundamental problem, and how it was made real through the deployment of a myriad of strategies and programmes.

Development as discourse shares structural features with other colonizing discourses, such as Orientalism, which, as Said argues:

> Can be discussed and analyzed as the corporate institution for dealing with the Orient – dealing with it by making statements about it, by teaching it, settling it, ruling over it; in short, Orientalism as a Western style for dominating, restructuring and having authority over the Orient. . . . My contention is that without examining orientalism as a discourse we cannot possibly understand the enormously systematic discipline by which European culture was able to manage – and even produce – the Orient politically, sociologically, ideologically, scientifically, and imaginatively during the post-Enlightment period.
>
> (Said 1978: 3)

Likewise, development has functioned as an all-powerful mechanism for the production and management of the Third World in the post-1945 period. The previous knowledge production system was replaced by a new one patterned after North American institutions and styles (Fuenzalida 1983, 1987; Escobar 1989). This transformation took place to suit the demands of the post-war development order, which relied heavily on research and knowledge to provide a reliable picture of a country's social and economic problems. Development disciplines and sub-disciplines – including development economics, the agricultural sciences, the health, nutrition and educational sciences, demography, and urban planning – proliferated.

The Third World countries thus became the target of new mechanisms of power embodied in endless programmes and 'strategies.' Their economies, societies and cultures were offered up as new objects of knowledge that, in turn, created new possibilities of power. The creation of a vast institutional network (from international organizations and universities to local develop-ment agencies) ensured the efficient functioning of this apparatus. Once consolidated, it determined what could be said, thought, imagined; in short, it defined a perceptual domain, the space of development. Industrialization, family planning, the 'Green Revolution,' macroeconomic policy, 'integrated rural development' and so on, all exist within the same space. All repeat the same basic truth, namely, that development is about paving the way for the achievement of those conditions that characterize rich societies: indus-trialization, agricultural modernization, and urbanization. Until recently, it seemed impossible to get away from this imaginary of development. Every-where one looked, one found the busy, repetitive reality of development: governments designing ambitious development plans, institutions carrying out development programmes in cities and countryside alike, experts study-ing development problems and producing theories *ad nauseam*, foreign experts everywhere and multinational corporations brought into the country in the name of development. In sum, development colonized reality, it became reality.

A critique of development as discourse has begun to coalesce in recent years (Mueller 1987a, 1987b; Ferguson 1990; Apffel Marglin and Marglin 1990; Sachs 1992). The critics aim to examine the foundations of an order of knowledge about the Third World, the ways in which the Third World is constituted in and through representation. Third World reality is inscribed with precision and persistence by the discourses and practices of economists, planners, nutritionists, demographers, and the like, making it difficult for people to define their own interests in their own terms – in many cases actually disabling them from doing so (Illich 1977). Development proceeded by creating abnormalities ('the poor,' 'the malnourished,' 'the illiterate,' 'pregnant women,' 'the landless') which it would then treat or reform. Seeking to eradicate all problems, it actually ended up multiplying them indefinitely. Embodied in a multiplicity of practices, institutions and structures, it has had a profound effect on the Third World: social relations, ways of thinking, visions of the future are all indelibly marked and shaped by this ubiquitous operator.

The view of development as a discourse differs significantly from analyses carried out from the perspective of political economy, modernization, or even 'alternative development.' Such analyses have generated proposals to modify the current regime of development: ways to improve upon this or that aspect, revised theories or conceptualizations, even its redeployment within a new rationality (for instance, socialist, anti-imperialist, or ecological). These modifications, however, do not constitute a radical positioning in relation to the discourse; they are instead a reflection of how difficult it is to imagine a truly different domain. Critical thought should help recognize the pervasive character and functioning of development as a paradigm of self-definition. But can it go further and contribute to the transformation or dismantling of the discourse?

First one must ask whether such a domain can be imagined. Philosophers have made us aware that we cannot describe exhaustively the period in which we happen to live, since it is from within its rules that we speak and think, and since it provides the basis for our descriptions and our own history (Benjamin 1969: 253–64; Foucault 1972: 130–1; Guha 1989: 215–23). We may be aware of regions or fragments of our era, but only a certain distance from it will enable us to attempt the critical description of its totality as an era which has ceased to be ours. We may be approaching this point in relation to the post-war order of development. The critiques of development by dependency theorists, for instance, still functioned within the same discursive space of development, even if seeking to attach it to a different international and class rationality. We may now be approaching the point at which we can delimit more clearly the past era. Perhaps we are beginning to inhabit a gap between the old order and a new one, slowly and painfully coming into existence. Perhaps we will not be obliged to speak the same truths, the same language, and prescribe the same strategies.

Inordinate care must be taken to safeguard this new discourse from attempts to salvage development through fashionable notions such as 'sustainable development,' 'grassroots development,' 'women and development,' 'market-friendly development,' and the like, or to restructure the Third World in line with the symbolic and material requirements of a new international division of labour based on high technology (Castells 1986; Harvey 1989; Amin 1990; López Maya 1991). Critical thought can rouse social awareness about the power that development still has in the present. It will also help in visualizing some possible paths along which communities can move away from development into a different domain, yet unknown, in which the 'natural' need to develop is finally suspended, and in which they can experiment with different ways of organizing societies and economies and of dealing with the ravages of four decades of development.

The number of Third World scholars who agree with this prescription is growing. Rather than searching for development alternatives, they speak about 'alternatives to development,' that is, a rejection of the entire paradigm. They see this reformulation as a historical possibility already underway in innovative grassroots movements and experiments. In their assessment, these authors share a number of features: a critical stance with respect to established scientific knowledge; an interest in local autonomy, culture and knowledge; and the defence of localized, pluralistic grassroots movements, with which some of them have worked intimately (Esteva 1987; Kothari 1987; Nandy 1987, 1989; Shet 1987; Fals Borda 1988; Rahnema 1988a, 1988b; Shiva 1989; Parajuli 1991; Sachs 1992). For these authors, as the links between development and the marginalization of people's life and knowledge become more evident, the search for alternatives also deepens. The imaginary of development and 'catching up' with the West is drained of its appeal. In sum, new spaces are opening up in the vacuum left by the colonizing mechanisms of development, either through innovation or the survival and resistance of popular practices.

What is at stake is the transformation of the political, economic and institutional regime of truth production that has defined the era of development. This in turn requires changes in institutions and social relations, openness to various forms of knowledge and cultural manifestations, new styles of participation, greater local autonomy over the production of norms and discourses. Whether or not this leads to significant transformations in the prevailing regime remains to be seen. The future cannot be predicted in this respect, nor can an explicit strategy be proposed. However, the grassroots initiatives of social movements may very well lead in this direction. Social movements constitute an analytical and political terrain in which the weakening of development and the displacement of certain categories of modernity (for example, progress and the economy), can be defined and explored. It is in terms of social movement discourse that 'development,' and its foundational role in the constitution of the 'Third World' and the post-war international economic order, can be put to the test.

SOCIAL MOVEMENTS AND THE TRANSFORMATION OF THE DEVELOPMENT ORDER

There is little point in speculating in the abstract about the character of a post-development era. If we accept that critical thought must be 'situated' (Haraway 1989; Fraser 1989), then a discussion of these issues should be practice-oriented, engaging with the politicized claims and actions of oppositional movements. In the long run, it is these movements which would largely determine the scope and character of any possible transformation. Hence it is important to link proposals for the transformation of development with the ongoing work of social movements.

Although there is disagreement about the nature and extent of today's social movements, it is clear to most analysts that 'there has begun a change in the structure of collective action . . . a new space for theory and social action' (Calderón and Reyna 1990: 19). Questions about daily life, democracy, the state, political practice, and the redefinition of development can be most fruitfully pursued in the context of social movements. But how are the practices of social movements to be studied? How can social science make visible the domain of popular practices and the inter-subjective meanings that underlie them? How can the self-interpretation of agents be accounted for? What is the field of meanings in which popular actions are inscribed and how has this field been generated by processes of domination and resistance, strategies and tactics, scientific knowledges and popular knowledges and traditions? What are the relations between cultural definitions of social life and political culture? How do collective actors build collective identities, and how do they create new cultural models?

The importance of daily life and its practices for the study of social movements is increasingly appreciated in Latin America. Reflection on daily life has to be located at the intersection of the micro-processes of meaning production, and the macro-processes of domination. Inquiry into social movements from this perspective seeks to restore the centrality of popular practices *without reducing the movements to something else*: the logic of domination or capital accumulation, the struggle of the working class or the labour of parties. This procedure privileges the value of everyday practices in producing the world in which we live. For if it is true that the majority of people live within structures of domination that are not of their own making, it is also true that they participate in these structures, adapting, resisting, transforming or subverting them through manifold tactics (de Certeau 1984; Scott 1985; Fiske 1989; Willis 1990; Comaroff and Comaroff 1991).

Much of the recent literature takes for granted that a significant social transformation has already taken place, perhaps the coming of a new period altogether. The 'old' is often yoked to analyses of modernization or dependency; to politics centred around traditional actors like parties, vanguards, and the working class who struggle for the control of the State; and to a view of society as composed of more or less immutable structures and class relations that only great changes (i.e. massive development schemes or revolutionary upheavals) can alter in a significant way. The 'new,' by contrast, is invoked in analyses based not on structures but on social actors; the promotion of democratic, egalitarian and participatory styles of politics; and the search not for grand structural transformations but rather for the construction of identities and greater autonomy through modifications in everyday practices and beliefs.

Social movement discourse thus identifies two orders – the old and the new – characterized by specific historical features. In the process, the many continuities between the two regimes – as well as the ways in which, for instance, old styles of politics are still pervasive among the new movements – are overlooked. Equally important, the past is endowed with features that are not completely accurate (for instance, the claim that all styles of politics in the past were clientilistic and non-participatory). To acknowledge the continuities existing between the two periods – both at the level of theories of politics, development, the economy, and that of popular practices – is important (Cardoso 1987; Mires 1987; Alvarez 1989). A more rigorous characterization of the nature of the change that is taking place is needed.

The demise of old models is arguably brought about by the failure of the developmentalist state to bring about lasting improvements, and of political mechanisms, on either Left or Right, to deal with that failure. Moreover, the untenability of the old models is reflected in the present crisis. This dual crisis of paradigms and economies is forcing a new situation, a 'social reconfiguration,' as Mires (1987) has aptly put it. For most observers, the crisis of the post-war models, centred around the agency of the modernizing bourgeoisie and the working class, necessarily entails the dissolution of the political discourses of these social actors thus paving the way for new voices and political manifestations. Beyond these general assertions, however, most talk about the crisis is imprecise at best. Crisis is conceptualized mostly in economic and political terms, but many questions remain: what, for instance, are the inherent contradictions of today's models? What specific problems of system control seem to be critical? What structures are being strained? How are legitimation, fiscal and economic crises interrelated in specific Latin American countries?

Still other questions are raised by the premise that culture and ideology are embedded in production and politics: what cultural features seem to pose limits to accumulation and the persistence of old political forms? Is the loosening of economic and political structures leading to new traditions and identities? What specific institutions are disintegrating? What groups of people feel their identity particularly threatened, and in what ways? If old systems of group identity are losing their integrating power, what are the new systems for identity formation? What new goals and values are being formulated? What new discourses are being put in circulation as the usual mechanisms for social and cultural discourse production are upset? These questions do not arise simply in relation to the very real and dramatic dislocations that Latin America suffers today. It is also necessary to probe deeper into the shifts and fluctuations in institutional arrangements, meanings and practices that result, in part, from the crisis.

The number and quality of studies of social movements in Latin America has grown steadily over the last decade. Amongst those studied are urban popular movements, Christian communities, peasant mobilizations, new types of workers' organizations and novel forms of popular protest (for basic needs and local autonomy, for example). Increasing attention is being paid to women's and ethnic movements and grassroots movements of various kinds; on the other hand, few studies exist of the gay (McRae 1990) and ecology movements (Viola 1987; García 1992). Human rights and defence of life issues, as well as youth forms of protest, have also attracted some attention.

The most complete study to date is the ten-country survey carried out by the Latin American Social Science Council (CLACSO) under the general direction of Fernando Calderón (1986). The study examines the relationships among crisis, movements and democracy, and the possible contribution of the movements 'to constructing new social orders, propitiating new models of development and promoting the emergence of new utopias' (Calderón 1986: 12). The existence of new tendencies towards greater autonomy and pluralism, less dependence on the state, and new values of solidarity and participation are also explored. In sum, the study purports to seek in the movements

> Evidence of a profound transformation of the social logic ... a new form of doing politics and a new form of sociality ... more importantly, a new form of relating the political and the social, the public and the private, in such a way that everyday practices can be included side by side with the politico-institutional.
>
> (Jelin 1986: 21)

One question that arises immediately is how this 'profound transformation,' will be conceptualized. Here we find a tautological proposition, since social movements are defined precisely in terms of what they supposedly bring about: new forms of politics and sociality whose definition in turn is left unproblematized. The 'new forms of doing politics' comprise not a new conception of politics but an expansion of the political domain to encompass everyday practices. Even the future of the movements is seen in relatively conventional terms: small organizations will branch out vertically and horizontally, non-party formations will give way to parties, short-term protest to long-term efforts. Similarly, social scientists see social movements as pursuing goals that look very much like conventional development objectives (chiefly, the satisfaction of basic needs). More radical questions about the redefinition of the political and the dismantling of development are thus overlooked. This is compounded by the fact that there is no agreement as to what counts as a 'movement' and what makes it 'new.'

Despite these difficulties, studies of social movements have clarified a number of macro issues. The relationships among crisis, social movements and democracy have been broadly defined. The reasons for the emergence of new actors have also been identified. These include the exclusionary character of development, increased fragmentation and precarious urbanization, general social decomposition and violence, the growth of the informal sector, loss of confidence in the government and political parties, the breakdown of cultural mechanisms, and so forth. Others have argued that the displacement of spaces and identities (from the working class to new actors, from the factory to the city, from the public sphere to the household, from the plaza to the neighbourhood) accounts for the new movements. Some of the movements have arguably achieved a transition 'from the micro to the macro, and from protest to proposal,' as they connect with each other in the building of coalitions and political movements, such as the Workers' Party in Brazil, the M-19 Democractic Alliance in Colombia, and the Cardenista movement in Mexico (Fals Borda 1992).

The CLACSO project also offers a typology of social movements based on their structures of participation; their temporality, including synchronic and diachronic aspects; their multiple and heterogeneous spread within a geographical space; their relation to the crisis and to other social forces; their effect on social relations; and their self-image and conception of everyday life. Finally, other studies have identified criteria for evaluating the democratizing potential of the movements: the extent to which they undermine prevalent authoritarian practices; the extent of their pluralizing and diversifying effects on dominant, homogenizing and reductionist forces; the relative weight of autonomy versus clientilism and dependence on the State and conventional political institutions; the possibility of bringing about new economic forms which transcend the rationality of the market; and the possible role played by Latin America within the international division of labour (Calderón 1986).

ISSUES IN SOCIAL MOVEMENT RESEARCH: KNOWLEDGE, POLITICS AND NEEDS

The intellectual and political challenges of social movements have provoked a significant academic reappraisal of civil society, the importance of the microsociology and politics of everyday life, the possibility for new types of pluralist democracies and alternative ways of satisfying basic needs (Calderón *et al.* 1992). There is, then, a sort of 'thematic renewal' which, despite conflicting demands and the existence of conservative tendencies (such as neoliberalism), is having a great impact on the social sciences in Latin America (López Maya 1991).

This does not mean that European and North American theories are not important. Post-structuralism, post-modernism and post-marxism have significantly influenced Latin American social movements theory. The most influential notions are Touraine's concept of historicity, Laclau and Mouffe's elaboration of articulation of identities and radical democracy, and Melucci's proposal of the social as a submerged network of practices and meanings. Touraine's and Melucci's work foreground the cultural aspects of collective mobilization. For the French sociologist, social movements struggle for the control of 'historicity,' defined as the 'set of cultural models that rule social practice' (Touraine 1988: 8). In other words, social movement actors recognize that there is a cultural project at stake, not merely a struggle for organizational control, services, or economic production. Melucci emphasizes the cultural character of contemporary collective action at an even deeper level. For this author, social movements have a very important symbolic function; collective actions 'assume the form of networks submerged in everyday life. . . . What nourishes [collective action] is the daily production of alternative frameworks of meaning, on which the networks themselves are founded and live from day to day' (Melucci 1988: 248). This also means that what we usually empirically observe as 'movements' is usually the manifestation of a larger, latent reality that involves continuous symbolic and cultural production.

For Laclau and Mouffe (1985), the collective identities that define a given movement (whether peasant, working-class, feminist, gay, ecologist, indigenous, or what have you) are never given from the start, but are the result of processes of 'articulation.' This process of articulation is always discursive, to the extent that it always entails a plurality of orientations and subject positions. The processes of negotiation, interaction, building of common interests, and relations to the social and political environment – like all social life – is endowed with and apprehended through meaning. From the dominant side, the process of discursive articulation results in a hegemonic formation; on the side of social movements, the logic of articulation can lead to radical democracy – groups and movements organizing in autonomous spheres, but also creating the possibility of articulations with other groups and movements, and, in the long run, the possibility of 'counter-hegemonic' formations.

While these works have been influential in Latin America, it should not be concluded that their application to that context, or that theory production, is a one-way street. Indeed as Calderón et al. (1992: 21) argue, one must question 'whether in spite of the richness of these foreign analyses there may not be something present in the social movements of the region impervious to the analytical categories provided by European theorists.' They conclude that Latin American researchers might actually be leading the way in the reformulation of social movements theory and methodology through continuous reflection on the practice of the movements. In sum, the belief that theory is produced in one place and applied in another is no longer acceptable practice. There are multiple sites of production and multiple mediations in the generation and production of theory. Social movement work provides a good example of 'travelling' theories and theorists in the post-colonial world (Said 1983; Clifford 1989). In the process, the West is partly reproduced as the site of enunciation, but also displaced and resisted. In important ways Third World intellectuals, while trying to extricate themselves from the West, remain bound to it in complex ways, sharing, to a greater or lesser extent, the theoretical imaginary of the West. Yet theoretical production cannot be seen in simple terms, as produced in one part and applied in another, but rather as a process of multiple conversations in a discontinuous terrain (Mani 1989).

Jean Cohen (1985) has introduced a useful distinction between those social movements primarily concerned with resource mobilization and those which emphasize struggles to constitute new identities as a means to open democratic spaces for more autonomous action. The focus of the identity-centered paradigm is primarily on social actors and collective action. This is true of the three most influential European conceptualizations of social movements already mentioned – those of Alain Touraine, Ernesto Laclau and Chantal Mouffe, and Alberto Melucci (see Escobar (1992b: 35–41) for a critique).

However, as Alvarez (1989) has remarked, disregard for the North American resource mobilization paradigm has had a high cost in Latin America. Many types of popular action have been crudely characterized in terms of groups 'reclaiming their identity' or searching for 'new ways of doing politics.' This leaves unexplained complex issues that impinge on the movements, such as organizational and institutional development, the role of external factors, constraints and opportunities *vis-à-vis* local or national politics, and so on. Some authors (Alvarez 1989; Tarrow 1988) argue that both paradigms should be combined for a more realistic portrayal of social movements in Latin America and elsewhere.

The recent work of the *Subaltern Studies* group of Indian historians provides rich insights for thinking 'the political' in a new manner. According to this group, conventional views of Indian politics, of the Right or the Left, are indelibly shaped by the institutions of colonialism, thus overlooking the existence of a whole different political domain:

> Parallel to the domain of elite politics there existed throughout the colonial period another domain of Indian politics in which the principal actors were not the dominant groups of the indigenous societies or the colonial authorities but the subaltern classes and groups. . . . This was an autonomous domain, for it neither originated from elite politics nor did its existence depend on the latter. . . . The experience of exploitation and labour endowed this politics with many idioms, norms and values which put it in a category apart from elite politics.
>
> (Guha 1988: 401)

Recognition of the existence of the subaltern domain of politics is the basis, according to Guha, for developing alternative conceptions of popular consciousness and mobilization, independent of conventional politics. In the case of peasant resistance in colonial India, for example, mobilization was achieved through horizontal rather than vertical integration; it relied on traditional forms such as kinship and territoriality; it sometimes grew out of outrage, even crime, to insurgency and uprisings; it was collective and often destructive and total; and it practised various styles of class, ethnic and religious solidarity. Nationalist leaders, on the other hand, tried to make the masses conform to a conventional politics with recognizable organizations and strategies (Guha 1983).

Much of the discussion of social movements in Latin America assumes a single political domain. Popular struggles sometimes resemble Guha's notion of subalternity but one of the effects of bourgeois hegemony has been the belief in a single political domain. Most analysts, in trying to get at the 'new forms of doing politics' or the 'new political culture,' still seem to be observing the movements from the conventional point of view. In other words, much of the discussion of social movements still relies on and shelters the political (and economic) culture of the West. With few exceptions, the possibility of a subaltern domain has been overlooked in Latin America even though scholars have tried to recuperate popular resistance as part of a theoretical and practical discussion of political practice and process. The conventional view of politics shapes any 'normal' understanding of the political, entrenched as it is in structures and everyday practices (including the state, interests groups, parties, forms of rationality and behaviour such as strikes, visible mobilizations, and so on). A redefinition of politics cannot occur without changing this political discourse. This is why, in Latin America, the question of the location of meaning is an essentially political question that has to be answered in the terrains of history, politics and dominant representations. Only by rethinking politics in this way will another space of history that registers popular experience be opened up (Lechner 1988; Quijano 1988).

Critical reflection on the politics of knowledge and the state is also crucial for transforming our understanding of social movements and development. Although social movements are usually thought of in terms of their connection to the state, they are also well beyond it. In the first place, relations of power exist outside the state, in a whole network of other relations (at the level of knowledge, the family, and so on) (Foucault 1980a, 1980b). Social movements may also hinder the consolidation of extra-social bodies such as the state. If the state is arborescent (characterized by unity, hierarchy, order), the new social movements are rhizomic (assuming diverse forms, establishing unexpected connections, adopting flexible structures, moving in various dimensions – the family, the neighbourhood, the region) (see Deleuze and Guattari 1987). Social movements are fluid and emergent, not fixed states, structures, and programmes. They might even be considered 'nomadic.' In perpetual interaction with the state and other megaforms like multinational corporations, they are irreducible to them.

A similar situation is found in the field of knowledge. State science and 'nomad' science coexist but the former is always trying to appropriate the latter. State science proceeds by territorializing, creating boundaries and hierarchies, producing certainties, theorems, and identities. Nomad (or popular) knowledge has a very different form of operation, opposed to that of the State and the economy, with its division of social space into rulers and governed, intellectual and manual labour. Nomad science stays closer to the everyday, seeking not to extract constants but to follow life and matter according to changing variables. While state science reproduces the world according to a fixed point of view, nomad science follows events and solves problems by means of real life operations, not by summoning the power of a conceptual apparatus or a pre-established form of intervention (Deleuze and Guattari 1987).

These features of new social movements – a certain independence of the state and the existence of a domain of popular knowledge – are hinted at in some of the literature. Fals Borda (1992), for instance, sees Latin

American social movements as fostering 'parallel networks of power' and a kind of 'neo-anarchism' resulting from the movements' search for greater autonomy from the state and conventional political parties. Some see today's social movements as 'nomad forms' which, although expanding the cultural and political terrain, may or may not coalesce into larger networks of action (Arditi 1988). Similarly, other systems of knowledge are invoked in the literature on alternatives to development (Apffel Marglin and Marglin 1990). Such alternative forms of knowledge are practised in the popular domain, particularly among women (Shiva 1989) and indigenous people. Participatory action research is based on this belief, focusing on the encounter between modern and popular forms of knowledge (Fals Borda 1988; Fals Borda and Rahman 1991).

As self-producing and self-organizing entities, social movements can be characterized as *autopoietic* (Maturana and Varela 1980, 1987). Through their own action they establish a distinct presence in their social and cultural environment. They produce themselves and affect the larger social order through their own organizing processes. It is necessary, then, to examine both the internal organization of these units and the history of their interaction with their environment. The result of this history of interactions is the creation of life worlds and social orders. Thus, movements would not merely be a reflection of the current crisis or any other principle, but would have to be understood in terms of their own rationality and the organization they themselves produce. Our knowledge of this history, of course, can only be fragmentary and dependent on our own systems of interpretation. Like hypotheses of the existence of subaltern domains of politics and knowledge, we need to be aware of the mediations that inevitably condition our perceptions of people's histories.

THE POLITICS OF NEEDS

The question of 'needs' is central to social movements analysis. The definition of needs presumes the knowledge of experts who certify 'needs,' and the institutionalization of 'social services' by the state. Needs discourses, as elaborated by development experts, universities, social welfare agents, and all kinds of professionals can be seen as 'bridge discourses, which mediate the relations between social movements and the state ... expert discourses play this mediating role by translating the politicized needs claimed by opposi- tional movements into potential objects of state administration' (Fraser 1989: 11). Most often, the interpretation of people's needs is taken as unprob- lematic, although it can easily be shown to be otherwise. There is an officially recognized idiom in which needs can be expressed: the means of satisfying 'needs' position people as 'clients' in relation to the state. Models of needs satisfaction are stratified along class, gender and ethnic lines. In other words, needs discourses constitute veritable 'acts of intervention' (Fraser 1989: 166) to the extent that the political status of a given need is an arena of struggle over how it is interpreted.

Social movements necessarily operate within dominant systems of need interpretation and satisfaction. But they do tend to politicize those interpretations by refusing to see needs purely as 'economic' or 'domestic.' This process contributes to the consolidation of alternative social identities by subaltern groups, especially if they manage to invent new forms of discourse for interpreting needs. As Fraser (1989: 171) argues, 'in oppositional discourses, needs talk is a moment in the self-constitution of new collective agents or social movements.' It is a problematic 'moment,' since it usually entails the involvement of the state and mediation by those who have expert knowledge. Whereas expert discourses (such as those of the agents of development) reposition groups as 'cases' for the state and the development apparatus, thus depoliticizing needs, popular actors challenge expert interpretations with varying degrees of success; for instance, rural development programmes may spawn movements for the recuperation of land.

In the Third World the process of needs interpretation and satisfaction is clearly and inextricably linked to the development apparatus. The 'basic human needs' strategy, pushed by the World Bank and adopted by most international agencies, has played a crucial role in this regard (World Bank 1975a; Leipziger and Streeten 1981). This strategy, however, is based on a liberal human rights discourse and on the rational, scientific assessment and measurement of 'needs.' Lacking a significant link to people's everyday experience, 'basic human needs' discourse does not foster greater political participation. This is why the struggle over needs interpretation is a key political arena of struggle for new social actors involved in redirecting the apparatuses of development and the state. The challenge for social movements – and the 'experts' who work with them – is to come up with new ways of talking about needs and of demanding their satisfaction in ways that bypass the rationality of development with its 'basic needs' discourse. The 'struggle over needs' must be practised in a way conducive to redefining development and the nature of the political. Finally, the language of 'needs' itself must be reinterpreted as one of the most devastating legacies of modernity and development, as Ivan Illich (1992) argues. These are open challenges that remain to be explored.

CONCLUSION

The possibility for redefining development rests largely with the action of social movements. Development is understood here as a particular set of discursive power relations that construct a representation of the Third World. Critical analysis of these relations lays bare the processes by which Latin America and the rest of the Third World have been produced as 'under-developed.' Such a critique also contributes to devising means of liberating

Third World societies from the imaginary of development and for lessening the Third World's dependence on the episteme of modernity. While this critical understanding of development is crucial for those working within social movements, awareness of the actions of the movements is equally essential for those seeking to transform development.

As regards social movement research, significant ambiguities and confusions still exist. A critical view of modernity, for instance, emphasizes the need to resist post-Enlightment universals (such as those of economy, development, politics and liberation); a reflection on historicity allows us to foreground the cultural aspects of the new movements; the discussion of meaning and background cultural practices provides a way to study the connection between cultural norms, definitions of social life and movement organization; this discussion also provides a conceptual tool for exploring the more profound effects of social movements, namely, those that operate at the level of life's basic norms.

Similarly, the notion of autopoiesis suggests that social movements are not merely a reflection of the crisis, but have to be understood in terms of the organization they themselves produce. They are, in important ways, self-producing, self-referential systems, even if their effects disseminate across large areas of economic, social and cultural life. In conceptualizing social movements as autopoietic entities, conventional definitions of the political, of knowledge, and of the relation between social movements and the state need to be scrutinized. Even if popular knowledge and politics are in continuous relation to the state, they nevertheless may have their own rationality and rules of operation.

To conclude, we may postulate the existence of three major discourses in Latin America with the ability to articulate forms of struggle. First, there is the discourse of the democratic imaginary (including the fulfilment of 'needs,' economic and social justice, human rights, class, gender and ethnic equality). Although it originates in the egalitarian discourses of the West, it does not necessarily have to follow the West's experience. This discourse offers the possibility of material and institutional gains and the emergence of more pluralistic societies. Second, there is the discourse of difference, which includes cultural difference, alterity, autonomy and the right of each society to self-determination. This possibility originates in a variety of sources: anti-imperialist struggles, struggles of ethnic groups and women, the challenge to European ethnocentrism and conventional epistemologies, revisions of history, and so on. The potential here is for the strategic release and furthering of some of these struggles. Third, there are anti-development discourses proper, which originate in the current crisis of development and the work of grassroots groups. The potential here is for more radical transformations of the modern capitalist order and the search for alternative ways of organizing societies and economies, of satisfying needs, of healing and living.

It should be clear by now that struggles in the Third World cannot be seen as mere extensions of the 'democratic revolution' or the consolidation of modernity. Although they may be necessary to help weather the precariousness of life conditions and to democratize social and economic life, the recent struggles in the Third World go well beyond the principles of equality, relations of production and democracy. They constitute arenas for redefining and recovering these terms. Even the possibility of building a 'counterhegemonic formation' (Laclau and Mouffe 1985) seems to be contrary to the movements' practices and would evince a type of rationality that popular movements may not share. This does not mean that alliances are impossible. They certainly do and must occur. In fact, networks of popular movements are appearing in several countries and internationally (in the case of indigenous peoples and women). But social movements are not ruled by the logic of all or nothing; they must consider the contradictory and multiple voices present in such experiences without reducing them to a unitary logic.

In the long run, new ways of seeing, new social and cultural self-descriptions, are necessary to displace the categories with which Third World groups have been constructed by dominant forces. As Ashis Nandy puts it:

> The recovery of the other selves of cultures and communities, selves not defined by the dominant global consciousness, may turn out to be the first task of social criticism and political activism and the first responsibility of intellectual stock-taking in the first decades of the coming century.
>
> (Nandy 1989: 265)

Perhaps social movements, as symbols of resistance to the dominant politics of knowledge and organization of the world, provide some paths in the direction of this calling, that is, for the re-imagining of the 'Third World' and a post-development era.

ACKNOWLEDGEMENT

An earlier version of this chapter appeared in *Social Text* 31 (1992). I am grateful to the publishers for permission to publish this version.

12

BLACK CONSCIOUSNESS AND THE QUEST FOR A COUNTER-MODERNIST DEVELOPMENT

Kate Manzo

Black Consciousness lives on more as an alternative vision than an organizationally active political party.

(Moodley 1992: 151)

INTRODUCTION

In February 1990, South African State President F.W. de Klerk unbanned all major opposition groups in South Africa and declared that his governm⌐ ⸱⸱ intended to 'normalize the political process in South Africa without jeopa⸱ ⸱ ising the maintenance of the good order' (*Southern Africa Report* 9 February 1990, p. 2). The 'good order' has been imperilled since then by political violence and intimidation and by 'a remarkable increase in industrial strikes around the country' (*The Star* 10 June 1994). And yet in a domestic context of political unrest, social disorder, and slowed economic growth, a South African indicator of business confidence rose in May 1994 to its highest level since December 1987. The 'peaceful transition to democracy' supposedly taking place in the country was cited by the South African Chamber of Business (*Business Day* 7 June 1994) as the reason for such optimism. If that one factor outweighs all others in business assessments of the future, it is because of the wide-ranging effects of a normalizing polity.

The idea of political normality in South Africa and elsewhere has become increasingly indissociable from that of competitive multi-party systems. World players anxious to either re-establish or initiate relations with the country have typically cast evidence of movement toward 'democracy' as the *sine qua non* of their own engagement. For some, the onset of negotiations was proof enough that South Africa was normalizing. Others waited until Nelson Mandela gave a political green light for the lifting of economic sanctions. For the rest, the announcement of a date for multi-party, multi-racial elections demonstrated once and for all the commitment of the regime

228

to the political emancipation of the majority, though US President Bill Clinton perhaps went further than most when he described the 1994 election as 'one of the new world miracles' (*The Star* 28 September 1993).

Clinton's depiction of a country that is reconstituting itself as politically normal and thus free to re-engage in normal international intercourse, is by no means unique. In the wake of meetings in Washington DC between the IMF/World Bank and a South African delegation (comprised mainly of representatives from the National Party government and the African National Congress), *The Star* reported with barely-concealed glee that both financial institutions had 'declared their eagerness to see vast sums invested in what is now being hailed as the world's newest emerging democracy' (*The Star* 29 September 1993). Once the election was over, the Group of Seven industrial countries vowed 'to provide further assistance to help strengthen economic and social development, in particular for the poorest groups' (*Business Day* 11 July 1994).

The reconstitution of South Africa as 'democratic' since the onset of a politics of negotiation cannot be separated from the increasingly global embracement of competitive elections as the solution to national political problems (Manzo 1992b: 253). Rob Nixon (1992) made a similar point when he faulted the Left for its tendency to view apartheid in isolation from 'shifts in global politics,' and argued that the current vogue for the norm of democracy across political spectrums worldwide (and not just in South Africa) reflects the collapse of the twin discourses of communism and anti-communism.

As the dawn of a golden democratic age in South Africa is anticipated with unabashed optimism in Washington and elsewhere, the normalization of the country's international relationships proceeds apace. Change is observable in the return of South Africa to international sport; in the return to international forums such as the UN, OAU and Commonwealth; in the restoration of ties to former investors and trading partners; and in the onset of a whole series of international relationships where none existed before.

Alongside presuppositions about the blessings of normalization in South Africa are analyses that caution against premature celebrations of the 'end of history.' According to Grant Farred (1992), for example, South African political identities have undergone transformations that call into question the notion of a peaceful transition to democracy. He points on the one hand to the emergence from negotiations with the National Party of a deracialized and thus unrepresentative ANC. He suggests also that the transformation of apartheid-produced cultural identities (such as that of Inkatha) into a 'sharp-edged political weapon' has been responsible for the 'horrendous' levels of political violence within the country.

Concerns have also been expressed about emergent international relationships, particularly those that threaten to subject South Africa more fully to 'that grim reaper, structural adjustment' (Nixon 1992: 248). While the IMF has promised to invest in 'peace and stability,' the World Bank stands ready to fund South Africa's Reconstruction and Development Programme (RDP). But the IMF and the World Bank are only the vanguard of what Statman (1992) calls the 'technocratic army of normalization.' Occupying a far larger space than the anti-apartheid activists of the 1980s are:

> The legion of economists and bankers; town planners and health specialists; educational consultants and conflict resolution gurus; IMF officials and World Bank technicians; USAID, Scandinavian, Japanese, EC, and Commonwealth careerists; academicians of all stripes; US, Taiwanese, and Nigerian businessmen; and God knows who else.
>
> (Statman 1992: 2)

Here it is worth repeating that political transformations and struggle in South Africa need to be understood as an effect of the changing global power relations within which the country is situated (Manzo 1992a and 1992b). Political normalization has noticeably shaped forms of struggle in South Africa through its production of new political identities and levels of violence, while international normalization undermines the credibility of 'the old solutions' to socio-economic problems and threatens to facilitate a wide array of technocratic interventions by 'the free market missionaries' (Nixon 1992: 249). What is perhaps less obvious and yet vital to recognize is how normalization has cleared the way for 'normal' development. It is not simply a case of the 'technocratic army' marching under development's banner, for that addresses only the international component of the normalization process. What both shifts together – from apartheid to democracy and from isolationism to engagement – have effected is South Africa's removal from the ranks of the global pariahs and its resituation within a different classificatory register. Put simply, it is through the normalization of South Africa's political system and international relationships that it is being reconstituted as a developing country.

South Africa's eligibility for World Bank loans to developing countries was contingent upon a handover of power to a representative government and the institution of an internal development programme. This is an example of how South Africa is reconstituted as underdeveloped; as captured by 'the imaginary of development' (Escobar, this volume). It could, of course, be argued that defining oneself as capital-poor and in need of World Bank loans is different from self-definition as underdeveloped; South Africans do not necessarily see themselves in terms of 'development' even as they participate in its rituals. Yet an underdeveloped identity has been fashioned out of the clay of negotiations as well as from the mud of conditionality.

Efforts by the former National Party government to have South Africa reclassified from a developed nation to a developing country within the GATT system so as to benefit from preferential trade arrangements were supported by the ANC and economists aligned to the organization. Such efforts have been accompanied by an unprecedented diplomatic openness to alliances and economic relationships with others in the Third World. In addition to Africa, upon which the gaze of foreign affairs and trade has long been fixed, China and the Middle East, India and Eastern Europe (particularly Russia) have been courted as much sought-after partners for various economic schemes. South Africa has 'firmly placed itself among developing countries' by becoming the 130th member of the 'Group of 77' in the United Nations. The spatial reorientation of South Africa from a bastion of anti-communism on the African continent into what F.W. de Klerk now calls 'the vanguard of the developing countries of the world' is but one effect of the imaginary of development (*South African Press Association* 19 April 1993).

This chapter argues that the transformation of South Africa into a 'developing country' reflects the global crisis in developmentalist discourse of which Arturo Escobar (this volume) and my own earlier work (Manzo 1992a, 1992b) has spoken. It is a crisis manifested in two simultaneous tendencies. First, there is the seeming hegemony of development due to its capacity to occupy ever larger perceptual domains and social spaces; those vacated, for example, by communist/anti-communist and apartheid/anti-apartheid commitments. This hegemony reflects in part the inability of major theoretical challenges, such as dependency, to detach development fully from its historic association with movement toward modernity (Manzo 1991). In order to understand how development establishes continuities with both colonialism and apartheid, it is necessary to situate it within the larger context of Western modernity (Escobar this volume; Manzo 1991, 1992b: 260).

The crisis is reflected, secondly, in the continued presence of resistance to modernist attitudes and practices. Discourses of Western modernity have circulated globally to justify colonialism in the name of development; but so too have counter-modernist challenges to them. Such challenges exist within the parameters of dependency theory for all its limitations. They are also to be found within the practices and attitudes of those social movements that draw inspiration from dependency, liberation theology, feminist ecology, and participatory action research (Manzo 1991).

One such social movement that confines its practices but not its intellectual inspiration to the South African context is the Black Consciousness Movement (or BCM). Before pronouncing the 'end of history' and political struggle in South Africa it is worth taking seriously the legacy of that movement. This involves asking critical questions about BCM's insights into the racial component of modern Western narratives; about its capacity to be aware of its own limits and willingness to engage in self-critique; and (perhaps most importantly) about its ability to withstand being drowned by the global tidal wave of normalization/modernization crashing across the political wasteland of South Africa.

The chapter begins with a discussion of development as a discourse of modernity, showing how the global circulation of modern narratives about man have been deployed to justify subjugation and domination in South Africa. The concept of counter-modernism is also explained in order to demonstrate how the BCM's critical attitude has been part of a global challenge to modernist thought despite its specifically South African and anti-apartheid orientation. The second part of the chapter considers Black Consciousness as a counter-modernist discourse. Drawing upon the writings of Steve Biko and other BCM activists, this section situates the movement within a global, as opposed to purely national, context. It does this by highlighting the similarities (as well as differences) between the BCM and other critical approaches like dependency, liberation theology, and particip-atory action research. The chapter points out the direct and indirect con-nections that link these various intellectual currents together through the practices of different social movements. South African society can be con-sidered a microcosm of the discursive relations between North and South, core and periphery, or white and black that exist on a global scale.

The third and final part of the chapter considers the extent to which the BCM's 'alternative vision' lives on within the two main political organ-izations to which many of its activists gravitated in the 1970s, namely the ANC and the Pan Africanist Congress (PAC). Although it is clear that the emphases of these organizations in their current policy manifestos are very different from those of the earlier BCM, it is also evident that the Black Consciousness philosophy continues to inform both cultural politics and community activism. Thus in South Africa, as globally, critical attitudes toward modernist development persist despite its seeming hegemony.

MODERNITY AND DEVELOPMENT IN SOUTH AFRICA

'The future of South Africa must be determined by the genuine leaders of our people in reasoned debate and not by emotionally charged crowds shouting slogans in our streets' (F.W. de Klerk, cited in *South African Press Association* 4 May 1993).

The epoch of modernity associated with the Enlightenment and the Protestant Reformation was characterized by a distinctive set of attitudes toward God, man, nature, and authority. At the centre of modernity stands the figure of secular man, who might achieve total knowledge, complete autonomy, and unlimited power through the development of his capacity to reason. Reason enables man to see himself, not God, as the origin of language, the maker of history, and the source of meaning in the world (Ashley 1989: 264–5). The spread of rational thought via the scientific method of Francis Bacon – which promised to conquer nature, to 'subdue her and shake her to her foundations' (Shiva 1989: 16) – provides the means to correct reasoning and to movement along the path of political emancipation and progess. Hence modernism is committed to reason, rationality, scientific knowledge, and democracy.

Among modernity's converts are the disciples of South African normalization as well as the 'modernization theorists' of mainstream development discourse. But while the faithful continue to celebrate the 'bright side' of reason and global emancipation, critical social theorists have sought to reveal the 'dark side of [modern] domination' (Escobar this volume). Following Michel Foucault's (1970) reading of Kantian philosophy as an exemplary break with classical epistemology, scholars such as Ashley (1989: 264) and Escobar (this volume) date the modern epoch from the end of the eighteenth century. However, the practice of colonialism and the routine justifications that were offered for it suggest that modernity's origins go back even further, and are to be found in other canonical texts of Western political theory.

Described by David Campbell as 'a polemic *for* science and the rationalism of the Enlightenment,' Thomas Hobbes' seventeenth-century text *Leviathan* cast reason as the antidote to civil and religious conflict (Campbell 1992: 67). Invoking a series of distinctions between science and superstition, man and beast, man and child, order and anarchy, sanity and madness, sense and nonsense, and reason and passion, Hobbes argued that rational man would restore sovereignty to the monarch in order to avoid sliding into a state of nature typified by brutality, poverty, evil and immanent death. As Campbell has pointed out, Hobbes' attempt to frighten the populace into submission to the sovereign power of the state depended upon the construction of a boundary between 'us' and 'them,' upon the conjuring up of an external realm that 'we' (in our right minds) could not possibly wish to resemble. In this regard, Hobbes' (1962: 101) reference to 'the savage people in many places of America' who 'live at this day in that brutish manner,' was integral to his overall project (Campbell 1992: 66).

Although Hobbes did not speak in the terms of underdevelopment, the logic of *Leviathan* provides an exemplary illustration of modern understandings of that term. To *be* underdeveloped was to exist in a condition of unreason, marked by the absence of industry, culture, navigation, trade, comfort, knowledge of the earth, time, art, letters, and society (Hobbes 1962: 100). For Hobbes, this mode of living in continual fear and danger was exemplified by children and 'savages' and attributable either to the absence of science ('ignorance') or to the presence of superstition and its reliance upon 'false rules' (Hobbes 1962: 45). To *become* underdeveloped was to slide back into a state of nature through a loss of reason and authority and thus to resemble 'them.' Whether through the onset of progress via scientific knowledge or the avoidance of regress through capitulation to the power of the sovereign, development signified movement away from the immature and the bestial.

In constructing 'America' as an external site of unreason, Hobbes spoke for the sovereignty of an emergent secular state in opposition to that of the universal church. While dependent upon the native American as Other, *Leviathan* sought to discipline the 'civilized' self within the boundaries of the modern state and was silent on the matter of Empire. Yet when lifted from their original spatial and temporal contexts, modern narratives have frequently been severed from the intentions of their authors and placed in the service of power relations they were never meant to justify. In this regard the Hobbesian state of nature is no exception.

To satisfy critics of land dispossession, extermination, and enslavement, colonial writings in South Africa and elsewhere spoke in part the language of self-defence. Referring to the 'bloody battles' and 'atrocities' of 'barbarous peoples' in their 'war of all against all,' Afrikaner narratives told a story of civilized man's struggle for survival in the state of nature (a natural state of war, for Hobbes) to which he had been transplanted (Du Toit and Giliomee, 1983: 212). Constructed as inhuman beasts rather than imperfect human children, the men and women of the San of southern Africa were routinely exterminated by hunting parties (de Villiers, 1987: 29). Such behaviour was not unusual in a context where European slavers described their victims as 'almost beasts in human form,' and where Europeans referred to Africans in general as 'bestial' and 'brutish' (Lauren 1988: 14, 18).

Colonial conquest in South Africa entailed 'bloody battles' between Africans and Afrikaners over land but it also involved Christian evangelization and British imperial rule. The discursive strategies deployed by earlier Dutch settlers to justify their own behaviour were now turned against them. Removed from the sovereignty first of the Dutch East India Company and then the British Empire, those who had trekked into the interior of the country were now 'lacking in civilization' and 'men without any idea of education, grown up in idleness, and in the unrestrained indulgence of the wild passions of nature' (Du Toit and Giliomee 1983: 96, 99). Unleashed from the discipline provided by European government and hard work, colonial man had lost his battle with nature and degenerated. He had slid backwards into the idle and brutish state of the Hottentots (Coetzee 1988: 3).

This particular narrative of underdevelopment served to locate responsibility for apartheid in the barbarous other, the Afrikaner, who in the course of his encounter with Africa became tribal, idiotic, the captive of irrational fears, and savage (Manzo 1992c: 117). Cleansed of responsibility for enslavement and genocide, Britain legitimized its annexation of the independent Afrikaner republics of the Transvaal and Orange Free State by heralding itself as the saviour of the white colonist in Africa. At the same time, what Coetzee (1988: 3) refers to as 'expansive imperialism' used the Afrikaners' own political vocabulary to justify the ongoing dispossession of African land and the project of enforced proletarianization.

Another set of arguments in routine circulation can be traced to a second exemplary modernist text, John Locke's *Second Treatise of Government*. The *Second Treatise* was an anti-monarchical tract. If forced to choose, reasoning man would exist in a state of nature before he would relinquish authority to absolute monarchs. 'Much better it is,' wrote Locke (1960: 317), 'in the State of Nature wherein Men are not bound to submit to the unjust will of another.' The natural condition of mankind was preferable to royal sovereignty because it was characterized by natural law and not anarchy, by idle play and not warfare, and by happy innocence and not brutality. To return to nature signified less a descent into Hell than a restoration of Eden.

Addressed to the urban merchants, tradesmen, artisans, and independent small gentry of England, Locke's *Treatise* has been described as a revolutionary manifesto within the context of seventeenth-century English politics (Ashcraft 1986). Its vision was post-monarchical rather than pre-monarchical because it advocated advance toward popular rule and not retreat to natural law. Regardless of how progressive Locke himself may have been, his work was bent easily to the service, in South Africa as elsewhere, of colonial relations of authority.

'God gave the world . . . to the use of the Industrious and Rational,' argued Locke (1960: 333) and 'in the beginning all the World was *America*' (Locke 1960: 343). Locke was as dependent as Hobbes on a temporal and spatial distancing of 'us' from 'them.' But he made the absence of labour and not the presence of warfare the key obstacle to progress. The improvement of land through labour transformed it from commonly held 'wast' into private property and enabled man to become rich 'in all the Comforts of Life.' Despite their natural endowments the 'several Nations of the *Americans*' lacked 'one hundreth part of the Conveniencies we enjoy' because they had failed to 'improv[e] it by labour' (Locke 1960: 338–9).

Lockean man was entitled to his colonial possessions because 'no real property rights to the land existed among its original inhabitants' (Du Toit and Giliomee 1983: 213). Since 'the Lord has created the earth to be lived upon and cultivated' (Du Toit and Giliomee 1983: 229), the Lord must approve of the emigration of industrious and rational men from modern Europe to various states of nature. In such terms was the dispossession of the common property of the 'idle' and their enforced proletarianization (supposedly for their own good) routinely legitimated. How, then, could Locke, who advocated majority government for his own people, be invoked to justify imperial rule?

Native American man provided, for Locke, a vision of all men 'in the beginning,' illustrating how far 'civilized' man had travelled, with the use of reason, from his own past and not where he was headed if he refused to submit to the monarch. In an age before Darwinian thinking suggested that humanity had evolved from apes, Locke's original man was more an underdeveloped human being than a brute; more akin to a child than a wild animal. When those in 'the imperfect state of Childhood' grew up by acquiring 'age and reason' they were to be freed from 'the bonds of this subjection' to their fathers in which human nature (listening to the word of God) had placed them. Unlike the 'Lunaticks and Ideots' who 'are never set free from the Government of their Parents,' children must one day be loosed from 'the

Father's Empire.' As 'Guardian of his Children,' a loving Parent would ensure correct development by means of education, itself enabled by 'restraint and correction; which is a visible exercise of Rule, and a kind of Dominion' (Locke 1960: 345–61).

As Ashis Nandy (1987) has pointed out, familiar colonial dichotomies such as white/black, civilized/uncivilized, European/native have typically been underpinned by a parent/child metaphor. Guardianship was long the principle that underpinned the behaviour of European colonizers in their relations with colonial peoples (Cowen and Shenton, this volume). In defending a lack of political rights for Africans in South Africa, for example, J.B.M. Hertzog argued:

> As against the European the native stands as an eight year old against a man of mature experience – a child in religion, a child in moral conviction; without art and without science; with the most primitive needs and the most elementary knowledge to meet those needs ... Differences exist in ethnic nature, ethnic custom, ethnic development and civilization and these differences shall long exist .... When he achieves his majority in development and civilization, and stands on an equal level with the white man, his adulthood will be acknowledged. Then the time will have come to take his claim to political rights into consideration, and further, to establish the relationship which he will have with the European.
>
> (Moodie 1975: 261)

In 1946, South African Prime Minister Jan Smuts expounded his United Party's plan to extend limited rights to blacks. He claimed that it was consistent with the idea and practice of guardianship to give Africans, as they 'developed,' certain political rights and, in so doing, the European could ensure his position in South African society (Lewsen 1988: 210–12). The National Party (NP), by contrast, treated Africans more like permanently underdeveloped 'lunaticks and ideots' than as normal growing children. Even now the NP resorts more to Hobbesian than Lockean logic when it cites 'black on black violence' in the townships as an example of what happens when the sovereign power of the state is lost.

Despite its apparent rejection of the possibility that Africans might one day develop to the same level as whites, apartheid was not an anti-modernist discourse. Throughout the modern age the 'advanced' world of 'civilized' man – with its supposedly higher material standard of life, its art, writing, science, and so on – has been placed in hierarchical opposition to the poor, tribal, 'underdeveloped' world of the 'barbarous.' Those defined solely by Europeans as inferior or 'primitive' to themselves are presumed to advance in direct proportion to their acquisition of European traits, so that normal development entails becoming, figuratively, white.

By pushing black South Africans literally and conceptually beyond the boundaries of the Republic, 'separate development' facilitated the construction of South Africa as a developed state; the 'whiter' it became through being scrubbed clean of 'black spots,' the more 'developed' it must be (Nixon 1993). Because they dismantle the barriers erected to keep the 'developed' from the 'underdeveloped,' the processes of normalization that transform 'resident aliens' into citizens and world economic actors also require a reconfiguration of the collective identity that is produced. It is in these terms, i.e. within the logic of modernist thought, that the reclassification of South Africa as a developing country can be understood.

'Separate development' was never overtly described, at least to its detractors, as a strategy of racial cleansing for the development of white society. As European states were forced by anti-colonial struggles to put the principle of guardianship into effect and grant to their colonies the political right of self-determination, the National Party tried to silence the critics of land dispossession, forced removal and disenfranchisement by insisting that its policies were consistent with that principle (Tapscott, in this volume). The NP claimed to recognize formally the incipient nationhood of black people and thus to be much more in tune with Western colonial practices than its predecessors had been. At the same time, the Afrikaans Press was quick to side with Europeans who insisted that communists would only take advantage of those who were unqualified for independence (Hugo 1988).

The synonym for 'development' in the post-colonial era is more often 'normalization' (or 'modernization') than 'civilization,' and overt references to black people as 'children' are certainly less common. The subjects of development are more likely to be states or communities than black individuals, which is why the National Party in South Africa now refers to 'social upliftment' when it speaks of black people's need for development. By the same token, the power relations that development is called upon to justify are typically those involving 'the disciplines of the IMF and the World Bank' (*The Economist* 29 March 1993, p. 4), meted out by national developmentalists rather than by imperialists or apartheid ideologues.

Despite these changes, there are important continuities between early and late-modern discourses of development. The idea of the modern West as a model of achievement, and the rest of the world as an inferior derivative, remains integral to the concept of development. The metaphor of a healthy adult continues to inform analysis of the 'modern' or 'core' world and that of a child the status of the 'traditional' or 'peripheral' (Manzo 1991). Both constitute an underdeveloped world inhabited primarily by people of colour as poor, lacking, and culturally inferior.

It is particularly problematic when the parent/child metaphor begins to underpin the self-understandings of the 'children' themselves. Gustavo Esteva (1987), for example, has described modern development and its effects in the following terms:

Development implies that one has started on a road that others know better, to be on one's way towards a goal that others have reached; to race up a one-way street. Development means the sacrifice of environment, solidarities, traditional interpretations, and customs, in the name of ever-changing expert advice. Development promises enrichment, although for the overwhelming majority, it has always meant the progressive modernization of their poverty; growing dependence on guidance and management.

(Esteva 1987: 144)

That the West (or European man) has so often recurred in modernist thought as the origin of truth and meaning in the world makes it easy to equate modernism with ethnocentrism, and anti-ethnocentrism with anti-modernism. Yet ethnocentrism is itself only one manifestation of the modernist procedure which Jacques Derrida (1978) calls 'logocentrism.' That term describes a tendency to impose hierarchy when encountering familiar and uncritically accepted distinctions between science and superstition, reason and passion, order and anarchy, industry and idleness, and so on. The first term in such oppositions is conceived as a higher reality, belonging to the realm of logos, or pure and invariable presence in need of no explanation. The other term is then defined solely in relation to the first, the sovereign subject, as an inferior or derivative form in need of correction. What distinguishes logocentrism for Derrida is a nostalgia for origins; for a foundational source of truth and meaning that is beyond doubt and criticism; and for a standpoint and standard supposedly independent of interpretation and political practice.

For Derrida a critical, counter-modernist attitude does not simply privilege one term over another, for example by insisting that truth and meaning reside with 'the primitive' and not with 'the advanced.' Such a move does not push beyond modern relations of domination and threatens to reinscribe them in their most violent form. For as Hobbes' and Locke's contrasting portrayals of 'Americans' demonstrate, 'the primitive' operates as a site for competing discourses about the self and is not a politically neutral description of an objective reality (see Torgovnick 1990). Efforts in the post-colonial world to reinvent a pre-colonial Eden that never existed in fact have been no less violent in their scripting of identity than those that practise domination in the name of development (Mudimbe 1988). From this perspective Gatsha Buthelezi may be more threatening, but is no less logocentric and modern in the leopard-skin robes of a 'traditional Zulu' than in a grey lounge-suit.

What is needed is a systematic and rigorous questioning of the history of modern concepts, which asks, for example, how a dichotomy such as advanced/primitive came to be produced in the first place. Counter-modernism also analyses what is politically at stake in the production and maintenance of such dichotomies. For Derrida the most effective critique of all is that which goes beyond the point at which most critical theory stops and critically evaluates *itself*, questioning in particular its ability to elude or even subvert logocentric reason (Derrida 1976, 1978). In this context, the challenge of Black Consciousness to modernist thinking, and the efforts of the movement to popularize ideas and practices consistent with a critical, counter-modernist attitude toward development, are of particular importance.

COUNTER-MODERNIST DISCOURSE IN THE PHILOSOPHY OF BLACK CONSCIOUSNESS

Steve Biko always insisted that Black Consciousness was 'an attitude of mind and a way of life' (Biko 1978: 91). As the writings of Biko and others demonstrate, such an attitude is consistent with counter-modernism because of the way that the BCM sought to *sub*vert, rather than simply *in*vert, an historically produced white/black dichotomy. Insisting that 'we have to examine and question old concepts, values and systems,' Biko (1978: 92) interrogated the ways in which the hierarchical placement of the category black in relation to white has been effected through culture, language, education, and religion. Given their centrality to both schools and churches, missionary societies were continually cited by Biko as major culprits during the colonial era. He argued, for example, that 'the traditional inferior-superior black–white complexes are deliberate creations of the colonialist. Through the work of missionaries and the style of education adopted, the blacks were made to feel that the white man was some kind of god whose word could not be doubted' (Biko 1978: 69).

Biko went on to implicate white liberals in the reproduction of the white/black dichotomy. He insisted that blacks must distance themselves politically from those 'who always knew what was good for the blacks and told them so' (Biko 1978: 20). This did not sit well with 'progressive' whites who saw Biko as a black racist. But as far as Biko was concerned, liberals have as much invested in the idea of blacks as 'children' as do more overtly racist whites. Throughout South African society, he argued, black people 'are being treated as perpetual under-16s' with 'the white a perpetual teacher and the black a perpetual pupil (and a poor one at that)' (Biko 1978: 21, 24). As long as the relationship between white and black continued to approximate that of parent and child, Biko argued that it would be better to have no relationship at all.

It is important to emphasize, in light of the charge of reverse racism, that Black Consciousness philosophy did not entail a rejection of white *people*. Biko (1978: 25) called on white liberals to 'realize that they themselves are oppressed if they are true liberals and therefore they must fight for their own freedom.' This is presumably what BCM activist C.R.D. Halisi (1992: 105) was alluding to when he said that the movement launched 'a call for a white consciousness movement that ran parallel to the activities of the BCM.' What the BCM called into question were the value-laden meanings attached to the concept 'white.' Biko (1978: 21, 49) argued, for example, that Black Consciousness 'seeks to demonstrate the lie that black is an aberration from the "normal" which is white.' Until black people themselves stopped believing in this lie, 'they will be useless as co-architects of a normal society where man is nothing else but man for his own sake.'

In continuing to talk about 'white people' and 'black people,' the BCM has been accused of accepting the legitimacy of colour as a marker (Moodley 1992: 146). This charge has some truth, but it is important to point out that, for Biko, 'race' signified attitudes, meanings, and assumptions as much as it did skin colour. He argued, for example, that 'being black is not a matter of pigmentation – being black is a reflection of a mental attitude' (Biko 1978: 48). This enabled him to subvert the National Party's racial classification scheme of white, coloured, Indian, and black by insisting that the latter three categories could theoretically be collapsed into one. Whether or not the oppressed in South Africa were one race or three depended not on the colour of their skin but on their willingness to 'hold their heads high in defiance rather than willingly surrender their souls to the white man.' Biko described those among the oppressed who aspired to whiteness as 'non-white' rather than black (Biko 1978: 49–50).

Black Consciousness refused to consider black an inferior or derivative form of white. In keeping with a counter-modernist attitude, the BCM interrogated the history of these concepts and showed how a superior/inferior, white/black hierarchy had been produced and sustained with the aid of missionaries and white liberals. It also showed how the uncritical acceptance of 'white' as a model to aspire to was implicated in the continued oppression of *all* people in South Africa, not just 'black' people. At stake politically is liberation, which cannot be effected as long as white people and black people relate to each other as parents and children.

Biko's (1978: 46–7) call for a re-evaluation of African value systems, cultures, religions, and outlooks on life was coupled with a rejection of 'the power-based society of the Westerner that seems to be ever concerned with perfecting their technological know-how while losing out on their spiritual dimension.' Both moves – as well as rejection of the white norm – affected the ways in which Black Consciousness understood and tried to instantiate the concept of development. But before turning to that discussion it is necessary to ask whether the BCM was guilty of falling (perhaps unwittingly) into the logocentric trap. Did the BCM, in other words, simply invert the white/black hierarchy and treat white as an aberration from black instead of the other way around?

According to BCM activist Mamphela Ramphele (1992: 178), 'a serious and costly error of the BCM was its failure to recognize that not all black people are necessarily committed to liberation and that the poor are not inherently egalitarian.' The ability of some blacks to exploit other blacks, she argues, was also insufficiently considered: 'The fact that black people were discriminated against as a group was regarded as a sufficient condition for their capacity and desire to identify with one another for the common good'

(Ramphele 1992: 171). A tendency to heroize 'the community' was therefore evident among some Black Consciousness activists, and identified by others in a spirit of ongoing self-evaluation and critique.

Yet it is difficult to trace the constitution of black people as a model of original and natural goodness to the philosophy of Black Consciousness itself. In the writings of Biko, for example, 'the black man' is described in the following terms:

> All in all the black man has become a shell, a shadow of a man, completely defeated, drowning in his own misery, a slave, an ox bearing the yoke of oppression with sheepish timidity.... The first step therefore is to pump back life into his empty shell, to infuse him with pride and dignity, to remind him of his complicity in the crime of allowing himself to be misused and therefore letting evil reign supreme in the country of his birth.
>
> (Biko 1978: 29)

Ramphele wants to remind the Movement that in Black Consciousness philosophy, black people can be both exploiter and exploited, hero and 'empty shell,' resilient fighter and collaborator. This is because black people are 'active agents of history' rather than 'those acted upon' (Ramphele 1992: 161). As full human beings, not children, blacks are capable of effecting their own liberation at the same time as they cannot avoid complicity in the perpetuation of the system that oppresses them. For Biko (1978: 49), the attainment by blacks of consciousness of themselves – warts and all – was necessary for the realization of 'the envisaged self which is a free self.'

Even if Black Consciousness did not simply replace a white model with a black one, there is another way in which the movement might not have avoided logocentrism. At issue is whether liberation is thought to entail the return to an heroic, unproblematic African past. If so then the BCM would be guilty of a nostalgia for origins, and of replacing European man with African man as a foundational source of wisdom and virtue. Citing Franz Fanon, Biko (1978: 69–70) argued that colonialism turns to the past of oppressed people 'and distorts it, disfigures, and destroys it.' As a consequence the oppressed must 're-write our history and describe in it the heroes that formed the core of resistance to the white invaders.' According to Shava (1989:105), in Black Consciousness 'the need to rewrite black history is taken as integral to the revival of black culture.'

In a 1971 essay, Biko at times endowed Africans with almost superhuman qualities:

> We are not a suspicious race. We believe in the inherent goodness of man. We enjoy man for himself. We regard our living together not as an unfortunate mishap warranting endless competition among us but as a deliberate act of God to make us a community of brothers and

sisters jointly involved in the quest for a composite answer to the varied problems of life. . . . We always refrain from using people as stepping stones. Instead we are prepared to have a much slower progress in an effort to make sure that all of us are marching to the same tune.

(Biko 1978: 42)

Clearly Biko did not always resist the temptation to identify with an heroic 'we' that never exploited other men. This created an unresolved tension in his writings because it was from amongst the ranks of the 'we' that the collaborating 'non-whites' were drawn. It is important to emphasize, however, that Biko's identification was generally with the African present, not the past, and with a 'modern African culture' that 'has used concepts from the white world to expand on inherent cultural characteristics' (Biko 1978: 45). Thus the task for the new generation is not to revert to 'tradition,' for the 'belief that traditional society can be devoid of hate and war is unrealistic and misleading' (Shava 1989: 131). It is instead to 'grapple with the problems of present-day industrial South Africa in the light of how the past has contributed in shaping those problems.' Black Consciousness sought to relearn the past in order to understand the present, not to revive a lost glory that never existed. This explains in part why the BCM's major political enemy in its early years was the Inkatha movement of Gatsha Buthelezi (Mzala 1988).

In devising strategies for liberation in South Africa, Black Consciousness turned for inspiration to the 'significant political discourse [that] was emerging worldwide at the end of the 1960s' (Halisi 1992: 108). Halisi (1992: 107) argues that 'the BCM was one of the few African political organizations to identify unashamedly with blacks in the diaspora by assimilating the international resistance against racial domination in its political outlook.' Black Consciousness incorporated three distinct traditions: black South African political thought; theories of colonialism and racial liberation developed in Africa and in the African diaspora, and New Left student radicalism (Halisi 1992: 101). These traditions are reflected in the influence exerted on the movement by the writings of Senghor, Memmi, Fanon, Cleaver, Carmichael, and Freire (Moodley 1992: 146).

In the light of its determination to question 'old concepts, values, and systems,' its commitment to social change, and its identification with critical social movements worldwide, it is hardly surprising that the BCM had something to say about the concept of development. Nor is it surprising that its approach shares common elements with other global, counter-modernist discourses such as the dependency school, liberation theology, and participatory action research.

BLACK CONSCIOUSNESS AND DEPENDENCY

Black Consciousness activists were heavily influenced by Julius Nyerere's Arusha Declaration of 1967, which committed Tanzania to *ujamaa* as a development philosophy (Ramphele 1992: 154–5). Instead of continuing to follow conventional development wisdom and pursue private investment, industrialization, Western-style education, and so on, Nyerere insisted that the ultimate objective was freedom. In a speech delivered in 1968, Nyerere described the freedom of Tanzania as 'a question of consciousness among all the people of the nation that they are free men who have something to defend' (Nyerere 1973: 59). In order to raise consciousness and foster a development that was *of the people*, *by the people*, and *for the people*, the TANU government was to promote 'leadership through education' and 'democracy in decision-making.' Central to national policy was commitment to 'creating socialist villages throughout the rural areas,' but Nyerere (1973: 66–7) insisted that this was 'not intended to be merely a revival of the old settlement schemes under another name.' *Ujamaa* villages were to be created from scratch by volunteer settlers and internally governed by their members; the intention was not to return to 'traditional society' but to achieve freedom and development within 'the organizations of a modern society' (Nyerere 1973: 64). The BCM was attracted to the 'person-centredness of this approach to life.' The idea of community-based autonomous development and co-operative decision-making also seemed consistent with the black South African principle of *ubuntu* (Ramphele 1992: 155). Its emphasis on black self-reliance and self-help – which 'progressive' whites and the ANC criticized as an abandonment of inter-racial politics – thus emerged from the context of African debates about development.

At a general level, both *ujamaa* and Black Consciousness aimed to remove black Africans from European control and influence. Self-reliance in both cases referred to the capacity of blacks to initiate, manage, and evaluate development efforts in light of their own perceived needs. The vision, in each case, was of 'a free self.' Thus it was under the influence of Nyerere's idea of self-reliance, according to Ramphele (1992: 149), that various Black Consciousness community projects explored ways in which blacks could become more self-supporting.

Despite the similarities, it is important to point out that concepts such as 'self-reliance' and 'dependency' acquired rather different meanings in South Africa. Whereas Nyerere was concerned about the dependence of the Tanzanian state on 'non-Tanzanians,' the BCM worried about the dependence of black South Africans on white charity organizations. Whereas dependency was a problem for Nyerere because it precluded national autonomy and self-sustaining growth, it was a problem for the BCM because it 'wrought havoc on the self-image of black South Africans, who lost self-confidence as a people' (Ramphele 1992: 156). In addition, the subject to be freed and thus the ultimate objective, differed in each case. For Nyerere the subject of freedom was Tanzania and the final aim was national development; for the BCM the subject to be freed was 'the black community' and the final aim was 'a normal society where man is nothing else but man for his own sake' (Biko 1978: 21).

A less obvious point of comparison is with Latin American writings on dependency. Although not directly influenced directly by them, BCM analyses of how dependency relations have been sustained and reproduced over time share something in common with these writings. In the best example of an historical–structural approach to dependency, Cardoso and Faletto (1979) undermined the internal/external, agent/structure dichotomies at the heart of much development theory. They refused to treat the people in once-colonial societies as either fully to blame for underdevelopment (the argument that 'traditional' attitudes and behaviours have impeded progress) or as the passive helpless victims of imperialism. Echoing Derrida on critical theory, they faulted analyses which merely invert the dichotomies and explain everything by reference to imperialist structures instead of to traditional agents, or to the external 'logic of capitalist accumulation' on a global scale instead of to national (internal) character traits (Cardoso and Faletto 1979: xv).

Cardoso and Faletto (1979) argued that in order to understand national dependency, it is necessary to consider the ways in which purely 'external' interests come to be internalized over time. A system of domination associated with external forces (multinational enterprises, international financial institutions, foreign states, and so on) is most effective when it 'reappears as an "internal" force, through the social practices of local groups and classes' (Cardoso and Faletto 1979: xvi). The question is not whether internal or external forces are more to blame, for the analytic distinction between inside and outside is false. It is instead a matter of investigating how the same global phenomena come to be articulated within, as well as transformed by, distinctively local contexts.

Black Consciousness also refused to consider oppressed people as powerless and hopeless victims. According to Ramphele (1992: 161), the movement was critical of 'dominant paradigms' which 'projected blacks as victims of racism and exploitation, while little attention was paid to the creativity and resilience which underpinned the stategies of survival blacks had elaborated over the years.' The BCM wanted the black community's capacity for social transformation to be recognized, at the same time as it was critical of those 'non-whites' who actively reproduced oppressive structures. Like Cardoso and Faletto (1979), the BCM emphasized both agency and structure, collaboration and resistance.

Another similarity is that the BCM rejected an internal/external dichotomy in its explanations for racial oppression. To the BCM it was obvious that a system of domination associated with an 'external force' such as European colonialism had become internalized – the European colonists never left South Africa or ceased oppressing the African people. Yet domination was not only considered internal to South Africa, and reproduced via the social practices of European settlers and their descendants. It was also internal to the black community, and sustained with the aid of collaborating 'non-whites.' Black Consciousness was always concerned about 'the growing distance between the poor and the black middle class, who modelled themselves on white liberals and thus perpetuated the myth of "white" as the norm and "black" as the aberration in all spheres of social relations' (Ramphele 1992: 156). The dependency school never spoke of social actors in the racial terms of the BCM. Yet its 'comprador bourgeoisie' – those who internalized and sustained external interests – can be seen as functionally equivalent to the 'non-white' collaborators in apartheid.

The dependency tradition is both modernist and counter-modernist in its thrust. Development tends to be equated with national autonomy and economic growth – even in Cardoso and Faletto (1979) – so that the objective remains the independence of the 'children' from their 'parents.' This repro-duces the parent/child dichotomy instead of calling it into question. At the same time, a series of other dichotomies are subverted and the 'victim' imagery of the Third World is problematized. Black Consciousness is therefore similar to, but not identical with, the dependency school because the BCM does not equate development with national autonomy and eco-nomic growth. This point becomes clearer in the BCM's understanding of 'community development.'

BLACK CONSCIOUSNESS AND LIBERATION THEOLOGY

Liberation theology has been described as a discourse that 'attempts to solve the problem of how the modern era can properly overcome its most debilitating characteristic: the dichotomy between *facts* and *values*' (Pottenger 1989: 3). The similarities between liberation theology and Black Conscious-ness are perhaps more obvious than they are between the BCM and dependency. 'From its inception,' according to Halisi (1992: 103), 'Black Consciousness philosophy had a humanist bent which allowed the develop-ment of a theological counterpart.'

Biko (1978: 59) described Black Theology as 'a situational interpretation of Christianity' which 'seeks to relate the present-day black man to God within the given context of the black man's suffering and his attempts to get out of it.' It 'shifts the emphasis from petty sins to major sins in a society, thereby ceasing to teach people to "suffer peacefully." Black Theology understands theology as a process that flows from the human endeavours of ordinary people, not from the 'erudite' writings of academic scholars or the 'esoteric' knowledge of priests. Consequently, Dwight Hopkins (1992: 199) has argued that theology becomes 'a popular activity bubbling forth out of the variegated, creative culture and politics of the folk.'

Biko (1978:94) argued that Black Theology is an 'important aspect of Black Consciousness, for quite a large proportion of black people in South Africa are Christians still swimming in a mire of confusion – the aftermath of the missionary approach.' In claiming a 'monopoly on truth, beauty, and moral judgement,' the missionaries denigrated 'native customs and traditions' in three ways (Biko 1978: 93–9). They argued that 'theirs was a scientific religion and ours a superstition;' they described those who defended indigenous religion as 'pagan;' and they summed up all native culture in a single word – barbarism. Biko rejected all three dichotomies – religion/superstition, Christian/pagan, and civilized/barbaric – that he considered inherent in missionary discourse in Africa. Why was African religion – which had its 'own community of saints through whom we related to our God' – constituted as pagan superstition? Biko (1978: 92–4) argued that the spread of the 'Christian message' was part of a larger system of discursive relations between white and black in which the former was cast as superior, expert, and in possession of universal truth. His case for Black Theology was thus part and parcel of a case against logocentrism in general.

Allan Boesak (1986: 265–71) has attributed the rise of liberation theology in South Africa to 'a human consciousness that we call black consciousness.' In 1977, he argued that people all over the Third World began to ask questions that they had never asked before. Why was blackness the reason for their oppression? Why had whiteness historically been tantamount to human 'beingness?' A black theology which asks what Jesus Christ has to say about a particular situation of oppression 'is also a theology of liberation.' Boesak suggested that the same themes – in particular the Exodus theme of escape from oppression and realization of liberation – linked together black theology in the United States, liberation theology in Asia and Latin America, and black liberation theology in South Africa. While not disagreeing with Boesak, I would argue that what the theological element in Black Consciouness shares most fundamentally with liberation theology is a critical, counter-modernist attitude. As does dependency, liberation theology tends to talk in terms of rich and poor as opposed to white and black. Yet it is evident, particularly in the highly influential work of Gustavo Gutierrez (1988), that the concerns expressed are the same ones found in Black Consciousness.

In a 1971 essay, Gutierrez (1988) suggested that the term 'development' can no longer synthesize the aspirations of poor peoples. This is because 'developmentalism came to be synonymous with reformism and modernization, that is to say, synonymous with timid measures, really ineffective in the long run and counterproductive to achieving a real transformation' (Gutierrez 1988: 17). The goal of history is the continuous creation of a new way to be human; this requires a conception of history as a process of human liberation from all the forces that oppress mankind. Only in the context of an historical vision in which humankind assumes conscious responsibility for its own destiny, and lives in communion with Christ, can development find its true meaning and accomplish something worthwhile.

Gutierrez (1988) argued that the possibility of enjoying a truly human existence – 'in the South as well as in the North, in the West as well as in the East, on the periphery and in the center' – cannot depend upon the poor countries modelling themselves after the rich ones:

> We must beware of all kinds of imitations as well as new forms of imperialism – revolutionary this time – of the rich countries, which consider themselves central to the history of humankind. Such mimicry would only lead the revolutionary groups of the Third World to a new deception regarding their own reality. They would be led to fight against windmills.
>
> (Gutierrez 1988: 18)

For Black Consciousness, as for liberation theology, progress is measured by the capacity of human beings to assume control of their own destiny, not by their acquisition of European traits. Since both discourses call into question the modern model of achievement as well as the supposed centrality of Europe to history, they tend to avoid the term development as a synonym for meaningful change and speak instead of liberation. Both have argued that theology must emerge out of the lives of the oppressed in their own encounter with biblical texts, and both have called for a reinterpretation of Scripture from the standpoint of contemporary situations of oppression and poverty. Both Black Consciousness and liberation theology, finally, have tried to 'conscientize' the poor and the black about their religious practices and situations of oppression.

BLACK CONSCIOUSNESS AND PARTICIPATORY ACTION RESEARCH

The notion of conscientization, derived from the work of Brazilian educator Paulo Freire, has come to underpin a diverse array of efforts to effect progressive change worldwide. According to Ramphele (1992: 155), 'Paulo Freire's conscientization approach in Latin America was found to have great relevance for the problems BC leaders identified amongst black people in South Africa.' The term *conscientización* came to refer (first in Latin America and then elsewhere) to a process by which poor people are brought to a critical consciousness of their situation, and then helped to come together as a community, to articulate their needs and to organize to effect change. The 'Freire method' aimed to convince people that they are active agents of history, and that grounding political practice in local histories, knowledge, and experiences is preferable to mimicry of the rich, the white, the powerful, the oppressive. As such it seemed to offer a way to break the psychological chains that prevent poor people from assuming control of their own destiny.

Systematized in an approach called Participatory Action Research (PAR), which combines techniques of adult education with social science research, political activism, popular theatre, and local grassroots initiatives, conscientization has been adopted not only by church people but also by leftist intellectuals and development practitioners in many Third World countries (Freire 1975; Escobar 1984; Fals-Borda 1988). As I have argued elsewhere, PAR displays a counter-modernist attitude in its rejection of the West as a model of achievement, and in its questioning of the divisions between research and practice, subject and object, and inside/outside so central to conventional conceptions of development (Manzo 1991: 28–9).

Under the influence of Nyerere and Freire, Black Consciousness exponents identified community development as an important strategy for liberation. In terms reminiscent of Gutierrez, they defined development as a process of empowerment which enables participants to assume greater control over their lives as individuals and as members of society. The goals of development were articulated, according to Ramphele (1992: 157), as 'the practical mani-festations of the Black Consciousness philosophy, which not only called for a critical awareness of social relations amongst the oppressed, but for the need to translate that awareness into active programmes for liberation from white domination.' The BCM argued 'that people who had known nothing but scorn and humiliation, needed symbols of hope to lift them out of despair and to empower them to liberate themselves.'

When Barney Pityana succeeded Steve Biko as president of the South African Students' Organization (SASO) in July 1970, he set about rewriting the SASO constitution and enlarging the executive, so as to accommodate action-oriented development programmes designed for political mobilization at the grassroots level (Buthelezi 1992: 122). Such programmes entailed the sorts of projects that the phrase 'community development' might invoke in mainstream development circles, i.e. health clinics, literacy campaigns, and home industries (Ramphele 1992). But since the object of development was a critical awareness amongst the oppressed of the ways that oppression has historically operated, far more was involved in the BCM's approach to development than efforts to improve material wellbeing.

Conscientization in the late 1960s often took the form of light-hearted, satirical, humorous plays (Moodley 1992: 149). The plays themselves appar-ently shared some common features with Poor Theatre, which evolved in Latin America and were used increasingly in conjunction with other efforts to promote participatory democracy, collaborative problem-solving, and mass mobilization (Mzamane, 1992: 188). As the state's initial tolerance of Black Consciousness gave way to violent attempts at suppression, satirical theatre remained vital as a method of political communication and resistance (Bunn and Taylor 1987: 273–304). One element of that theatre – performance poetry – took root within the trade-union movement and 'carved a role for itself in our [Black South African] culture of liberation' (Mzamane 1992: 191).

Poetry, drama, and fiction were not valued by the BCM for their capacity to disguise politics as culture. The revival of cultural forms – like performance poetry – that played important roles in pre-colonial African societies was intended to counteract 'the cultural imperialism that had sapped the spirit and being of the racially oppressed.' Just as culture can be oppressive, it can also be liberating. This means that culture is always political, and cultural struggle of vital political importance (Mzamane 1992: 193).

Cultural imperialism was understood by the BCM to entail two interconnected elements. On the one hand, it entailed 'the Eurocentric assumption that literature could not exist without writing' (Mzamane 1992: 191). Black Consciousness thus objected to the logocentric creation of a literate/illiterate dichotomy – itself predicated upon the presumed superiority of the written over the spoken word – and the concomitant relegation of oratory to the margins of literary, cultural and political respectability. On the other hand, cultural imperialism involved the acceptance of this logocentric procedure by the oppressed themselves, who came to believe that societies which performed poetry in social contexts were culturally inferior to those which wrote it for individual consumption. For example, Mzamane (1992: 190–1) has argued that African poetry 'bowed before the supremacy of print and the influence of Western models to express itself in an art form written and read by individuals.'

Black Consciousness exponents advocated a return to traditional forms of social expression as part of its cultural offensive. But this signified less a celebration of the past than it did a subversion of the norm of whiteness in the present. In addition to using culture as a tool, Black Consciousness sought to conscientize through social science research. Activists complained that blacks were mainly objects of research, and expressed frustration at the absence of attempts to draw them into the process of formulating the questions to be posed and responding to analyses of the results of such enquiries. The first task identified in the research field was publication of an annual survey of events, to counter the one issued by the liberal South African Institute of Race Relations (SAIRR). The SAIRR's survey was criticized mainly for its portrayal of blacks as victims, a portrayal that contributed to the poor self-image blacks had of themselves and reinforced their sense of powerlessness. Before 'victimhood' became a self-fulfilling prophecy, it was necessary to conduct and publish research into the positive efforts of blacks to cope with their social disabilities (Ramphele 1992: 161).

While waging a continual struggle for the right to publish in the face of a growing number of political restrictions, the BCM managed to produce four issues of *Black Review* – from 1973 to 1977 – as well as a few issues of *Black Viewpoint* and one issue of *Black Perspectives*. The impact of *Black Review*, according to Ramphele (1992: 162–3), was immeasurable. She argues that it inspired black organizations and facilitated networking between community workers; it also changed forever the format and focus of the SAIRR's annual survey. The other two publications ran into problems, for a variety of reasons. But the sheer effort put into all three by Biko and others illustrates the importance attached by Black Consciousness to grounding development in popular knowledge and experiences.

Given the similarities between PAR and the 'community development' of the BCM, it is relevant to ask whether the movement managed to avoid the same dangers that face all of the disciples of Paulo Freire. In particular, did Black Consciousness activists assume that only the consciousness of the poor needed to be raised because they themselves were already fully conscious of how oppression operated? If, as Rahnema (1990: 205) has pointed out, the 'change agents' end up treating the poor as people in need of 'truth' then 'the scenario is hardly different from that of the conventional developers, and their coactors are hardly more independent in their acting than the extras participating in development projects.'

Ramphele (1992: 164) insists that 'no one within the BC ranks claimed the monopoly over "truth" with regard to strategies and tactics for liberation.' At the same time, it appears that the priorities for action of community workers rarely coincided with those of the communities to which they were sent. The voice of 'the community' was hardly ever heard, and the result was that 'the poor and the women voted with their feet, and did not show up for voluntary activities beyond the initial stages' (Ramphele 1992: 172).

As I have tried to show, a critical, counter-modernist attitude was absolutely central to Black Consciousness philosophy. Yet in its quest for liberation the BCM still occasionally reproduced the practices and assumptions that it identified as facilitators of oppression. That the BCM was unable to avoid in practice what it rejected in principle – namely the treatment of blacks as objects – demonstrates the difficulty for even the most radically critical and self-aware social movements of shedding completely the discourse they seek to contest. It also helps to explain why the questioning of modernity on a global scale has failed to dislodge the West as a model of achievement and progress.

POST-APARTHEID DEVELOPMENT AND THE LEGACY OF BLACK CONSCIOUSNESS

On the margins of contemporary South African politics are the organizations – such as the Azanian People's Organization (AZAPO), its armed wing the Azanian National Liberation Army (AZANLA), and the Black Consciousness Movement of Azania (BCMA) – which claim to be the only true heirs of Black Consciousness. Such a claim has been disputed. In their recent book, Pityana et al. (1992: 10) assert that 'the legacy of Black Consciousness spans the whole political spectrum: ANC, PAC, Unity Movement, and many individuals not aligned to any particular political organization.'

The legacy of Black Consciousness is apparently widespread for a couple of different reasons. First, many BCM activists did not end up in AZAPO or the BCMA: 'they gravitated in many political directions; they joined the ANC or PAC, trade unions and civic associations, formed new black community organizations or were forced into prison or exile' (Halisi 1992:

101). Second, in the demand it made on blacks to acknowledge their active agency in history, the BCM gave 'oxygen and new life' to oppositional politics. The legacy of defiance associated with Black Consciousness 'infused the liberation movement as a whole with a new energy' (Pityana et al. 1992: 11). As Wilson (1992: 76) puts it, 'the Biko generation inspired the culture of fearlessness.'

While emphasizing the continued influence of the BCM, Pityana et al. (1992: 11) bemoan the fact that the victim mentality has crept back into oppositional discourse; that self-respect and pride in oneself have largely disappeared in the struggle for political turf; and that political tolerance has been thrown overboard as the struggle for political turf intensifies. What also seems to have disappeared, if the discourse of the ANC and the PAC is any indication, is the BCM's counter-modernist understanding of the concept of development.

In the ANC's Reconstruction and Development Programme (RDP), priority is given to economic growth rather than human freedom and its agents are foreign investors and local capitalists instead of black communities themselves. At the same time as it remains committed to the ultimate alleviation of poverty and inequality, the ANC promises to create a climate favourable to foreign investors who wish to 'act upon' deprived areas by setting up plants, joint-ventures and 'export processing zones.' In a clash between the old solutions to problems (such as industrial action) and the entailments of development, the Mandela government has already sided with the latter against the former.

The ultimate aim of the PAC is 'a political order that will represent the economic and political interests of all Africans in Azania' (Shabalala 1991: 6). The PAC conceives of its mission as emancipation, rather than the promotion of economic growth. Yet in its strategies for emancipation the PAC appears even more modernist than the ANC. Among other things it advocates 'self-sufficiency in food production and other basic necessities' and suggests that 'innovative technological development will receive massive state support.' More ominously, the PAC argues that 'the country will be one of scientists and technical people, not philosophers' (Shabalala 1991: 6–7). The model of autonomous, scientific man is thus central to the thinking of a PAC which sees no need for the continued philosophical questioning of old values, concepts, and systems.

Businessmen, who once worried that an ANC election victory would mean the 'familiar ANC line' on mass action, wealth distribution, and affirmative action, are now cautiously optimistic. As the imaginary of development fills up the spaces vacated by anti-apartheid activity, 'Radical reconstruction . . . remains for the present in the realm of promise rather than performance' (The Star 23 June 1994). Many South Africans await the delivery of development's promises in the same way that they once awaited white charity. Yet it is certainly not the case that Black Consciousness philosophy

251

has been supplanted completely by modernist thinking. The counter-modernism of the BCM lives on in 'community development' projects and in cultural sites that continually restage 'acts of dissent' (Farred 1992: 231).

CONCLUSION

All over the so-called Third World, people have been questioning values, assumptions, and meanings in a manner that has been described here as counter-modernist. In rejecting the equation of progress with whiteness, and in calling into question the logocentric production of a white/black dichotomy, Black Consciousness drew inspiration from a number of other critical discourses. It is thus appropriate to consider Steve Biko's BCM as a social movement situated within a wider context of discursive relations between white and black, rich and poor, and oppressive and oppressed that exist on a global scale.

The centrality or marginality of any critical discourse cannot depend only upon the balance of forces within the specific local site that it inhabits. Embedded within a larger system of meanings, values, and concepts, it must rely for sustenance on the dissemination of particular attitudes and practices worldwide. Black Consciousness is marginal in South Africa because the philosophy it expresses remains marginal worldwide. In a sense then, it is the centrality of modernist thought and not the centrality of the ANC, that accounts for the weakened legacy of the BCM.

13

POST-MODERNISM, GENDER AND DEVELOPMENT

Jane L. Parpart

Advocating the mere tolerance of difference between women is the grossest reformism. It is a total denial of the creative function of difference in our lives. For difference must be not merely tolerated, but seen as a fund of necessary polarities between which our creativity can spark like a dialectic.

(Lorde 1981: 99)

INTRODUCTION

The post-modern critique, with its attention to difference and discourse, and its attack on the universalizing truths of Enlightenment thinking, has much to offer those who are critical of development theory and practice. Some Third World and Western scholars have drawn on this perspective to challenge the assumption that modernization is necessarily possible or desirable. They have questioned the belief that Third World development and westernization/modernization are synonomous and that Western political, social and economic institutions and practices (whether liberal or socialist) hold the answers to the Third World's development problems (Escobar 1984; Ferguson 1985, 1990; Moore 1992).

Drawing on postmodernist conceptions of power and knowledge, particularly the role of discourse in the construction of power/knowledge systems (Foucault 1972, 1980b), these scholars argue that development discourse is embedded in the ethnocentric and destructive colonial (and post-colonial) discourses designed to perpetuate colonial hierarchies rather than to change them. It has defined Third World peoples as the 'other,' embodying all the negative characteristics (primitive, backward and so forth) supposedly no longer found in 'modern,' Westernized societies (Said 1978; Escobar 1984; Spivak 1990b). This representation of Third World realities has provided the rationale for development experts' belief in modernization and the superiority of the values and institutions of the North, and has legitimated a process that 'entails an incomprehensible amount of destruction or, at the least, discrediting and subordination of local techniques, knowledges, practices,

253

and lifestyles.' While not all of this loss is lamentable, much of it is (Du Bois 1991:21). This critique asserts a people's right to their own culture, history and world view, and argues for a new form of development, one that is based on the knowledge and needs of peoples in the South rather than the so-called 'expertise' of Northern (and some Third World) development agents. It challenges received wisdom about Third World realities and the nature and goals of development. While accepting the importance of economic development, it rejects the adoption of mindless modernity, calling instead for a creative synthesis of tradition and modernity, drawing on local knowledge and culture (Edwards 1989).

Does a postmodern feminist perspective offer similar insights into the women and development enterprise? I believe it does. After all, the criticisms levelled against Western feminists' representations of Third World peoples (especially women) also apply to Women in Development (WID) practitioners. Much of the WID and Gender and Development (GAD) literature represents Third World women as benighted, overburdened beasts, helplessly entangled in the tentacles of regressive Third World patriarchy. The recent economic crisis has not helped matters as it has legitimized representations of women in the South as the vulnerable 'other,' victimized by the reactionary traditions and economic ineptitude of Third World societies. The poor Third World woman remains truly the 'other' to her development expert sisters (CIDA 1987; CEGWSA 1989; Wiltshire 1988).

DIFFERENCE AND THE THIRD WORLD 'OTHER'

One of the most appealing aspects of postmodernism to many feminists has been its focus on difference. The notion that women have been created and defined as 'other' by men has long been argued and explored by feminists, most notably Simone de Beauvoir (1952) in her book *The Second Sex*. She challenged male definitions of woman and called on women to define themselves outside the male/female dyad. Women, she urged, must be the subject rather than the object (other) of analysis. This concern was echoed and expanded by other feminists, particularly those calling for the recovery of women's voices and the development of knowledge from the standpoint of women (Harding 1987, 1992). However, the concern with women as 'other' emanated largely from the writings of white middle-class women in the North, whose generalizations were grounded for the most part in their own experience. Feminist theory 'explained' women as if this reality applied to women from all classes, races and regions of the world. Feminist concern with female 'otherness' ignored the possibility of differences among women themselves (Gilligan 1982; Spelman 1990).

Not surprisingly, the postmodern focus on difference has offered ammunition to women who felt excluded by this approach. Black and Native women in North America and Europe became increasingly vocal about their unique problems, and the need to incorporate race and culture as well as class and gender into feminist analysis. While minority feminists have been arguing for some time for a racially and ethnically specific feminism (Lorde 1984; Anzaldua 1990; Anthias and Yuval-Davis 1990; Minh-Ha 1989), post-modernism has provided a space which legitimizes the search for 'the voices of displaced, marginalized, exploited and oppressed black people' (hooks 1984: 25). bell hooks eloquently argues for a black postmodernism where difference and otherness can be used to explore the realities of the black experience in North America and the connection between that experience and critical thinking. Only then, she argues, will feminism truly incorporate difference into its analysis (hooks 1984, 1991; see also Collins 1989). Julia Emberley (1995) makes a similar point for Native Canadian women.

A number of feminists in the South have taken up this argument as well. They have accused Western scholars of creating a colonial discourse which represents women in the South (and minority women in the North) as an undifferentiated 'other', oppressed by both gender and Third World under-development. Chandra Mohanty analyses the writings about Third World women by a number of Western feminists and concludes that they:

> Colonize the material and historical heterogeneities of the lives of women in the third world, thereby producing/re-presenting a compos-ite, singular 'third-world woman' – an image which appears arbitrarily constructed but nevertheless carries with it the authorizing signature of western humanist discourse.... assumptions of privilege and ethno-centric universality on the one hand, and inadequate self-consciousness about the effect of western scholarship on the 'third world' in the context of a world system dominated by the west on the other, characterize a sizable extent of western feminist work on women in the third world.
>
> (Mohanty 1988: 62–3)

Third World women are presented as uniformly poor, powerless and vulner-able, while Western women are the referent point for modern, educated, sexually liberated womanhood. This analysis both distorts women's multiple realities and reduces the possibility of coalitions among (usually white) Western feminists and working-class and feminist women of colour around the world. While recognizing the important contribution of postmodern feminists such as Luce Irigaray and Helene Cixous, who have revealed the peripheral nature of women in Western humanist discourse, Mohanty quite rightly reminds us that Western feminists have equally ignored and mar-ginalized Third World women in their own discourse (Mohanty 1988; Mohanty et al. 1991). Aihwa Ong (1988: 90) puts it rather more bluntly, arguing that 'for feminists looking overseas, the non-feminist Other is not so much patriarchy as the non-Western woman' (see also Ong 1990; Barriteau Foster 1992).

The tendency to essentialize Third World women does not just occur in the writings of women in the North. It is also pronounced in some of the work of Third World women trained in Western institutions, particularly when writing for a Western audience. For example, Marnia Lazreg (1988) discovered that both Western and Western trained scholars writing on Algeria often uncritically adopt Western stereotypes about Arab peoples and culture, particularly the primacy of Islam, which is seen as a self-contained and flawed belief system impervious to change. Arab women are represented as passive pawns, trapped in a world dominated by hopelessly outdated and retrogressive religious traditions. The Islamic world is characterized as in inexorable decline: progress for Arabic women can only come from the adoption of modern Western values. Feminist theory, when applied to Arab women, is often seen as an opportunity to enlarge liberal feminist knowledge rather than as a chance to explore the variety of modes of being female. Lazreg calls for a new approach, one that recognizes difference and accepts the need to explore the concrete, lived experiences of women in different cultures. It requires studies of Third World women which reveal women's lives 'as meaningful, coherent and understandable instead of being infused "by us" with doom and sorrow' (Lazreg 1988: 98). To keep difference from becoming mere division, she believes indigenous scholars must take on a double burden. They must work toward an epistemological break with the prevailing paradigm while also re-evaluating the structure of gender relations in their own societies (see also Schick 1990; Spivak 1990b).

POSTMODERNIST FEMINISM AND DEVELOPMENT

Does this debate have anything to offer theorists and practitioners concerned with the problems of women's development in the Third World? I think it does, particularly for those who write and work within the dominant development paradigm. The liberal approach to development matured in the post-war period of the 1940s, when Western thinkers and policy makers assumed that Third World peoples could gradually, but steadily, be helped to achieve political and economic systems similar to those in the industrialized world. Development was seen as a linear process, in which a nation or people proceeded from underdevelopment, which was characterized as backward/traditional/primitive, to full development, which was identified as modern/rational/industrialized. The rationale for proceeding along this path was provided by European imperialism, which had fostered the assumption that everything European was superior to any aspect of Third World life. This sense of difference and superiority was enshrined in a colonial discourse that compared Third World peoples and cultures unfavourably with 'progressive' Western societies (Curtin 1974; Said 1978), and equated global modernization with world development.

The problem of development became one of bringing 'backward' colonial peoples into the modern (i.e. developed) world. This was seen more as a logistical problem – how to go about it, how quickly it could be achieved and so on – than as a goal that might itself be questioned. Economists like Rostow (1960) developed models explaining the 'how' of development, and development experts set about trying to bring Third World societies to the 'take off' stage where modernization could be assured (Escobar 1984; Moore 1992). The validity of the development project – making the world modern – was never in dispute. Third World women, if they were considered at all, were seen as an impediment to development.

Colonial discourse represented Third World women 'as exotic specimens, as oppressed victims, as sex objects or as the most ignorant and backward members of "backward" societies' (de Groot 1991: 115). During the colonial period, missionaries, colonial officials, and settlers put forward a blend of information, imagination, pragmatic self-interest and prejudice to explain why Third World women were inferior beings, bound by tradition – either unable or unwilling to enter the modern world. Development planners adopted these assumptions uncritically, regarding Third World women as an important block to modernity and thus to development. Consequently, development theory and practice in the first two post-colonial decades (the 1950s and 1960s) ignored women on the assumption that they would eventually be forced to adopt a more 'progressive' stance towards development once the modernization process had been set in motion and Third World men had learned how to organize their societies along modern lines (Afshar 1991).

As a result, development plans were designed on the assumption that productive work was performed entirely by men. Women as workers, owners or entrepreneurs were totally ignored, apart from the gratuitous admission that women produced future labourers and thus had a role to play in population policy (Hirschman 1958). These assumptions reflected Western patriarchal patterns of ownership, work and control, which, although assuming the modernity of Western women, in the sense of developed, still relegated them to a subordinate role in society, particularly in regard to economic and political matters.

By the late 1960s some economists began to realize that development was not taking place as easily as they had hoped; a number of scholars were particularly concerned by the continuing underdevelopment of Third World women. Ester Boserup's landmark 1970 study, *Woman's Role in Economic Development*, proved that development schemes, rather than improving the lives of Third World women, had often deprived them of economic opportunities and status. Modernization had displaced women from their traditional productive functions, particularly in agriculture, where they had generally played a crucial role as food producers. Boserup appealed to development planners and policy makers to recognize and account for women's roles in economic development. Only then, she argued, would development occur in the Third World. Other development experts joined her, and in 1973 the Percy Amendment to the United States Foreign Assistance Act enshrined the principle that US development assistance should try to improve the status of Third World women by integrating them into the development process (Tinker 1990).

A new subfield emerged entitled 'Women in Development' (WID), with its own language and preoccupations. Drawing on liberal feminist thinking in the North, with its call to integrate women into male power structures, women in development specialists initially sought women's equality through improved access to education, employment and material benefits such as land and credit. While tolerated by male development planners, these early WID experts were fairly marginal to the development process (Moser 1989; Buvinic et al. 1983).

WID's status within the development community was enhanced by a number of global changes. In the 1970s, poverty and overpopulation continued to expand in the South, lending support to the assertion that liberal trickle-down development had failed. In response to the growing criticisms by Third World economists such as Andre Gunder Frank (1978) and Samir Amin (1974) – who advocated Third World disengagement and self-reliance – liberal development specialists shifted their policies to address the basic needs of the poor. Since women were disproportionately represented in this group and played a key role in population control, they became a central concern of policy makers and development planners. The United Nations declared 1976 to 1985 the 'Decade for the Advancement of Women,' which was inaugurated by a world meeting in Mexico City. Women were put on the development agenda, and obstacles to women's progress were identified. Research on women increased and professional WID experts gained status in the development bureaucracy, although most development agencies did not have full-time professional WID officers until the mid-1980s (Tinker 1990; Mueller 1987b).

Alternative development approaches and practices emerged around the critiques of Frank, Amin and others. They blamed underdevelopment on the machinations of Northern elites (who were largely male) and their (again largely male) collaborators in the South. They called for self-reliant development, free from the self-interested 'assistance' of capitalist elites and their indigenous henchmen. This thinking, along with the radical feminist critique of patriarchy (Gilligan 1982), inspired a new approach to the development of women. Activists in the North and South who were influenced by these perspectives began calling for small-scale, women-only projects that would be designed to avoid male domination, both metropolitan and indigenous. This approach, known as Women and Development (WAD), found considerable support in non-governmental organizations (NGOs) and became a staple of NGO activities. Some mainstream development agencies supported this approach through their NGO programmes (Parpart 1989; Rathgeber 1990).

258

WID policy responded to these critiques by modifying mainstream development policy for women. Concern with equality between women and men fell by the wayside as planners emphasized basic human needs, particularly for health, education and training. WID specialists argued that this approach would increase women's effectiveness and productivity at work, thus assisting both economic development and women's lives. Reduced fertility would be a side benefit. Planners also called for more credit, greater access to land, legal reform and for more female involvement in development planning. As a USAID report on women in development put it, 'a focus on the economic participation of women in development is essential' (USAID 1982: 1). The possibility that fundamental change might be required was rarely discussed (Kandiyoti 1990; Mueller 1987b; Moser 1989, 1993).

However, development practitioners working within both traditions still largely accepted the ideas that development for Third World women meant becoming more modern. For it is clear that development thought, in both its Marxist and liberal expressions, has been (and continues to be) the embodiment of Enlightenment thinking, and therefore does not reject the modernist paradigm. Thus, development practitioners from both the mainstream and alternative approaches have for the most part continued to frame policies and programmes on the assumption that Northern expertise holds the answers to Third World women's developmental problems. Moreover development specialists, whatever their perspective, have generally accepted the idea that Third World women's problems and solutions bear little relation to the struggles and concerns of Western feminists, which were seen as entirely unrelated to the more practical concerns of development. This gap was exacerbated by clashes between feminists from the North and South at the 1975 meeting to launch the UN Decade for Women, and by conflict between some American feminists and the Third World women who attended the 1977 Wellesley College conference on women and international development. This confrontation undermined many Western feminists' sanguine assumptions about global feminist solidarity and reinforced development specialists' belief in their unique capacity to analyse and solve the problems of Third World women (Maguire 1984; Papanek 1984; Stamp 1989).

This division narrowed somewhat during the 1980s, however. The 1985 Nairobi Conference to celebrate the end of the UN Decade for Women encouraged contacts and better understanding between feminists world-wide. It provided a springboard for South–South linkages among women, including the creation of an international organization, Development Alternatives with Women for a New Era (DAWN), which grew out of discussions in India before the conference. DAWN has continued to organize and deliberate on development issues of concern to Third World women. The group has published a book which emphasizes the importance of listening to and learning from women's diverse experiences and knowledge, and to maintaining a commitment to long-range strategies dedicated to breaking down the structures of inequity between genders, classes and nations. Its political goals thus maintain a belief in modernization, but with greater sensitivity to cultural differences and Third World agendas (Sen and Grown 1987).

The Nairobi conference also facilitated dialogue between some feminists in the South and the North, particularly those working within the socialist feminist framework. Links between these groups had been growing rather tentatively in North America, and rather more surely in Europe (most notably at Sussex University and The Institute of Social Sciences in the Hague), but they received a boost at Nairobi. DAWN members had been influenced by the writings of socialist feminists (Beneria and Sen 1981) and the deliberations of the Sussex workshops on the subordination of women (Young *et al.* 1981). This perspective, with its commitment to understanding class and gender inequalities in a global context, provided an intellectual meeting point for some Western and Third World feminists. The resulting dialogue has enriched our understanding of comparative women's studies and produced some important texts (Robertson and Berger 1986; Afshar 1987, 1991; Stichter and Parpart 1988; Agarwal 1989). Gender and development (GAD) research and training has grown as well, for instance at Sussex and at a number of Canadian institutes supported by the Canadian International Development Research Centre (IDRC).

The GAD perspective focuses on gender rather than women, particularly the social construction of gender roles and gender relations. 'Gender is seen as the process by which individuals who are born into biological categories of male or female become the social categories of men and women through the acquisition of locally-defined attributes of masculinity and femininity' (Kabeer 1991: 11). The possibility of transforming gender roles is thus established. This approach also emphasizes the importance of examining the gender division of labour in specific societies, particularly the more invisible aspects of women's productive and reproductive work, and the relation between these labour patterns and other aspects of gender inequality. It looks at the issue of power as it relates to gender and at strategies for empowering women and challenging the structures and ideas maintaining gender hierarchies (Kabeer 1991).

While this approach has had considerable influence on academic development discourse, its willingness to consider fundamental social transformation does not sit well with the large donor agencies who prefer government-to-government aid, with its respect for the sovereign rights of member states. Although some government development agencies (most notably the Scandinavians, Dutch and Canadians) and some non-governmental development organizations have adopted a more gender-oriented approach to women's development, adding gender analysis training to established WID training programmes, this approach has only rarely been integrated into development planning (Moser, 1989).

Most development practitioners have seen no reason to alter their comfortable belief that development requires the adoption of Western values and institutions. This position has been buttressed by the economic crisis which has enveloped much of the South over the past ten years. The crisis has reinforced stereotypes about Third World underdevelopment and incompetence. Colonial discourse, with its emphasis on Third World inferiority, has re-emerged in the language of the international development agencies. The paternalism of colonial rule, with its rhetoric of 'uplifting the natives,' has been replaced by development discourse which emphasizes the vulnerability and helplessness of Third World peoples, especially women. This language provides the rationale for the neo-classical market-oriented 'reforms' which the World Bank and the International Monetary Fund have foisted on Third World governments, promising that they will 'reconstruct' their hapless economies. These structural adjustment plans (SAPs) are designed to establish neo-liberal economies in the South and are based on the assumption that capitalist economic and political (i.e. modern) systems should be the model for all nations (Bernstein 1990; Elson 1991; World Bank 1989a, 1990a, 1991a, 1992, 1993). The recent collapse of most socialist economies has further strengthened this position (Fukuyama 1989).

Concern to mitigate some of the harsher aspects of these programmes, particularly among the poor, has led to renewed concern for women's welfare. SAP's disproportionately severe impact on the poor or 'vulnerable groups,' which consist largely of women, children, the disabled and the elderly, has preoccupied UNICEF and other UN agencies (Cornia *et al.* 1987). A report of the Commonwealth Expert Group on Women and Structural Adjustment (CEGWSA 1989), advocated special support programmes for women, increased participation of women in decision-making processes and the creation of socially responsible structural adjustment programmes that address social equity as well as economic growth and efficiency. While raising important issues, this language reinforces WID discourse, which represents Third World women as helpless victims trapped by tradition and incompetence in an endless cycle of poverty and despair. The possibility that women (and men) in the South might have skills and strategies to protect themselves rarely surfaces. Third World women are characterized as uniformly poor, inadequately prepared to cope with the current economic crisis and desperately in need of salvation through foreign expertise (see also Kandiyoti 1990; de Groot 1991).

As a result, WID policy and projects have continued to focus on the poorest of the poor. Of late, some of the more enlightened development agencies, most notably those of Scandinavia and Canada, have directed some of their aid to encouraging women's entrepreneurial talents, supporting women's studies at post-secondary institutions and strengthening women's ministries. While a salutary and progressive trend, the amount of money and human resources deployed in this direction pales when compared with traditional development projects, such as roads, agricultural projects, and so on, which

261

continue to be designed for the most part by men for men. Moreover, most development projects for women are still aimed at the poorest women, and are carefully designed with very small-scale goals. Women are taught to make baskets or sew clothes while men are trained to use machinery. Women's projects often seem determined to increase the productivity of women within subsistence agricultural production rather than providing alternative economic activities which might offer women more economic and personal autonomy (Kandiyoti 1990). They frequently support economic initiatives while ignoring the need to empower women through collective action (World Bank, 1989b). The preoccupation with women's maternal and domestic roles overshadows most efforts to develop skills that might enhance women's economic autonomy. The focus remains on women's poverty and vulnerability, which reinforces the image of Third World women as helpless victims, while sustaining the reality that feeds such an image.

POSTMODERN FEMINISM AND THE PRACTICE OF DEVELOPMENT

The possibility that women in the South (and minority women in the North) have anything to offer to development specialists has been buried in the glib assurance that modernization (and Westernization) holds all the answers. Yet it is clear that there is much to learn from these women. Development plans based on inadequate knowledge of women's lives and attitudes have consistently failed (Rogers 1980; Moser 1989). A postmodern feminist focus, with its critique of the modern and its focus on localized, subjugated knowledge/power systems, would encourage development planners to pay more attention to the concrete circumstances of Third World and minority women's lives. The desire to understand the lived realities of women in the South (and North) would encourage a search for previously silenced women's voices, particularly their interpretation/representation of their life experiences, their successes and failures and their desires for change. The goals and aspirations of Third World women would be discovered rather than assumed, and strategies for improving their lives could be constructed on the basis of lived experiences and expressed needs.

The attention to difference encouraged by postmodernist feminist thinking also reminds women from the North that women in the South (and North) cannot be lumped into one undifferentiated category. The notion of Third World homogeneity, especially among women, may be comforting to WID practitioners, but it is damaging to both understanding and practice. It ignores the intersection of class and gender in the Third World and the need to evaluate that intersection in concrete historical circumstances. As in the North, elite women in the South share some experiences with their less advantaged sisters, particularly the constraints of patriarchal values, but many have opportunities to circumvent patriarchy in ways unavailable to the poor.

Circumstances can of course change over time – formerly prosperous women in Africa and Latin America are now bearing much of the brunt of the current economic crisis – but the experience of that crisis still varies by class, and those variations need to be recognized.

WID practitioners have generally dealt with class by ignoring it and focusing on poor women. The socialist feminist critique of development has been more sensitive to the importance of gender and class, but it has emphasized women's work experience, particularly their unrecorded reproductive labour. While paying lip service to the importance of gender ideology in a materialist analysis, the socialist feminist perspective has not been able to provide the tools for investigating the construction of meaning and its dissemination through language which the postmodernist feminist approach can offer. I am not suggesting that we abandon the socialist feminist concern with gender and class, but rather that we add a postmodern feminist analysis of discourse, knowledge/power relations and difference to materialist analysis.

This more nuanced approach to the construction of gender ideology in Third World societies would assist development planners and scholars to understand how gender ideology shapes and limits women's access to knowledge and power in particular societies. Postmodernist feminist critiques not only call into question received ideas about modernization, but also offer new insights into women's lived experiences, particularly the way societies define women's sense of themselves and the limitations of that sensibility for social change and development. Unless we achieve this level of understanding, women and development studies will remain another area of colonization of the South by the North (Wiltshire 1988). The postmodernist feminist focus on difference and discourse offers the possibility of understanding and transcending both Western and Third World patriarchal ideologies, without abandoning the search for a more gender equitable world.

This approach to gender and development has implications for the practice as well as the analysis of development. WID practitioners' concentration on vulnerable women permits them to ignore the existence of many highly skilled indigenous women in the Third World, women who have much to offer the development process in their own countries (Staudt 1985). By denying this reality, development practitioners perpetuate the belief that they alone can save women in the South.

Morever, many WID experts know little about women's organizations and development writings in the South, nor do they see the need to learn about them. Many are unaware that DAWN is a well-organized, articulate movement calling for new approaches to development. While somewhat utopian in approach, the DAWN document reminds us that the development concerns of women in the South are less bound to gender struggle and more embedded in issues of global redistribution and underdevelopment than feminist struggles in the West (Sen and Grown 1987). This is an important lesson for Western feminists, and for development experts, who are apt to equate their own priorities with those of women everywhere in the world.

Right now there is a lot of talk about consulting women from the South about development plans and programmes, but very little practice. Development practitioners from the North offer consultancies to Third World women and then all too often ignore their suggestions. Moreover, most local consultancies in the South go to men, who rarely see gender as an important issue. The use of trained women experts in the South has only increased of late, and even then they are mainly concentrated among the more gender sensitive development agencies such as the Scandanavians and Canadians (Wiltshire 1988; CIDA 1991, 1992). The belief that real expertise can only come from the North, meanwhile, afflicts not only Northern WID practitioners, but many indigenous experts who have been trained in the North; they too accept this dictum and often ignore indigenous knowledge unless clothed in the rhetoric and assumptions they learned in Northern institutions. This attitude blinds many female (and male) experts from the South (and North) to the developmental potential of poor women and consequently often leads them to ignore and undermine the self-confidence and self-reliance so necessary to the development of autonomous self-reliant and equitable societies.

CONCLUSION

The 'crisis' of development has led some scholars to question the validity of the entire development project; some critics of women and development activities make similar allegations (Edwards 1989; Mueller 1987b). Rather than 'throw the baby out with the bathwater,' I believe much can be gained by adding a postmodernist feminist perspective to both the analysis and practice of women's development in the South. This perspective questions the equation between modernity and development so dear to most development specialists. It rejects a purely top–down approach to development, arguing for development planning based on a closer, more localized and contextualized examination of women's strategies for survival, both in the South and the North. Third World women become the subject rather than the object of development theory and practice.

This approach to development recognizes the connection between knowledge, language and power, and seeks to understand local knowledge(s), both as sites of resistance and power. It provides a more subtle understanding of Third World women's lives, one that questions development discourses that represent Third Women as the vulnerable 'other.' It reminds development practitioners (and theorists) that women's realities can only be discovered by uncovering their voice(s) and subjugated knowledge(s), and that once that is understood, women's 'vulnerability' in the South is neither so clear nor so pervasive. This approach provides new insights into the lives and behaviour of Third World women, it undermines the authority of Northern development experts, and hopefully will lead to development theory and practice which is grounded in the multiple realities of women's lived experiences in the South.

However, it is important to recognize the pitfalls of an overemphasis on difference. A postmodernist feminist approach to women's development must encourage global, regional and national solidarity among women over issues of mutual concern. An uncritical extreme postmodernism encourages a view of gender as endlessly multiple, 'inherently unstable and continually self-deconstructing' (Bordo 1990:134). The danger of dissolving into relativity and political paralysis is very real, but a materialist postmodernist feminist approach can provide 'practical spaces for both generalist critique (suitable when gross points need to be made) and attention to complexity and nuance' (Bordo 1990:153). A synthesis of this kind, which incorporates postmodernist critiques without losing touch with the insights and political goals of feminist thinking, holds out the possibility of deepening our global understanding of women's multiple realities while remaining committed to the struggle to improve the lives of women throughout the world.

ACKNOWLEDGEMENTS

This chapter is based on an earlier article which appeared in *Development and Change* 24 (1993): 439–64 (copyright holder: Institute of Social Studies, The Hague). I am grateful to the Institute of Social Studies, The Hague, for giving me permission to include sections of that article.

14

BECOMING A DEVELOPMENT CATEGORY

Nanda Shrestha

> History, despite its wrenching pain,
> Cannot be unlived, and if faced
> With courage, need not be lived again.
> (Maya Angelou 1993)

'Colonial domination,' claimed Fanon, 'manages to disrupt in spectacular fashion the cultural life of a conquered people ... [T]he intellectual throws himself in frenzied fashion into the frantic acquisition of the culture of the occupying power and takes every opportunity of unfavorably criticizing his own national culture' (1967: 236–7). Mesmerized by the glamorous notion of development, I was mentally slow to scale its ideological contours, to comprehend how development ideology is produced and reproduced, how it is propagated across space and through time, how it conquers the minds of native elites, and how it paves the path for a monolithic culture of materialism which stigmatizes poverty and the poor. Increasingly, it has dawned on me that my own development odyssey served as an autopsy of how the imported discourse of development had possessed the mind of a national ruling class, and how such a mindset had, in turn, played a major role in deepening the social roots of poverty – all, of course, in the name of development.

This chapter is an account of the process of my own seduction. This is a self-reflective narrative, a wrenching dialogue with myself, based on my encounter with development as a young student aspiring to join the ranks of educated elites and the well-to-do. However, my objective here is *not* to write my own personal biography; this is rather a post-mortem of the body of development by a colonized mind, designed to serve as a research method. Even though such a methodology is uncommon in academic research, it is valuable in exposing the experience of most elites – whether self-made like myself or those born and raised in elite families. This personal narrative reveals how and why the discourse of development, with the help of foreign aid, solidifies the colonial mindset in the post-imperial world, crafting cultural values, thinking, behaviour, and actions. This is how, under the guise of development, the culture of imperialism is methodically reproduced in order to maintain continued Western dominance over the myriad of nation-states which have emerged since the downfall of the formal colonial–imperial order. As Edward Said (1993: 25), describing the lingering legacy of imperialism, points out: 'Westerners may have physically left their old colonies in Africa and Asia, but they retained them not only as markets but also as locales on the ideological map over which they continued to rule morally and intellectually.'

As a *garib* (poor) boy growing up in a rustic town of Pokhara in Central Nepal more than 40 years ago, I had few possessions of material value. My aspirations were limited to an occasional desire to have enough food and some nice clothes. Based on the contemporary measure of poverty, the World Bank and its agents would have labelled my family extremely poor. Indeed, the 1992 *World Development Report* shows Nepal as the fifth poorest country in the world. I grew up in a tiny house with a leaky roof. My family had about 1.5 acres of non-irrigated land. Along with some vegetables, we usually grew maize and millet. My mother sometimes brewed and sold millet liquor, known locally as *raksi*. This is how my family eked out a meagre existence. Life was always hand-to-mouth, a constant struggle for survival. It was not unusual at all for me to go to school hungry, sometimes three or four days in a row.

I specifically recall one Dashain – the biggest Hindu festival which is celebrated with a great deal of fanfare for ten consecutive days. It signifies a celebration of victory of good over evil, namely the victory won by Goddess Durga. During this festival, most temples are littered with blood from sacrificed animals (uncastrated goats, roosters, ducks, and buffaloes). The smell of blood and raw meat is everywhere. Large quantities of meats are consumed during this festival. Even the poorest are expected to eat some meat, one of the very few times during the year that most poor families get to do so. Dashain is not just a religious celebration; it is equated with status. There is immense pressure on every family, rich and poor, to celebrate the festival with as much pomp and show as possible. Parents are expected to get brand new clothes and other material items for their children. As a consequence, each year countless families plunge deep into debt. Many mortgage, if not sell outright, whatever little land or other assets (e.g. gold) they have to raise money for celebration. The festival is very expensive, with many households never recovering from debt. My father used to call Dashain *dasha* (misery) or the 'Festival of Sorrow.'

That particular Dashain, I was 8 years old. My family had no money to acquire any of the necessities for the Dashain. It was the eighth day of the Dashain, two days before its culmination. On the eighth or ninth day, families are supposed to sacrifice animals. We had not even a rooster to worship Goddess Durga. We all sat in the house the whole day, huddled around and feeling sad, not knowing what to do. My parents could not get me even one new shirt, let alone a complete outfit. Even today, the memory of that Dashain brings tears to my eyes. Because of that bitter memory, I have never been able to enjoy any festival. Finally, on the morning of the ninth day of the Dashain, I received a small sum of money from my brother-in-law, for whom I had done some work. The money saved that Dashain, and my family was just able to ward off a social embarrassment.

To my innocent mind, poverty looked natural, something that nobody could do anything about. I accepted poverty as a matter of fate, caused by bad *karma*. That is what we were repeatedly told. I had no idea that poverty was largely a social creation, not a bad karmic product. Despite all this, it never seemed threatening and dehumanizing. So, poor and hungry I certainly was. But underdeveloped? I never thought – nor did anybody else – that being poor meant being 'underdeveloped' and lacking human dignity. True, there is no comfort and glory in poverty, but the whole concept of development (or underdevelopment) was totally alien to me and perhaps to most other Nepalis.

There is a word for development in the Nepali language: *bikas*. Following the overthrow of the Rana autarchy in 1951, the word began to gain currency. A status divide emerged between the *bikasi* and the *abikasi*. Those who had acquired some knowledge of so-called modern science and technology identified themselves as *bikasis* (developed), supposedly with a 'modern' outlook, and the rest as *abikasis* or *pakhe* (uncivilized, underdeveloped, or backward). There was money in *bikas*, and the funding for *bikas* projects, mostly through foreign aid, was beginning to swell. Development was thus no longer just a concept. It became a practice which fortified, and even exacerbated, the existing class hierarchy. The wealthy, the powerful, the more educated embraced *bikas*, becoming *bikasis*. The *garib* (poor) were *abikasis*. As the logic went, the poor became poor because they were *abikasi*; they impeded *bikas*.

Bikas was generally associated with objects such as roads, airplanes, dams, hospitals, and fancy buildings. Education was also a key component, essential to build human capital. Education could salvage the *abikasi* mind, but only if it was 'modern,' emphasizing science, technology, and English, the language of *bikas*. Sanskrit, previously the language of the learned, was a deterrent to *bikas*. There was tension in the family. Educated children were viewed as future agents of *bikas*, and our parents were usually seen as *abikasis*. True, there were things our parents did that had little scientific basis or made any logical sense. But there were also many things they did that had more practical values than the theoretical 'science' we were learning at school. Yet, in the eyes of *bikasis*, whatever human capital, productive forces or knowledge our parents had accumulated over the years did not count for much. Many students felt ashamed to be seen in public with their parents. The new education gave us the impression that our parents' manual labour was antithetical to *bikas*. So we sneered at manual work, thinking that it was something only an *abikasi* or intellectually 'underdeveloped' mind would do. It was not for the high-minded *bikasis*. The new educational system was producing a whole new way of thinking about the value of labour. *Bikas* meant, to apply Ivan Illich's (1992) logic, denying as well as uprooting the existing labour use system, traditional bonds, and knowledge base, rather than building on them.

Before development, hard manual labour was a common way of life. The vast majority of people did it from early childhood, from the time they were 7 or 8 years old. Now the delusionary vision of *bikas* had made it an anathema. The new attitude toward labour created a backlash against education in general. My father opposed my education although I always did manual labour. Many children were actually pulled out of their schools by their parents before completing their elementary education. In an agrarian society like Nepal, children formed a vital source of labour or economic assets, but they had developed an aversion to manual work as a result of education. So what good was their education if it meant depriving the family of much needed family labour and potential supplementary income the children would generate when hired by others? Such a calculation was particularly important among the poor parents who did not see much prospect for their educated children's employment in the civil service – the principal source of salaried employment for the educated. To most poor parents, their children's education did not mean an investment in future prosperity; rather it entailed, at least in the short run, lost labour and potential income.

The devaluation of manual labour was hardened by our observation of Westerners whom we considered educated, developed, sophisticated, civilized. We rarely observed any of the growing contingency of Westerners in Nepal doing manual work. They all had at least one maid; some had two or three. Even meagrely paid Peace Corps Volunteers (PCVs) had personal cooks or maids. Many lived a life of luxury. They saw themselves as advisers and exhibited an aura of superiority. We thought that their life-style represented that of a modern, educated *bikasi*. Consequently, local educated people began to emulate them and aspire to the 'good life' the Westerners enjoyed and represented. Development was the fountain of good life.

Not all parents resented 'Western' education, however. For the elites, the architects of the national culture, modern education was the umbilical chord between themselves and the West. Since they cherished such linkage and wanted to be associated with *bikas*, educating their children in 'modern' schools and in the West was very important for them. Within Nepal they preferred to send their children to St Mary's School for girls and St Xavier's School for boys, run by Christian missionaries, mostly from England. Several wealthy families in Pokhara sent their children to these two schools, both located in Kathmandu Valley. When these children, some of whom lived in my neighbourhood, came back home during breaks, we could hear them speak fluent English. They would have little contact with us, and sometimes treat us like *pakhes* (uncivilized). Educationally, we felt very deficient in front of those elite children. The new education was preparing a new generation who not only controlled the rapidly expanding bureaucracy, but also dominated the development enterprise, thereby reaping the lopsided benefits of *bikas*. Education and *bikas* both not only displayed a distinct class character, but also accentuated the prevailing class biases of colonial society. Most educated people shunned hard work and looked for work in the civil service sector where they could boss their juniors around. They wore two disparate faces: one looking meek and saying *hajur, hajur* (yes sir, yes sir) to those above them and another stern and rude, treating those below them as worthless subhumans.

By the mid-1950s, the idea of *bikas* had been firmly transplanted in the Nepalese psyche. Whether *bikas* was actually occurring did not matter. It had permeated almost every Nepalese mind, from peons to the prime minister and the king. The higher the bureaucratic authority, the louder the voices of *bikas*. *Bikas* was regarded as a secret passage to material paradise. The myth of *bikas* projected materialism as human salvation, the sole source of happiness, emancipation, and redemption from hunger and poverty (Ullrich 1992: 275). Materialism appeared to have replaced a traditional Hindu conception of *bhakti* (devotion) and *dharma* (duty, good deeds) as a channel of *moksha* (salvation). Not that Hinduism is devoid of material values; it has always played hide-and-seek with materialism. *Laxmi* (the goddess of wealth) is actually highly revered. But this new form of materialism was much more pronounced and had quickly emerged as a new deity.

I believe it was 1951 when the first group of British Christian missionaries arrived in Pokhara (missionary activity had started in Nepal much earlier). Although they probably were not the first white people to come to Pokhara, they are the first ones I remember. Because of the British policy of Gorkha (Gurkha) recruitment, many recruits from the surrounding hills had already served the British. While the citizens of other colonies were exploited as slaves, indentured plantation workers, and coolies, Britain's exploitation of young and able Nepalis was somewhat unique, raw material for the war machine of the British imperial army. Although their bodies belonged to Nepal, their labour belonged to the British. In this sense, the dance of British imperialism was already in full swing across Nepal.

The missionaries' 'civilizing mission' brought Christianity and modern medical facilities to the town of Pokhara as they set up a small hospital called the Shining Hospital. While the hospital seemed to have brought medical miracles as patients often responded faster to their (Western) medicines than to local medical practices, it also undermined local medical knowledge. Missionaries mocked our local medical practices, and made us feel ashamed of them. Even more important, however, their presence led to a total psychological metamorphosis in our perception of whites. Almost everybody, regardless of their socioeconomic status in the community, started addressing white missionaries, or for that matter all whites, as *sahib* or *sab* for males and *mimsab* for females (master, boss, teacher, or sir/madam

race

depending upon the context). Although the word *sahib* is a fairly common honorific term, it clearly has connotations of dominance and subordination. Whites, called *sahib* from the very start, were thus accorded a dominant position. The *sahib* culture became engraved in the Nepalese mind, a culture in which whites were placed at the apex, with the Nepalis looking up to them in the way devotees look up to the statues of their gods, begging for blessings or waiting at the end of the table for crumbs to fall. This, in turn, accentuated whites' pre-existing feeling of superiority and, in their own minds, justified their treatment of us as uncivilized and inferior or as needing salvation.

Previously, white people were often referred to as monkeys (in appearance). The Hindu caste codes regarded whites as *mlaksha*, the polluted, the untouchable, and hence relegated to the bottom of the caste hierarchy. If any high-caste individual (Brahman, Kshatriya, and Vaishay) touched a white person, that person would be considered unclean, and thus required to undergo a cleansing ritual. In fact, as late as the 1940s, all Nepalese recruits serving in the British imperial army and those who had crossed any of the oceans were, upon their return home, subjected to such ritual, for they were presumed to have come in physical contact with whites. Now whites were no longer viewed as monkeys or as *mlakshas*. Instead, they were beautiful, the *sahibs*, the masters, a super caste, even to the highest ranking caste group: the Brahmans. Even the most sacred of the Hindu social codes was no longer sacrosanct when it came to applying them to white people. Here was a fundamental transformation of Nepalese culture, attitude, and behaviour towards whites. It was hard to fathom why whites had been elevated so quickly to the top of the social hierarchy. The oppressive and archaic caste system had simply been rearranged to accommodate the emerging *sahib* culture and nascent *bikas* enterprise; caste relations had been transformed into power relations in our dealings with whites, the latter occupying the position of power and prestige.

The hospital was a sign of *bikas*, the first such symbol in Pokhara. It was brought by white people, the harbingers of *bikas*. To us, they were obviously economically superior. They spoke the language of *bikas*; they knew the modern science and technology of *bikas*. They embodied *bikas*. Being associated with them, learning their language, and imitating them became important attributes of *bikas*, attributes that all *bikasis* were expected – and wanted – to possess.

Shortly after the arrival of British missionaries came an airplane, an old DC3. When some people heard the roaring sound of an approaching airplane, it caused an incredible commotion in Pokhara and surrounding villages. The serenity of bucolic Pokhara surrounded by hills and mountains was disrupted by that noisy machine. When the airplane landed, pandemonium broke out throughout the town. Almost everybody flocked to see it. We were clamouring to touch it as if it were a divine creation, sent to us by God. Some wondered how something so big could fly. Others searched in their Hindu religious tradition to see if they could identify some divine figure resembling an airplane. They did find one: the Garuda, the eagle-looking Hindu mythical bird, the heavenly vehicle of Vishnu, who in the Hindu trinity of Brahma, Vishnu, and Shiva is the universal god of preservation, the Saviour. The airplane was the talk of the town for several days. We had seen another facet of *bikas*. Not only could *bikas* cure the sick, but it could fly like the Garuda, carrying *bikasis* around the country. We adapted this *bikas* symbol to our own Hindu tradition. *Bikas* was justified.

Then came a used jeep, flown in by the mechanical Garuda. The jeep was brought in pieces, along with a foreign mechanic to assemble it. In that jeep, some saw the chariot driven by Lord Krishna during the epochal war called the *Mahabharat*, the war fought for justice between the Pandavas and the Kauravas, brothers from two different mothers. In that war, the chariot carried Arjuna, who led the five Pandavas' forces representing justice and ultimately defeated the evil forces represented by the one-hundred Kauravas. The jeep was later followed by bicycles and oxen-driven carts. Such was the order of transportation development in Pokhara and in many parts of Nepal: a retrogressive order. This was quite symptomatic of the whole process of development, everything backwards. What we were observing was imported *bikas*, not true progress from within. We had achieved very little on our own. *Bikas* was our new religion. Various material objects represented the pantheon of *bikas* gods and goddesses. The symbolism of *bikas* and Hinduism were uncannily alike.

The first wave of *bikas* was encapsulated in the first five-year development plan launched in 1956, and almost entirely financed by foreign aid. As this plan institutionalized the development enterprise, the march of *bikas* was now official though few knew where it would lead. Following the advice of Western experts, Nepalese *bikasis* advocated industrial growth. Some actually built factories, even before embarking on the path of agricultural improvement and setting up infrastructure. Merchants in Pokhara established a match factory, but the venture collapsed because of the absence of marketing networks and transportation facilities. Such a regressive trend continued to mar the national development horizon. North Atlantic consumer culture penetrated, unchecked, every nook and corner of Nepal, rapidly generating previously non-existent wants and hence scarcities, a situation which only aggravated poverty. The local production system remained incapable of meeting the demands of this rising consumerism. So, *bikas* had arrived in Pokhara (and in Nepal in general) in many forms, represented by various objects, most of which had little use value for the general public. Excitement filled the air even though few outside the *bikas* circle climbed the ladder of progress. The jeep was symptomatic of Nepal's *bikas*: second-hand and out of reach of the masses.

In 1962 the first group of PCVs arrived in Pokhara, most of them as instructors to teach different subjects. I was in the sixth grade at that time.

Before their arrival, a high school was constructed with financial aid from the United States. Our high school was chosen as one of the first multi-purpose schools in Nepal. Along with regular courses, it offered vocational education in trade and industry (carpentry and rudimentary drafting and electric wiring), home economics (cooking, sewing, and knitting), agriculture, and commerce (typing and some shorthand writing and bookkeeping). Vocational education was designed to produce a pool of skilled workers, to build human capital, needed for development, because our existing knowledge and skills were presumed worthless. So we, the vocational students, were expected to fill the knowledge and skill void and play a big role in national *bikas*. We were subsumed by this tide of *bikas*. We were its recipients, groomed as its agents.

In order to carry out the vocational training plan, fancy chairs, desks, and tables were flown in from overseas as part of the aid package. All sorts of tools and equipment for various vocational fields came from the United States which planned and funded the whole project. The headmaster and three vocational teachers went to the United States for training. We had no idea that our school, Pokhara, Nepal were the fulfilment of President Truman's grand plan for the 'poor, underdeveloped' peoples. Through the Peace Corps initiative, President Kennedy took the Truman plan to new levels, placing his own stamp on it. The Peace Corps plan was the least expensive yet most effective mechanism of intensifying American influence and countering communism. Perhaps, most PCVs were not aware of the grand plan either. There was a good mix of volunteers. There were some who had joined the Peace Corps (PC) for an idealistic purpose: the do-gooders. Some had joined the PC, to avoid being drafted for the Vietnam war, and others did it because they were indulging in the hippie movement or alternative lifestyles. They were going overseas, as PC volunteers, to 'exotic' countries, some in search of cheap marijuana and hashish and others in search of cultural relief from the material opulence of stale suburban life. Nepal was viewed as a mecca for such relief. How ironic that many volunteers, sent to promote American values and materialistic development, were themselves yearning for reprieve from that very same material life in a culture that was described as backward and poverty-stricken.

We sought ways to be close to Westerners, for we viewed them as the messiahs of development. Since the PC policy presented the best opportunity to be close to whites, we hailed it. PCVs were usually friendly and accessible unlike most high-flying diplomatic types and so-called development advisers. PCVs lived and socialized with local people, and rarely demonstrated the religious zeal of the missionaries. We constantly hung around the PCVs, and fantasized about going to America with them. We neither knew nor cared about the motives and hidden agenda of the Truman/Kennedy plan. The degrading spectre of colonialism appeared to have vanished like a shadow. The vituperative language of colonial hegemony and racial superiority had been replaced by a new language with a neutral tone. A euphemistic lexicon of American partnership and collaboration for development emerged. It proved to be a potent seductive force in the modern diplomacy of domination. So I was sold on *bikas*.

Bikas seemed to be spreading: a brand new school with a corrugated tin roof that had nice windows and blackboards, fine furniture and tools, objects beyond our imagination, and of course an ever increasing horde of Westerners. For those who grew up going to school in an open field or in open sheds made of bamboos and thatch, who used to play football (soccer) with unripe grapefruits, the school looked like a castle in a fairy tale. I had never dreamed of such things; now they were part of our daily reality. Our school even had a generator to produce electricity and operate fancy equipment. *Bikas* looked glistening and sumptuous, at least on the outside and at school. A little bit of US educational aid had done wonders. So we thought. We felt like we were taking a giant leap to the top of the stairway. We did not even have to work, let alone work hard. *Bikas* could bring things instantly, and we did not have to work hard to acquire what we wanted. But we were all bewitched. Foreign aid had become our sole medium of material nirvana. Pride in self-achievement and self-reliance was conspicuously absent.

Bikas solidified the colonial notion that we were incapable of doing things for ourselves and by ourselves. The colonial 'civilizing mission' was resurrected as the mission of development. These Western 'civilizers' first undermined our relative self-sufficiency and self-reliance, and then categorized us as inferior and poverty-stricken. Closely interwoven with nature and its cyclical rhythm, our way of life was certainly different, but not inferior. True, it was not prepared to bring nature under large-scale human subjugation. But our relatively harmonious coexistence with nature was interpreted as a sign of backwardness and primitiveness. Development was measured in terms of the distance between humans and nature. The greater the distance between the two, the higher the level of development. The distance between the two definitely increased – in some cases literally, as poor Nepalese village women walked further and further every year in search of fire wood and animal fodder.

In hindsight, I see a great deal of sadness in the glitter of *bikas*. While we saw *bikas* at school, there was no change at home, at least not for most poor families. *Bikas* had done nothing to reduce our hunger. Life at school and at home were an ocean apart. Every morning we went to school excited, ready to enjoy our new chairs and work with fancy tools. After school many of us returned home to face the same old hunger. Nonetheless our expectations had been raised. Disappointment became more frequent as the gap between the promise and the reality widened. Since wants were rising, poverty had grown a new face. It had a much deeper materialistic undertone than ever before. Poverty was never so frightening and degrading in the past. We did not help ourselves either. Self-reliance and cooperation gave way to despondency and dependency. In the past, if a trail was damaged, the villagers from the surrounding villages organized a work force and repaired it. Now the villagers felt that somebody else, a foreign donor or government agency would come and fix it. Nowadays, nothing moves without foreign aid.

Before the onslaught of *bikas*, the poor and poverty were rarely stigmatized. Despite the oppressive feudalistic social structure that existed in Nepal, the rich seemed to bear some sense of shared moral responsibility toward the poor. Patron–client relations, though onerous in many ways, offered some economic cushion for the poor (Brass 1990). Poverty in the past was padded with a modicum of security; now it meant total insecurity. The principles of *bikas* denigrated traditional behaviours. Everything was defined in stark economic terms. Those who disregarded these principles were labelled irrational. Development categories were being constantly invented and reinvented, used and reused.

The national ruling elites internalized the new civilizing mission of development. As Nandy (1992: 269) has observed elsewhere: 'When, after decolonization, the indigenous elites acquired control over the state apparatus, they quickly learnt to seek legitimacy in a native version of the civilizing mission and sought to establish a similar colonial relationship between state and society.' As envisioned and practised, development legitimized the ruling elites' authority. Well-accustomed to the Western way of life, irrespective of their political ideology, they subscribed to the mistaken belief that Western-style development was the only way out of poverty. They also managed to project themselves as the champions of the poor. Prevailing modes of life were vilified by development fetishism acquired from the West.

When I reflect on my own development experience and journey, it is transparent that my mind had been colonized. I was proud of my contact with PCVs. Being able to speak a little bit of broken English was a big thing. I viewed my PCV contact and English-speaking ability as my *bikas* ladder to the summit of modernity. I acquired American values, copied their habits. In my mind, I thought like an American although I had no idea what that really meant. I believed that if a person spoke English, they were very bright, *bikas*-minded, and sophisticated. That person also gained respect from others. At school, I decided to pursue vocational education because it was an American initiative. We were told that if we passed the national high school matriculation examination in first class, we would receive a full scholarship to go overseas to study. Such a prospect had a magnetic appeal to my colonized mind. Since foreign education was deeply cherished, many students aspired to go to America and Europe to study. America was the most preferred destination, followed by England and other countries.

I passed the examination in first class. But no scholarship came my way. A sense of betrayal surrounded me. With my *bikas* hopes and dreams dashed, there seemed a big void in my life. I felt that *bikas* had failed to deliver on its promise. With nothing left to look forward to, I became a primary school teacher, attended college in the morning, and stayed active in student politics.

275

Then, in 1971, my life suddenly took a new turn. I received a letter from a Peace Corps friend who had returned to the USA in 1968. Thanks to his efforts, I obtained a full college scholarship in Minnesota. *Bikas* had at last arrived. Such was the development odyssey of my colonized mind. In recent years though, I have come full circle. I am no longer the passionate subscriber to Western development that I once was. The more I observe what is happening in countries like Nepal, especially the social, political, and economic outcomes of their booming enterprise of development rooted in Western materialism, the more I question its value.

These days, I am frequently haunted by the many diverse images I have encountered over the years – all victims of development in one respect or another – some struggling to survive, some going hungry, and others rejoicing in their financial success and ostensive material acquisitions. In my quiet moments, many muttering voices fill my ears, with a sense of both ecstacy and deep pain. 'We have been seduced by the goddess of development, by the voracity of the North Atlantic material culture,' pronounce these voices. Yes, I too have been seduced; we have all been seduced. There has been a structural violence of our psyche. But who caused it and how can it be repaired?

I am not trying to suggest that whatever was old was good and desirable and that every aspect of our lost heritage should be reclaimed. Nor am I implying that the old social structure should be revived in its entirety and that we should adopt an exclusionary position and advocate 'nativism.' Such a fundamentalist position is neither possible nor acceptable. Nobody should be oblivious to the many tyrannical practices of our feudal–religious heritage. My contention is that the indigenous economic system and values were generally self-reliant, self-sufficient, sustainable, and far less destructive of humanity as well as nature. At least, it served as a hedge against total deprivation. But now in the name of *bikas*, the dignity and humanity of the poor were questioned, while poverty itself deepened. Yet, this seemed to matter little. We had already developed a blind faith in *bikas* and its objects. We accepted development as a *fait accompli*. We seemed to have convinced ourselves that more *bikas* meant less poverty. What a fallacy!

In this self-reflective narrative, I have recounted the development journey of my own colonized mind. In doing so, I have attempted to show how the culture of imperialism transfused Nepalese society, how the colonial mindset was created among its elites, how manual labour and indigenous economic activities were devalued. In all of this, foreign aid played a critical role, captivating minds and actions. Many still claim that foreign aid is being used to achieve economic development for all citizens. We still insist that the poor need the kind of development we have practised since the early 1950s. Although the poor were never asked if they wanted to be helped or preferred Westernized development at all, now they too seem to have been intoxicated by the brew called foreign aid.

A cruel choice confronts us all. The underlying logic of this narrative dictates that we reframe our mindset and take a hard look at the seductive power of development. Even if we can gather enough strength and determination to navigate a relatively self-reliant path, our efforts should not be guided by what Edward Said (1993) calls 'nativism' – a twisted nationalistic tendency often rooted in religious fundamentalism, which is no less dangerous than the seductive power of 'Westernism.' The way I see it, the elites – whether self-made like myself or born and raised like those from elite families – are at the root of most social and economic problems haunting Nepal. In the name of development, we pursued our own interests, both individually and as a class. We incarnated ourselves as domestic *sahibs*, denigrating the poor and their labour. In our attempt to look and become Westernized, we have created a monster out of developmentalism, lost touch with our social consciousness and humanity, and surrendered our national dignity and culture. We trust Westerners more than ourselves, virtually in every respect. We learned how to seize the currents of international development, propelled by the World Bank, US Agency for International Development (USAID), and other prominent development agencies. We turn their fads into overriding national concerns, instantly churning out reports to corroborate our claims. When they were concerned about deforestation and other environmental problems, we suddenly discovered our deforestation, soil erosion, and many other environmental ills.

Let us get serious and have enough moral courage first to challenge our own elitism and vested interests. Let us free ourselves from the trappings of Westernized development fetishism; let us *un*learn the Western values and development thinking which have infested our minds. However, *un*learning is not complete without *re*learning. So let us relearn. All of this, of course, requires that we consciously deconstruct our colonial mindset. This is a colossal battle against the entrenched culture of imperialism. If it is to succeed, it needs to be fought on two fronts. First, the battle is waged at the personal front to decolonize individually our colonized minds. Second, the battle is fought at the societal front. This demands a collective force to deconstruct the colonial mindset that pervades Nepalese society. The outcome of the second battle will depend on the degree of success achieved at the personal front. If we muster enough moral courage to wage these battles and win them, we can then consciously demystify the seductive power of development. I am fully cognizant that this is very bitter medicine, but we have few other choices if we want to create a future of human dignity and relative economic autonomy.

BIBLIOGRAPHY

Abdullah, A. (ed.) (1990) *Al-Jaysh wa-l-dîmuqrâtîya fî misr*, Cairo: Dâr Sînâ li-l-Nashr.

Abelove, H., Blackmar, B., Dimock, P., and Schneer, J. (eds.) (1983) *Visions of History*, Manchester: Manchester University Press.

Abu-Lughod, J. (1978–79) 'The shape of the world system in the thirteenth century', *Studies in Comparative International Development* 22, 4: 1–25.

Adams, W. (1990) *Green Development: Environment and Sustainability in the Third World*, London: Routledge.

—— (1993) 'Sustainable development and the greening of development theory', in F. Schuurman (ed.) *Beyond the Impasse: New Directions in Development Theory*, London: Zed Books.

Adams, W. and Thomas, D. (1993) 'Mainstream sustainable development: the challenge of putting theory into practice', *Journal of International Development* 5, 6: 591–604.

Adelman, I. (1984) 'Beyond export-led growth', *World Development* 12, 9: 937–949.

Adorno, T. (1993) [1951] *Minima Moralia*, London: Verso.

Afshar, H. (ed.) (1987) *Women, State and Ideology*, London: Macmillan.

—— (ed.) (1991) *Women, Development and Survival in the Third World*, London: Longman.

Agarwal, B. (ed.) (1989) *Structures of Patriarchy*, London: Routledge.

Ahmad, A. (1992) *In Theory: Classes, Nations, Literatures*, London: Verso.

Alderman, H. and von Braun, J. (1984) *The Effects of the Egyptian Food Ration and Subsidy System on Income Distribution and Consumption*, Research Report No. 45, Washington D.C.: International Food Policy Research Institute.

Alexander, J. and Alexander, P. (1992) 'Protecting peasants from capitalism: the subordination of Japanese traders by the colonial state', *Comparative Studies in Society and History* 33, 2: 370–394.

Alexander, Y. (1966) *Technical Assistance Experts: A Case Study of the UN Experience*, New York: Praeger.

Allen, R. (1980) *How to Save the World: Strategy for World Conservation*, London: Kogan Page.

Alvares, C. (1992a) *Science, Development and Violence: The Twilight of Modernity*, Delhi: Oxford University Press.

—— (1992b) 'Science', in W. Sachs (ed.) *The Development Dictionary: A Guide to Knowledge as Power*, London: Zed Books.

Alvarez, S. (1989) 'Conceptual problems and methodological impasses in the study of contemporary social movements in Brazil and the Southern Cone', paper presented at the XV International Conference of the Latin American Studies Association, Miami, Florida.

BIBLIOGRAPHY

Amin, S. (1974) *Accumulation on a World Scale: A Critique of the Theory of Underdevelopment*, 2 volumes, New York and London: Monthly Review Press.
—— (1985) 'Apropos the "green" movements', in H. Addo *et al.* (eds.) *Development as Social Transformation: Reflections on the Global Problematique*, Sevenoaks: Hodder and Stoughton.
—— (1989) *Eurocentrism*, trans. R. Moore, New York and London: Monthly Review Press.
—— (1990) *Maldevelopment: Anatomy of a Global Failure*, London: Zed Books.
Anderson, B. (1992) 'The new world disorder', *New Left Review* 193: 3–14.
Anderson, B.R. (1990) 'Reflections on the ecology of Southeast Asian studies in the United States, 1950–1990', paper presented at Conference on Southeast Asian Studies and the Social and Human Sciences, Wingspread Conference Center.
Anderson, D. (1984) 'Depression, dustbowl, demography and drought: the colonial state and soil conservation in East Africa during the 1930's', *African Affairs* 83: 321–342.
Anderson, D. and Throup, D. (1989) 'The agrarian economy of Central Province, Kenya, 1918 to 1938', in I. Brown (ed.) *The Economies of Africa and Asia in the Inter-War Depression*, London: Routledge.
Anderson, P. (1992) *Zone of Engagement*, London: Verso.
Anthias, F. and Yuval-Davis, N. (1990) 'Contextualising feminism – gender, ethnic and class divisions', in T. Lovell (ed.) *British Feminist Thought: A Reader*, Oxford: Basil Blackwell.
Anzaldua, G. (ed.) (1990) *Making Face, Making Soul/Haciendo caras: Creative and Critical Perspectives by Women of Color*, San Francisco: Aunt Lute Books.
Apffel Marglin, F. and Marglin, S. (eds.) (1990) *Dominating Knowledge: Development, Culture and Resistance*, Oxford: Clarendon Press.
Appadurai, A. (1990) 'Disjuncture and difference in global culture and economy', *Public Culture* 2, 2: 1–24.
Appiah, A. (1992) *In My Father's House: Africa in the Philosophy of Culture*, New York: Oxford University Press.
Apter, D. (1987) *Rethinking Development: Modernization, Dependency and Post-Modern Politics*, London: Sage Publications.
Apthorpe, R. (1970) *People, Planning and Development Studies*, London: Frank Cass.
—— (1986) 'Development policy discourse', *Public Administration and Development* 6: 377–389.
Arditi, B. (1988) 'La sociedad a Pesar del Estado', in F. Calderón (ed.) *La Modernidad en La Encrucijada Postmoderna*, Buenos Aires: CLACSO.
Armstrong, W.R. and McGee, T.G. (1985) *Theatres of Accumulation: Studies in Asian and Latin American Urbanisation*, London: Methuen.
Arndt, H. (1981) 'Economic development: a semantic history', *Economic Development and Cultural Change* 29: 457–466.
Asenerio, G. (1985) 'A reflection on developmentalism: from development to transformation', in H. Addo *et al.* (eds.) *Development as Social Transformation: Reflections on the Global Problematique*, Sevenoaks: Hodder and Stoughton.
Ashcraft, R. (1986) *Revolutionary Politics and Locke's Two Treatises of Government*, Princeton: Princeton University Press.
Ashforth, A. (1990a) 'Reckoning schemes of legitimation: on commissions of inquiry as power/knowledge', *Journal of Historical Sociology* 3, 1: 1–22.
—— (1990b) *The Politics of Official Discourse in Twentieth-Century South Africa*, Oxford: Oxford University Press.
Ashley, R. K. (1989) 'Living on border lines: man, poststructuralism, and war', in J. Der Derian and M.J. Shapiro (eds.) *International/Intertextual Relations: Post-modern Readings of World Politics*, Lexington, Mass.: Lexington Books.

279

Atkinson, P. (1990) *The Ethnographic Imagination: Textual Constructions of Reality*, London: Routledge.

Aveling, H. (ed./trans.) (1975) *Contemporary Indonesian Poetry*, St. Lucia: University of Queensland.

Baillie, J. (1950) *The Belief in Progress*, London: Oxford University Press.

Baker, R. (1974) 'Famine: the cost of development?', *The Ecologist* 4, 5: 170–175.

Bantu Investment Corporation (1975) *Homelands: The Role of the Corporations in the Republic of South Africa*, Johannesburg: Chris van Rensburg Publications.

Bardhan, P. (ed.) (1989) *The Economic Theory of Agrarian Institutions*, Oxford: Clarendon Press.

Barnes, R. (1992) 'Reading the texts of theoretical economic geography', in T. Barnes and J. Duncan (eds.) *Writing Worlds: Discourse, Text and Metaphor in the Representation of Landscape*, London: Routledge.

Barnes, T. and Duncan, J. (eds.) (1992) *Writing Worlds: Discourse, Text and Metaphor in the Representation of Landscape*, London: Routledge.

Barriteau Foster, E.V. (1992) 'The construct of a postmodernist feminist theory for Caribbean social science research', *Social and Economic Studies* 41, 2: 1–43.

Bates, R. (1989) *Beyond the Miracle of the Market*, Cambridge: Cambridge University Press.

Beckman, B. (1990) 'Empowerment or repression?: The World Bank and the politics of African adjustment', Department of Political Science, Stockholm University.

Beinart, W. (1984) 'Soil erosion, conservationism, and ideas about development in southern Africa', *Journal of Southern African Studies* 11, 2: 52–83.

—— (1989) 'Introduction: The politics of colonial conservation', *Journal of Southern African Studies* 15, 2: 143–162.

Beinart, W. and Bundy, C. (1987) *Hidden Struggles in Rural South Africa: Politics and Popular Movements in the Transkei and Eastern Cape, 1890–1930*, London: James Currey.

Bell, C. (1989) 'Development economics', in J. Eatwell, M. Milgate, and P. Newman (eds.) *The New Palgrave: Economic Development*, London: Macmillan.

Bello, W., Kinley, D., and Elinson, E. (1982) *Development Debacle: The World Bank and the Philippines*, San Francisco: Institute for Food and Development Policy.

BENBO (Bureau for Economic Research re Bantu Development) (1976) *Black Development in South Africa*, Johannesburg: Perskor.

Beneria, L. and Sen, G. (1981) 'Accumulation, reproduction and women's role in economic development: Boserup revisited', *SIGNS* 7, 2: 279–298.

Benjamin, W. (1969) 'Theses on the philosophy of history', in W. Benjamin (ed.) *Illuminations*, New York: Schocken Books.

Berman, B. and Dutkiewicz, P. (eds.) (1993) *Africa and Eastern Europe: Crises and Transformations*, Kingston: Centre for International Relations, Queen's University.

Berman, M. (1982) *All That is Solid Melts into Air: The Experience of Modernity*, New York: Simon and Schuster.

Bernstein, H. (1971) 'Modernisation theory and the sociological study of development', *Journal of Development Studies* 7: 141–160.

—— (1990) 'Agricultural "modernisation" and the era of structural adjustment: observations on SubSaharan Africa', *Journal of Peasant Studies* 18, 1: 3–35.

Bernstein, R. (1983) *Beyond Objectivism and Subjectivism: Science, Hermeneutics and Praxis*, Oxford: Basil Blackwell.

Berry, L. and Townshend, J. (1973) 'Soil conservation policies in the semi-arid regions of Tanzania: a historical perspective', in A. Rapp, L. Berry, and P. Temple (eds.) *Studies of Soil Erosion and Sedimentation in Tanzania*, Research Monograph

No. 1, Dar es Salaam: Bureau of Resource Assessment and Land Use Planning, University of Dar es Salaam.

Berry, S. (1989) 'Social institutions and access to resources', *Africa* 59, 1: 41–55.

Berthoud, G. (1990) 'Modernity and development', *European Journal of Development Research* 2, 1: 22–35.

Beukes, E.P. (1990) *Chairman's Report for the Period 1988–1990*, Johannesburg: Development Society of Southern Africa.

Bhabha, H. (1984) 'Of mimicry and man: the ambivalence of colonial discourse', *October* 28: 125–133.

—— (1991) '"Race", time and the revision of modernity', *Oxford Literary Review* 13, 1–2: 193–220.

—— (1994) *The Location of Culture*, London: Routledge.

Bhatia, B.M. (1991) *Famines in India 1860–1990*, Delhi: Konark Publishers.

Bierstecker, T. (1992) 'The "triumph" of neo-classical economics in the developing world', in J. Rosneau and E-O. Czempiel (eds.) *Governance Without Government: Order and Change in World Politics*, Cambridge: Cambridge University Press.

—— (1993) 'Evolving perspectives on international political economy', *International Political Science Review* 14, 1: 1–29.

Biko, S. (1978) *I Write What I Like*, London: Heinemann.

Binder, L. (1986) 'The natural history of development theory', *Comparative Studies in Society and History* 28: 3–33.

Binswager, H. (1986) 'Agricultural mechanization: a comparative historical analysis', *World Bank Research Observer* 1, 1: 27–56.

Blaikie, P. (1985) *The Political Economy of Soil Erosion in Developing Countries*, London: Longman.

—— (1987) 'The SADCC countries' historical experience of soil conservation and people's participation in it', in *History of Soil Erosion in the SADCC Region*, Report No. 8, SADCC Soil and Water Conservation and Land Utilization Programmes, Maseru, Lesotho.

—— (1989) 'Environment and access to resources in Africa', *Africa* 59, 1: 18–40.

Blaikie, P. and Brookfield, H. (1987) *Land Degradation and Society*, London and New York: Methuen.

Blaut, J.M. (1970) 'Geographic models of imperialism', *Antipode* 2, 1: 65–85.

—— (1973) 'The theory of development', *Antipode* 5, 3: 22–26.

—— (1975) 'Imperialism: the Marxist theory and its evolution', *Antipode* 7, 1: 1–19.

Boardman R. (1981) *International Organisations and the Conservation of Nature*, Bloomington, Ind.: Indiana University Press.

Bode, B. (1989) *No Bells to Toll: Destruction and Creation in the Andes*, New York: Charles Scribners.

Boeke, J.H. (1953) *Economics and Economic Policy of Dual Societies*, New York: Institute of Pacific Studies.

Boesak, A. (1986) 'Liberation theology in South Africa', in D.W. Ferm (ed.) *Third World Liberation Theologies: A Reader*, Maryknoll, New York: Orbis Books.

Bohle H-G. (ed.) (1993) *Worlds of Pain and Hunger: Geographical Perspectives on Disaster Vulnerability and Food Security*, Freiburg Studies in Development Geography, No. 5, Saarbrücken:Verlag Breitenbach.

Bookchin, M. (1979) 'Ecology and revolutionary thought', *Antipode* 10, 3/11, 1: 21–32.

Booth, D. (1985) 'Marxism and development sociology: interpreting the impasse', *World Development* 13, 7: 761–787.

Bordo, S. (1990) 'Feminism, postmodernism, and gender-scepticism', in L. Nicholson (ed.) *Feminism/Postmodernism*, London: Routledge.

Boserup, E. (1970) *Woman's Role in Economic Development*, New York: St. Martin's Press.

Boshoff, J.L. (1970) 'The implications of homeland development for white South Africans', *BANTU* December 1970.

Bourdieu, P. (1988) *Homo Academicus*, Cambridge: Polity Press.

Bradley, A.C. (1904) *Shakespearean Tragedy: Lectures on Hamlet, Othello, King Lear, Macbeth*, London: Macmillan.

Brand, S.S. (1984) 'Introduction', *Development Southern Africa* 1, 1: 1–2.

Brandt, W. (1980) *North-South: A Programme for Survival*, London: Pan.

—— (1983) *Common Crisis North-South: Cooperation for World Recovery*, London: Pan.

Brass, T. (1990) 'Class struggle and the deproletarianisation of agricultural labour in Haryana (India)' *Journal of Peasant Studies* 18, 1: 36–67.

Braudel, F. (1972) *The Mediterranean and the Mediterranean World in the Age of Philip II*, 2 volumes, New York: Harper and Row.

Bray, F. (1986) *The Rice Economies: Technology and Development in Asian Societies*, Oxford: Basil Blackwell.

Breckenridge, C. and van der Veer, P. (eds.) (1993) *Orientalism and the Postcolonial Predicament*, Philadelphia: University of Pennsylvania Press.

Breitbart, M. (1981) 'Peter Kropotkin: the anarchist geographer', in D. Stoddart (ed.) *Geography, Ideology and Social Concern*, Oxford: Basil Blackwell.

Brookfield, H. (1975) *Interdependent Development*, London: Methuen.

Brouwer, E. (1982) 'Social science, development assistance and evaluation', in Australian Development Assistance Bureau, *Summaries and Reviews of Ongoing Evaluation Studies, 1975–80*, Canberra: Australian Government Publishing Service.

Brown, N. (1991) *Apocalypse and/or Metamorphosis*, Berkeley: University of California Press.

Brundtland, H. (1987) *Our Common Future*, Oxford: Oxford University Press, for the World Commission on Environment and Development.

Brunner, E. (1948) *Christianity and Civilisation: Part One*, London: Nisbet.

Buchanan, K. (1970) *The Transformation of the Chinese Earth: Perspectives on Modern China*, London: C. Bell and Sons.

—— (1973) *The Geography of Empire*, London: Spokesman Books.

Buck-Morse, S. (1989) *The Dialectics of Seeing: Walter Benjamin and the Arcades Project*, Cambridge, Mass.: MIT Press.

Bunn, D. and Taylor, J. (eds.) (1987) *From South Africa: New Writing, Photographs and Art*, Chicago and London: University of Chicago Press.

Burchell, G., Gordon C., and Miller P. (eds.) (1991) *The Foucault Effect: Studies in Governmentality*, London: Harvester Wheatsheaf.

Buthelezi, S. (1992) 'The emergence of black consciousness: an historical appraisal', in B. Pityana *et al.* (eds.) *Bounds of Possibility: The Legacy of Steve Biko and Black Consciousness*, London: Zed Books.

Buttel, F. (1991) 'Environmentalism, origins, processes, and implications for rural social change', Presidential address to annual meeting of the Rural Sociological Society, Columbus, Ohio, 18 August.

Buttel, F. and Taylor, P. (1992) 'How do we know we have global environmental problems?' *Geoforum* 3: 405–416.

Buttel, F., Hawkins, A., and Power, A. (1990) 'From limits to growth to global change: constraints and contradictions in the evolution of environmental science and ideology', *Global Environmental Change* 1, 1: 57–66.

Buvinic, M., Lycette, M., and McGreevey, W. (1983) *Women and Poverty in the Third World*, Baltimore, Maryland: Johns Hopkins University Press.

Caccia, C. (1990) 'OECD nations and sustainable development', in D. Angell, J.

BIBLIOGRAPHY

Comer and M. Wilkinson (eds.) *Sustaining Earth: Response to Environmental Threats*, Basingstoke: Macmillan.

Calderón, F. (ed.) (1986) *Los Movimientos Sociales ante la Crisis*, Buenos Aires: CLACSO/UNU.

Calderón, F. and Reyna, J.L. (1990) 'La irrupción incubierta', *David y Goliath* 57: 12–20.

Calderón, F., Piscitelli, A., and Reyna, J.L. (1992) 'Social movements: actors, theories, expectations', in A. Escobar and S. Alvarez (eds.) *The Making of Social Movements in Latin America: Identity, Strategy and Democracy*, Boulder, Colorado: Westview Press.

Caldwell, J. (1976) 'Towards a restatement of demographic transition theory', *Population and Development Review* 2.

Campbell, D. (1992) *Writing Security: U.S. Foreign Policy and the Politics of Identity*, Minneapolis, Minnesota: University of Minnesota Press.

Canguilhem, G. (1966) *Le Normal et le Pathologique*, Paris: Vrin.

CAPMAS (Central Agency for Public Mobilization and Statistics) and UNICEF (1988) *The State of Egyptian Children*, Cairo: CAPMAS.

Cardoso, F. H. and Faletto, E. (1979) *Dependency and Development in Latin America*, Berkeley and Los Angeles: University of California Press.

Cardoso, R.C.L. (1987) 'Movimientos sociais na América Latina', *Revista Brasileira de Ciências Sociais* 3, 1: 27–37.

Castel, R. (1991) 'From dangerousness to risk', in G. Burchell, C. Gordon, and P. Miller (eds.) *The Foucault Effect: Studies in Governmentality*, London: Harvester Wheatsheaf.

Castells, M. (1986) 'High technology, world development, and structural transformation: the trends and the debate', *Alternatives* 11, 3: 297–344.

Castoriadis, C. (1991) 'Reflections on rationality and development', in C. Castoriadis (ed. D.A. Curtis) *Philosophy, Politics, Autonomy*, New York: Oxford University Press.

CEGWSA (Commonwealth Expert Group on Women and Structural Adjustment) (1989) *Engendering Adjustment for the 1990s*, London: Commonwealth Secretariat.

Chadwick, O. (1957) *From Bossuet to Newman: The Idea of Doctrinal Development*, Cambridge: Cambridge University Press.

Chakrabarty, B. (1992) 'Jawaharlal Nehru and planning, 1938–1941', *Modern Asian Studies* 26, 2: 275–287.

Chambers, R. (1969[1844]) *Vestiges of the Natural History of Creation*, New York: Humanities Press.

Chandra, B. (1991) 'Colonial India: British versus Indian views of development', *Review* xiv, 1: 81–167.

Chatterjee, P. (1993) *The Nation and its Fragments: Colonial and Postcolonial Histories*, Princeton: Princeton University Press.

Chaudhuri, K.N. (1990) *Asia Before Europe: Economy and Civilisation of the Indian Ocean from the Rise of Islam to 1750*, Cambridge: Cambridge University Press.

Chesneaux, J. (1989) *La Modernité Monde*, Paris: Harmattan.

CIDA (Canadian International Development Agency/ Women in Development) (1987) 'WID Policy Framework', Ottawa: CIDA.

—— (1991) 'CIDA's Women in Development Program: Evaluation Assessment Report', Ottawa: CIDA.

—— (1992) 'WID Policy Framework', Ottawa: CIDA.

Claasz, G. (1979) 'Social factors in overseas aid projects', unpublished paper, presented to Sociological Association of Australia and New Zealand Conference, Canberra.

Clegg, S. (1976) 'Power, theorizing and nihilism', *Theory and Society* 3: 65–87.

283

Cleland, D. and King, W. (1968) *Systems Analysis and Project Management*, New York: McGraw-Hill.

Clifford, J. (1989) 'Notes on theory and travel', *Inscriptions* 5: 177–187.

Clifford, J. and Marcus, G. (1986) *Writing Culture: The Poetics and Politics of Ethnography*, Berkeley: University of California Press.

Clough, M.S. (1990) *Fighting Two Sides: Kenyan Chiefs and Politicians, 1918–1940*, Niwot, Colorado: University Press of Colorado.

Clough, P. and Williams, G. (1987) 'Decoding Berg: the World Bank in rural Nigeria', in M. Watts (ed.) *State, Oil and Agriculture in Nigeria*, Berkeley: University of California Press.

Coetzee, J.M. (1988) *White Writing: On the Culture of Letters in South Africa*, New Haven, Connecticut: Yale University Press.

Cohen, J. (1985) 'Strategy or identity: new theoretical paradigms and contemporary social movements', *Social Research* 52, 4.

Cohen, J. and Arato, A. (1992) *Civil Society and Political Theory*, Cambridge, Mass.: MIT Press.

Colclough, C. and Manor, J. (eds.) (1991) *States or Markets? Neo-Liberalism and the Development Policy Decade*, Oxford: Clarendon Press.

Collins, P. (1989) 'The social construction of black feminist thought', *SIGNS* 14, 4: 745–773.

Comaroff, J. and Comaroff, J. (1991) *Of Revelation and Revolution*, Chicago: University of Chicago Press.

Commander, S. (1987) *The State and Agricultural Development in Egypt Since 1973*, London: Ithaca Press, for the Overseas Development Institute.

Commoner, B. (1972) *The Closing Circle: Nature, Man and Technology*, New York: Knopf.

Comte, A. (1875[1851]) *System of Positive Polity*, vol.1, London: Longmans, Green, and Co.

Cooper, F. (1989) 'From free labor to family allowance', *American Ethnologist* 14, 4: 745–766.

Cooper, F. and Packard, R. (1992) 'Development knowledge and the social sciences', unpublished manuscript, University of Michigan, Department of History, Ann Arbor.

Cooper, F. and Stoler, A. (eds.) (1989) 'Tensions of empire: colonial control and visions of rule', *American Ethnologist* 16, 4: 609–621.

Coquery-Vidrovitch, C., Hemery, D., and Piel, J. (eds.) (1988) *Pour une Histoire du Développement*, Paris: Harmattan.

Corbridge, S. (1986) *Capitalist World Development: A Critique of Radical Development Geography*, London: Macmillan.

—— (1989) 'Marxism, post-Marxism and the geography of development', in R. Peet and N. Thrift (eds.) *New Models in Geography*, Volume 1, London: Unwin Hyman.

—— (1990) 'Post-marxism and development studies: beyond the impasse', *World Development* 18: 623–639.

—— (1991) 'Third World development', *Progress in Human Geography* 15, 3: 311–321.

—— (1993) 'Marxisms, modernities, and moralities: development praxis and the claims of distant strangers', *Society and Space* 11: 449–472.

Cordell, D. and Gregory, J. (eds.) (1987) *African Population and Capitalism: Historical Perspectives*, Boulder, Colorado: Westview Press.

Cornia, G.A., Jolly, R., and Stewart, F. (eds.) (1987) *Adjustment with a Human Face, Volume I: Protecting the Vulnerable and Promoting Growth*, Oxford: Oxford University Press.

Cosgrove, D. (1990) 'Environmental thought and action: pre-modern and post-modern', *Transactions of Institute of British Geographers* 15: 344–358.

Cotgrove, S. (1982) *Catastrophe or Cornucopia: The Environment, Politics and the Future*, Chichester: Wiley.

Coulson, A. (1978) 'Agricultural policies in mainland Tanzania', *Review of African Political Economy* 10: 74–100.

Cowen, M. (1981) 'Commodity production in Kenya's Central Province', in J. Heyer, P. Roberts and G. Williams (eds.) *Rural Development in Tropical Africa*, London: Macmillan.

Cowen, M. and Shenton, R. (1991) 'The origin and course of Fabian colonialism in Africa', *Journal of Historical Sociology* 4, 2: 143–174.

— — (1993) 'The theology of development', paper presented to the Historicizing Development Workshop, Emory University, Atlanta, December.

Crehan, K. and von Oppen, A. (1988) 'Understandings of "development": an arena of struggle', *Sociologia Ruralis* XXVIII, 2/3: 111–145.

Crummey, D. (ed.) (1986) *Banditry, Rebellion and Social Protest in Africa*, London: James Currey.

Crush, J. (1991) 'The discourse of progressive human geography', *Progress in Human Geography* 15, 4: 395–414.

— — (1994) 'Post-colonialism, de-colonization and geography', in A. Godlewska and N. Smith (eds.) *Geography and Empire*, Oxford: Blackwell.

Culwick, A.T. (1943) 'New beginning', *Tanganyika Notes and Records* 15: 1–6.

Cunningham, V. (1994) *In the Reading Gaol: Postmodernity, Texts, and History*, Oxford: Blackwell.

Cuny, F.C. (1983) *Disasters and Development*, Oxford: Oxford University Press.

Curtin, P. (1974) *The Image of Africa*, Madison: University of Wisconsin Press.

Cutter, S.L. (1993) *Living with Risk: The Geography of Technological Hazards*, London: Edward Arnold.

Dahlberg, K. (1992) 'Renewable resource systems and regimes: key missing links in global change studies', *Global Environmental Change* 2, 2: 128–152.

Dalby, S. (1992) 'Ecopolitical discourse: "environmental security" and political geography', *Progress in Human Geography* 16, 4: 503–522.

Daly, H. (1977) *Steady-State Economics: The Economics of Biophysical Equilibrium and Moral Growth*, New York: F.H. Freeman.

Dankelman, I. and Davidson, J. (1988) *Women and Environment in the Third World*, London: Earthscan.

Dasmann, R. (1979) 'Nature reserves in global perspective', in J. Nelson, R. Needham, S. Nelson, and R. Scace (eds.) *The Canadian National Parks: Today and Tomorrow, Conference II: Ten Years Later*, Waterloo: University of Waterloo Faculty of Environmental Studies.

Davis, I. (ed.) (1981) *Disasters and the Small Dwelling*, Oxford: Pergamon Press.

de Beauvoir, S. (1952) *The Second Sex*, trans. H. M. Parshley, New York: Bantam.

de Certeau, M. (1984) *The Practice of Everyday Life*, Berkeley: California University Press.

de Groot, J. (1991) 'Conceptions and misconceptions: the historical and cultural context of discussion on women and development', in H. Afshar (ed.) *Women, Development and Survival in the Third World*, London: Longman.

de Janvry, A., Sadoulet, E., and Thornbecke, E. (1992) 'State, market and civil organizations: new theories, new practices and their implications for rural development', unpublished paper, University of California, Berkeley, Department of Agricultural Economics.

de Villiers, M. (1987) *White Tribe Dreaming*, New York: Penguin.

de Waal, A. (1989) *Famine that Kills: Darfur, Sudan, 1984–1985*, Oxford: Clarendon Press.

de Wet, C. (1989) 'Betterment planning in a rural village in Keiskammahoek, Ciskei', *Journal of Southern African Studies* 15, 2: 326–345.

Dean, M. (1991) *The Constitution of Poverty*, London: Routledge.

Dear, M. (1989) 'Survey 16. Privatisation and the rhetoric of planning practice', *Environment and Planning D. Society and Space* 7: 449–462.

Defert, D. (1991) 'Popular life and insurance technology', in G. Burchell, C. Gordon, and P. Miller (eds.) *The Foucault Effect: Studies in Governmentality*, London: Harvester Wheatsheaf.

Deleuze, G. (1988) *Foucault*, Minneapolis: University of Minnesota Press.

Deleuze, G. and Guattari, F. (1977) *Anti-Oedipus: Capitalism and Schizophrenia*, New York: Viking Press.

—— (1987) *A Thousand Plateaus: Capitalism and Schizophrenia*, Minneapolis: University of Minnesota Press.

Derrida, J. (1976) *Of Grammatology*, Baltimore, Maryland: Johns Hopkins University Press.

—— (1978) *Writing and Difference*, London: Routledge & Kegan Paul.

Dethier, J-J. (1989) *Trade, Exchange Rate, and Agricultural Pricing Policies in Egypt*, Volume 1, *The Country Study*, World Bank Comparative Studies, the Political Economy of Agricultural Pricing Policy, Washington D.C.: World Bank.

Devall, B. and Sessions, G. (1985) *Deep Ecology: Living as if Nature Mattered*, Salt Lake City: Peregrine Smith.

Development Society of Southern Africa (1985) 'Background paper: aims and functions', Johannesburg: Development Society.

Dietz, J. and James, D. (eds.) (1990) *Progress Toward Development in Latin America*, Boulder, Colorado: Westview Press.

Dobby, E.G.H. (1950) *Southeast Asia*, London: London University Press.

Dreze, J. and Sen, A.K. (1990) *The Political Economy of Hunger*, Volumes 1–3, Oxford: Clarendon Press.

Drinkwater, M. (1989) 'Technical development and peasant impoverishment: land use policy in Zimbabwe's Midlands Province', *Journal of Southern African Studies* 15, 2: 287– 305.

Du Bois, M. (1991) 'The governance of the Third World: a Foucauldian perspective on power relations in development', *Alternatives* 16: 1–30.

Du Pisani, J.A., Lombard, J.A., Olivier, G.C., and Vosloo, W.B. (1980) *Alternatives to the Consolidation of kwaZulu: Progress Report*, Special Focus No. 2, Bureau for Economic Policy Analysis, University of Pretoria.

Du Toit, A. and Giliomee, H. (1983) *Afrikaner Political Thought, Volume 1 (1780–1850)*, Cape Town: David Philip.

Dubow, S. (1989) *Racial Segregation and the Origins of Apartheid in South Africa*, Basingstoke: Macmillan.

Dupuis, J. (1960) *Madras et le Nord du Coromandel: Etude des Conditions de la Vie Indienne dans un Cadre Géographique*, Paris: Adrien-Maison-Neuve.

Eagleton, T. (1990) 'Defending the free world', in R. Miliband and L. Panitch (eds.) *Socialist Register*, London: Merlin.

Eatwell, J., Milgate, M., and Newman, P. (eds.) (1989) *The New Palgrave: Economic Development*, London: Macmillan.

Economist (1993) 'The final lap: a survey of South Africa', 30 March: 3–26.

Edwards, M. (1989) 'The irrelevance of development studies', *Third World Quarterly* 11, 1: 116–135.

Ehrlich, P. (1972) *The Population Bomb*, London: Ballantine.

Ehrlich, P. and Ehrlich, A. (1970) *Population, Resources and Environment: Issues in Human Ecology*, New York: W.H. Freeman.

EIU (Economist Intelligence Unit) (1989) *Egypt: Country Report No. 4, 1989*, London: Economist Intelligence Unit.

—— (1993) *Country Profile 1993/4: Egypt*, London: Economist Intelligence Unit.

El-Rafie, M., Hassouna, W.A., Hirschhorn, N., Loza, S., Miller, P., Nagaty, A., Nasser, S., and Riyad, S. (1990) 'Effect of diarrhoeal disease control on infant and child mortality in Egypt', *The Lancet* 335: 334–338.

Elson, D. (1991) 'From survival strategies to transformation strategies: women's needs and structural adjustment', in L. Beneria and S. Feldman (eds.) *Economic Crises, Household Strategies and Women's Work*, Boulder, Colorado: Westview Press.

Emberley, J. (1995) 'The inner text of colonial violence: Anne Wheeler's film *Loyalties*', in M. Marchand and J. Parpart (eds.), *Feminism/Postmodernism/ Development*, London: Routledge.

Enzensberger, H. (1974) 'A critique of political ecology', *New Left Review* 8: 3–32.

ERA 2000 Inc. (1979) *Further Mechanization of Egyptian Agriculture*, New York: ERA 2000 Inc.

Erikson, K.T. (1976) *Everything in its Path: Destruction of Community in the Buffalo Creek Flood*, New York: Simon and Schuster.

Erskine, J. (1985) 'Rural development: putting theory into practice', in L. A. Van Wyk (ed.) *Development Perspectives in Southern Africa*, ABEN, Research Paper 85–1, Potchefstroom: Potchefstroom University for Christian Higher Education.

Escobar, A. (1984) 'Discourse and power in development: Michel Foucault and the relevance of his work to the Third World', *Alternatives* 10: 377–400.

—— (1988) 'Power and visibility: the invention and management of development in the Third World', *Cultural Anthropology* 4, 4: 428–443.

—— (1989) 'The professionalization and institutionalization of "development" in Colombia in the early post-World War II period', *International Journal of Educational Development* 9, 2: 139–154.

—— (1991) 'Anthropology and the development encounter', *American Ethnologist* 18, 4: 658–682.

—— (1992a) 'Reflections on "development"', *Futures* 24, 5: 411–436.

—— (1992b) 'Imagining a post-development era? critical thought, development and social movements', *Social Text*, 31/32: 20–54.

—— (1992c) 'Culture, economics and politics in Latin American social movements: theory and research', in A. Escobar and S. Alvarez (eds.) *The Making of Social Movements in Latin America: Identity, Strategy and Democracy*, Boulder, Colorado: Westview Press.

—— (1992d) 'Planning', in W. Sachs (ed.) *The Development Dictionary: A Guide to Knowledge as Power*, London: Zed Books.

—— (1995) *Encountering Development. The Making and Unmaking of the Third World*, Princeton: Princeton University Press.

Escobar, A. and Alvarez, S. (eds.) (1992) *The Making of Social Movements in Latin America: Identity, Strategy and Democracy*, Boulder, Colorado: Westview Press.

Esteva, G. (1987) 'Regenerating people's space', *Alternatives* 10, 3: 125–152.

—— (1992) 'Development', in W. Sachs (ed.) *The Development Dictionary: A Guide to Knowledge as Power*, London: Zed Books.

Evans, P. (1991) 'Predatory, developmental and other apparatuses: a comparative political economy perspective on the third world state', unpublished paper, Department of Sociology, University of California, Berkeley.

Fairclough, N. (1992) *Discourse and Social Change*, Cambridge: Polity Press.

Faithfull, G. (1982) 'Urban project planning and development policy', paper presented to the Symposium on Australian Development Practice in Asia and the Pacific: Experience and Issues, Planning Research Centre, Sydney.

Fals Borda, O. (1988) *Knowledge and People's Power*, Delhi: Indian Social Institute.

—— (1992) 'Social movements and political power in Latin America', in A. Escobar and S. Alvarez (eds.) *The Making of Social Movements in Latin America: Identity, Strategy and Democracy*, Boulder, Colorado: Westview Press.

Fals Borda, O. and Rahman, A. (eds.) (1991) *Action and Knowledge: Breaking the Monopoly with Participatory Action-Research*, New York: Apex Press.

Fanon, F. (1967) *Black Skin, White Masks*, New York: Grove Press.

—— (1968) *The Wretched of the Earth*, trans. C. Farrington, New York: Grove Press.

FAO (Food and Agriculture Organization of the United Nations) (1986) *Inter-Country Comparisons of Agricultural Production Aggregates*, FAO Economic and Social Development Paper 61, Rome: Food and Agricultural Organization.

—— (1988) *Yearbook 1987: Production*, volume 41, Rome: Food and Agriculture Organization.

—— (1989a) *Yearbook 1988: Production*, volume 42, Rome: Food and Agriculture Organization.

—— (1989b) *The State of Food and Agriculture*, Rome: Food and Agriculture Organization.

—— (1993) *Yearbook 1992: Production*, volume 46, Rome: Food and Agriculture Organization.

Farred, G. (1992) 'Unity and difference in black South Africa', *Social Text* 31/32: 217–234.

Feeley-Harnik, G. (1992) *Green Estate*, Washington, D.C.: Smithsonian Press.

Ferguson, J. (1985) 'The bovine mystique: power, property and livestock in rural Lesotho', *MAN* 20, 4: 647–674.

—— (1990) *The Anti-Politics Machine: 'Development', Depoliticisation and Bureaucratic Power in Lesotho*, Cambridge: Cambridge University Press.

Financial Mail (1979) 'Supplement on the urban foundation', 16 February.

Fisher, C.A. (1964) *Southeast Asia: A Social, Economic and Political Geography*, London: Methuen.

Fisher, J. (1954) *The Anatomy of Kikuyu Domesticity and Husbandry*, Nairobi and London: Department of Technical Cooperative.

Fisher, W.B. (1971) *The Middle East: A Physical, Social and Regional Geography*, London: Methuen.

Fiske, J. (1989) *Understanding Popular Culture*, Boston: Unwin.

Fitter, R. (1978) *The Penitent Butchers*, London: Collins.

Folch-Serra, M. (1989) 'Geography and postmodernism: linking humanism and development studies', *Canadian Geographer* 133, 1: 66–75.

Forbes, D.K. (1984) *The Geography of Underdevelopment*, Baltimore: Johns Hopkins University Press.

Foucault, M. (1965) *Madness and Civilisation*, trans. R. Howard, New York: Mentor Books.

—— (1970) *The Order of Things*, New York: Vintage Books.

—— (1972) *The Archaelogy of Knowledge and the Discourse on Language*, New York: Tavistock Publications.

—— (1979) 'What is an author?', in J. Harari (ed.) *Textual Strategies: Perspectives in Post-structuralist Criticism*, Ithaca, New York: Cornell University Press.

—— (1980a) *The History of Sexuality*, New York: Random House.

—— (1980b) *Power/Knowledge: Selected Interviews and Other Writings, 1972–1977*, C. Gordon (ed.), New York: Pantheon Books.

—— (1984) *Discipline and Punish: The Birth of the Prison*, Harmondsworth: Penguin.

—— (1985) *The Use of Pleasure*, New York: Vintage Books.
—— (1986) 'Two lectures', in *Power/Knowledge: Selected Interviews and Other Writings 1972–1977 by Michel Foucault*, C. Gordon (ed.) New York: Pantheon Books.
—— (1990) *The History of Sexuality: An Introduction*, New York: Vintage Books.
Fourie, P.C. and de Vos, T. (1986) *Bevordering van Totale Ontwikkeling in Landelike Gebiede van Suid-Africa*, Institute for Social and Economic Development, University of the Orange Free State, Bloemfontein.
Fowler, A. (1992) 'Distant obligations: speculations on NGO funding and the global market', *Review of African Political Economy* 55: 9–29.
Frank, A.G. (1978) *Dependent Accumulation and Underdevelopment*, London: Macmillan.
—— (1983) 'Introduction' and 'Crisis and transformation of dependency in the world system', in R.H. Chilcote and D.L. Johnson (eds.) *Theories of Development: Mode of Production or Dependency*, Beverly Hills: Sage.
Fraser, N. (1989) *Unruly Practices: Power, Discourse and Gender in Contemporary Social Theory*, Minneapolis: University of Minnesota Press.
Freire, P. (1972) *Pedagogy of the Oppressed*, Harmondsworth: Penguin.
—— (1975) *Cultural Action for Freedom*, Harmondsworth: Penguin.
Friberg, M. and Hettne, B. (1985), 'The greening of the world: towards a non-deterministic model of global processes', in H. Addo *et al.* (eds.) *Development as Social Transformation: Reflections on the Global Problematique*, Sevenoaks: Hodder and Stoughton.
Friedmann, H. (1993) 'The political economy of food: a global crisis', *New Left Review* January/February: 29–57.
Friedmann, J. and Alonso, W. (eds.) (1964) *Regional Development and Planning: A Reader*, Cambridge, Mass.: MIT Press.
Frisby, D. (1992) *Simmel and Since*, London: Routledge.
Fromm, E. (1966) *Marx's Concept of Man*, New York: Frederick Ungar Publishing Company.
Fry, G. and Martin, G. (1991) *The International Development Dictionary*, Santa Barbara: ABC-CLIO.
Fryer, D.W. (1970) *Emerging Southeast Asia*, London: Philips and Sons.
Fuenzalida, E. (1983) 'The reception of 'scientific sociology' in Chile', *Latin American Research Review* 18, 1: 95–112.
—— (1987) 'La reorganización de las Instituciones de Enseñanza Superior e de Investigación en América Latina entre 1950 y 1980 sus Interpretaciones', *Estudios Sociales* 52: 115–137.
Fukuyama, F. (1989) 'The end of history?', *The National Interest* Summer: 3–18.
—— (1992) *The End of History and the Last Man*, New York: Free Press.
Galâl, Dînâ (1988) *Al-ma'ûna al-amrîkîya li-man: Misr am Amrîkâ*, Cairo: al-Ahram.
Galois, B. (1976) 'Ideology and the idea of nature: the case of Peter Kropotkin', *Antipode* 8: 1–16.
García, M., del Pilar (1992) 'The Venezuelan ecology movement: symbolic effectiveness, social practice, and political strategies', in A. Escobar and S. Alvarez (eds.) *The Making of Social Movements in Latin America: Identity, Strategy and Democracy*, Boulder, Colorado: Westview Press.
Geertz, C. (1963) *Peddlers and Princes: Social Change and Economic Modernization in Two Indonesian Towns*, Chicago: University of Chicago Press.
—— (1988a) *Works and Lives: The Anthropologist as Author*, Stanford: Stanford University Press.
—— (1988b) 'Being there, writing here', *Harpers*, March: 32–38.

Geggus, C. (1986) *University Training and Career Opportunities 1985*, Guidance Series GS-5, Pretoria: Human Sciences Research Council.

Gendreau, F., Meillasoux, C., Schlemmer, B., and Verlet, M. (eds.) (1991) *Les Spectres Malthus: Déséquilibres Alimentaires, Déséquilibres Démographiques*, Paris: ED, ORSTOM, CEPED.

Gendzier, I. (1985) *Managing Political Change: Social Scientists and the Third World*, Boulder, Colorado: Westview Press.

George, E. (1984) *Ill Fares the Land: Essays on Food, Hunger and Power*, New York: Penguin Books.

George, S. (1990) 'Conscience "planétaire" et "trop nombreux" pauvres', *Le Monde Diplomatique* 434: 18–19.

George, S. and Sabelli, F. (1994) *Faith and Credit: The World Bank's Secular Empire*, Boulder: Westview Press.

Georgescu-Roegen, N. (1971) *The Entropy Law and the Economic Process*, Cambridge, Mass.

—— (1976) *Energy and Economic Myths*, New York.

Gerschenkron, A. (1992) [1952] 'Economic backwardness in historical perspective', reprinted in M. Granovetter and R. Swedberg (eds.) *The Sociology of Economic Life*, Boulder, Colorado: Westview Press.

Giddens, A. (1979) *Central Problems in Social Theory: Action, Structure and Contradiction in Social Analysis*, London: Macmillan.

—— (1984) *The Constitution of Society*, Berkeley: University of California Press.

—— (1990) *The Consequences of Modernity*, Stanford: Stanford University Press.

Giliomee, H. (1982) *The Parting of the Ways: South African Politics 1976–1982*, Cape Town: David Philip.

Gillespie, K. (1989) 'The key to unlocking sustainable development', mimeo, Washington, D.C.: World Bank.

Gilligan, C. (1982) *In A Different Voice*, Cambridge, Mass.: Harvard University Press.

Ginsburg, N. (ed.) (1958) *The Pattern of Asia*, Englewood Cliffs, New Jersey: Prentice Hall.

—— (ed.) (1960) 'Essays on geography and economic development', Research Paper No. 62, Department of Geography, University of Chicago, Chicago.

—— (1990) *The Urban Transition: Reflections on the American and Asian Experiences*, Hong Kong: Chinese University Press.

Godlewska, A. and Smith, N. (eds.) (1994) *Geography and Empire*, Oxford: Blackwell.

Goodland, R. and Ledec, G. (1984) *Neoclassical Economics and the Principles of Sustainable Development*, Washington, D.C.: World Bank Office of Environmental Affairs.

Gottman, J. (1961) *Megalopolis: The Urbanized Northeastern Seaboard of the United States*, New York: Kraus International Publications.

Goulet, D. (1971) *The Cruel Choice: A New Concept in the Theory of Development*, New York: Atheneum.

Goureau, P. (1953) *L'Asie*, Paris: Hachette.

Graber, L. (1976) *Wilderness as Sacred Space*, Washington, D.C.: Association of American Geographers.

Gramsci, A. (1971) *Selections from Prison Notebooks*, London: Lawrence and Wishart.

Gran, P. (1978) 'Modern trends in Egyptian historiography: a review article', *International Journal of Middle East Studies* 9, 3.

Greenberg, S. (1987) *Legitimating the Illegitimate: State, Markets and Resistance in South Africa*, New Haven, Connecticut: Yale University Press.

Gregory, D. (1990) 'Chinatown, Part Three? Soja and the missing spaces of social theory', *Strategies* 3: 40–104.

BIBLIOGRAPHY

—— (1994) *Geographical Imaginations*, Oxford: Blackwell.

Gronemeyer, M. (1992) 'Helping', in W. Sachs (ed.) *The Development Dictionary: A Guide to Knowledge as Power*, London: Zed Books.

Grove, R. (1987) 'Early themes in African conservation: the Cape in the nineteenth century', in R. Grove and D. Anderson (eds.) *Conservation in Africa: People, Policies and Practice*, Cambridge: Cambridge University Press.

—— (1989) 'Scottish missionaries, evangelical discourses and the origins of conservation thinking in Southern Africa 1820–1900', *Journal of Southern African Studies* 15, 2: 163–187.

—— (1990a) 'Threatened islands, threatened earth: early professional science and the historical origins of global environmental concerns', in D. Angell, J. Comer and M. Wilkinson (eds.) *Sustaining Earth: Responses to Environmental Threats*, London: Macmillan.

—— (1990b) 'The origins of environmentalism', *Nature* 345, 6270: 11–14.

—— (1990c) 'Colonial conservation, ecological hegemony and popular resistance: towards a global synthesis' in J. Mackenzie (ed.) *Imperialism and the Natural World*, Manchester: Manchester University Press.

Guha, R. (1983) *Elementary Aspects of Peasant Insurgency in Colonial India*, Delhi: Oxford University Press.

—— (1988) 'On some aspects of the historiography of colonial India', in R. Guha and G. Spivak (eds.) *Selected Subaltern Studies*, Delhi: Oxford University Press.

—— (1989) 'Dominance without hegemony and its historiography', in R. Guha (ed.) *Subaltern Studies*, Delhi: Oxford University Press.

Gutierrez, G. (1988) *A Theology of Liberation: History, Politics, and Salvation*, Maryknoll, New York: Orbis Books.

Guyer, J. (1990) 'Food regulation in Britain and Nigeria', unpublished manuscript, Department of Anthropology, Johns Hopkins University.

Haas, P. (1989) 'Do regimes matter? Epistemic communities and Mediterranean pollution control', *International Organization* 43, 4: 377–403.

Hadjor, K. (1993) *Dictionary of Third World Terms*, London: Penguin.

Haggard, S. (1990) *Pathways from the Periphery: The Politics of Growth in Newly Industrializing Countries*, Ithaca: Cornell University Press.

Halisi, C.R.D. (1992) 'Biko and black consciousness philosophy: an interpretation', in B. Pityana *et al.* (eds.) *Bounds of Possibility: The Legacy of Steve Biko and Black Consciousness*, London: Zed Books.

Halliday, F. (1991) 'An encounter with Fukuyama', *New Left Review* 193: 89–95.

Hannerz, U. (1990) 'Cosmospolitans and locals in world culture', in M. Featherstone (ed.) *Global Culture: Nationalism, Globalization and Modernity*, London: Sage.

Haraway, D. (1989) *Primate Visions: Gender, Race and Nature in the World of Modern Science*, London: Routledge.

—— (1992) 'Promises of monsters', in L. Grossberg, C. Nelson, and P. Treichler (eds.) *Cultural Studies*, London: Routledge.

Harding, S. (1987) *The Science Question in Feminism*, Milton Keynes: Open University Press.

—— (1992) 'Subjectivity, experience and knowledge', *Development and Change* 23, 3: 175–194.

Harik, I. (1977) 'Decentralization and development in rural Egypt: a description and assessment', mimeo, prepared for USAID, Egypt.

Harley, J. B. (1992) 'Deconstructing the map', in T. Barnes and J. Duncan (eds.) *Writing Worlds: Discourse, Text and Metaphor in the Representation of Landscape*, London: Routledge.

Harris, G. (1989) *The Sociology of Development*, London: Longman.

Harrison, R. and Livingstone, D. (1982) 'Understanding in geography: structuring the subjective', in D. Herbert and R. Johnston (eds.) *Geography and the Urban Environment: Progress in Research and Applications*, volume V, London: John Wiley and Sons.

Hart, G. and Watts, M. (1992) 'Recycling populisms in development discourse', unpublished manuscript, University of California, Berkeley, City and Regional Planning Department.

Hart, K. (1973) 'Informal income opportunities and urban employment in Ghana', *The Journal of Modern African Studies* 11: 61–89.

Hartmann, B. (1987) *Reproductive Rights and Wrongs: The Global Politics of Population Control and Contraceptive Choice*, New York: Harper and Row.

Hartmann, B. and Boyce, J.K. (1983) *A Quiet Violence: View from a Bangladesh Village*, London: Zed Books.

Harvey, D. (1974) 'Population, resources and the ideology of science', *Economic Geography* 50: 256–277.

—— (1985) *The Urbanization of Capital*, Baltimore: Johns Hopkins University Press.

—— (1989) *The Condition of Postmodernity: An Enquiry into the Origins of Cultural Change*, Oxford: Basil Blackwell.

—— (1993) 'From place to space and back again', in J. Bird *et al.* (eds.) *Mapping the Futures: Local Cultures, Global Change*, London: Routledge.

Haynes, D. and Prakash, G. (1992) *Contesting Power: Resistance and Everyday Social Relations in South Asia*, Berkeley: University of California Press.

Hays, S. (1959) *Conservation and the Gospel of Efficiency: The Progressive Conservation Movement 1890–1920*, Cambridge, Mass.: Harvard University Press.

Hepple, L. (1992) 'Metaphor, geopolitical discourse and the military in South America', in J. Barnes and J. Duncan (eds.) *Writing Worlds: Discourse, Text and Metaphor in the Representation of Landscape*, London: Routledge.

Herbert, C. (1991) *Culture and Anomie*, Chicago: University of Chicago Press.

Hettne, B. (1990) *Development Theory and the Three Worlds*, London: Methuen.

Hewison R. (1987) *The Heritage Industry*, Andover: Methuen.

Hewitt, K. (1983) 'The idea of calamity in a technocratic age', in K. Hewitt (ed.) *Interpretations of Calamity: From the Viewpoint of Human Ecology*, London: Allen and Unwin.

—— (1987) 'Risks and emergencies in Canada: a national overview', *Ontario Geography* 29: 1–36.

—— (1992) 'Mountain hazards', *Geojournal* 27: 47–60.

—— (1994a) '"When the great planes came and made ashes of our city...": towards an oral geography of the disasters of war', *Antipode* 26, 1: 1–34.

—— (1994b) 'Hidden damages, shadow risks: making the social space of disasters visible', in *Proceedings: Seminario Internacional Sociedad y Prevención de Desastres*, Coordinación de Humanidades, Autonomous University, Mexico City.

Heyer, J. (1981) 'Agricultural development in Kenya from the colonial period to 1975', in J. Heyer, P. Roberts, and G. Williams (eds.) *Rural Development in Tropical Africa*, London: Macmillan.

Heyer, J., Maitha, J., and Senga, W. (eds.) (1976) *Agricultural Development in Kenya: an Economic Assessment*, Nairobi: Oxford University Press.

Heyer, J., Roberts, P., and Williams, G. (eds.) (1981) *Rural Development in Tropical Africa*, London: Macmillan.

Hill, P. (1986) *Development Economics on Trial: The Anthropological Case for a Prosecution*, Cambridge: Cambridge University Press.

Hinnebusch, R.A. (1993) 'Class, state and the reversal of Egypt's agrarian reform', *Middle East Report*, 184: 20–23.

Hirschman, A.O. (1958) *The Strategy of Economic Development*, New Haven, Connecticut: Yale University Press.

—— (1967) *Development Projects Observed*, Washington D.C.: Brookings Institution.

—— (1981) *Essays in Trespassing*, Cambridge: Cambridge University Press.

Hobart, M. (ed.) (1993) *An Anthropological Critique of Development*, London: Routledge.

Hobbes, T. (1962) *Leviathan: Or the Matter, Forme and Power of a Commonwealth Ecclesiasticall and Civil*, New York: Collier.

Hobsbawm, E.J. (1968) *Industry and Empire*, London: Weidenfeld and Nicolson.

—— (1988) *The Age of Revolution: Europe 1789–1848*, London: Cardinal.

Holden, C. (1987) 'World Bank launches new environmental policy', *Science* 236: 769.

Hont, I. (1983) 'The "rich country-poor country" debate', in I. Hont and M. Ignatieff, *Wealth and Virtue: the shaping of political economy in the Scottish Enlightenment*, Cambridge: Cambridge University Press.

hooks, b. (1984) *Feminist Theory: from Margin to Center*, Boston, Mass.: South End Press.

—— (1991) *Yearning: Race, Gender, and Cultural Politics*, Boston: South End Press.

Hopkins, D. (1992) 'Steve Biko, black consciousness, and black theology', in B. Pityana *et al.* (eds.) *Bounds of Possibility: The Legacy of Steve Biko and Black Consciousness*, London: Zed Books.

Hopper, W. D. (1957) 'The economic organization of a village in North Central India', unpublished Ph.D. thesis, Cornell University.

Horesh, E. (1985) 'Labelling and the language of international development', *Development and Change* 16: 503–514.

Hoselitz, B. (1952) 'Non-economic barriers to economic development', *Economic Development and Cultural Change* 1: 8–21.

Hosle, V. (1991) 'The Third World as a philosophical problem', *Social Research* 59, 2: 227–262.

HSRC (Human Sciences Research Council) (1976) *Institutes of Higher Learning in South Africa*, Pretoria: Human Sciences Research Council.

Huang, P-W. (1975) *The Asian Development Bank: Diplomacy and Development in Asia*, New York: Vantage Press.

Hugo, G.F. (1975) 'Population mobility in West Java', unpublished Ph.D. thesis, Australian National University.

Hugo, P. (1988) 'Towards darkness and death: racial demonology in South Africa', *Journal of Modern African Studies* 24, 4: 567–590.

Hunt, D. (1989) *Economic Theories of Development: An Analysis of Competing Paradigms*, New York: Simon and Schuster.

Husayn, 'Adil (1985) *Nahw fikr 'arabî jadîd: al-Nâsirîya wa-l-tanmîya wa-l-dîmuqrâtîya*, Cairo: Dar-al Mustaqbal al-'arabî.

Ibrahim, S.A. (1979) 'Social mobility and income distribution', Egyptian Income Distribution Research Project, No. 4, mimeo.

IDS (1992) *Institute of Development Studies Bulletin*, special issue on flexible specialization in the Third World.

IGBP (International Geosphere-Biosphere Programme) (1992) *Global Change: Reducing Uncertainties*, Sweden: International Geosphere-Biosphere Programme, International Council of Scientific Unions.

Iggers, G. (trans.) (1972[1829]) *The Doctrine of Saint-Simon: An Exposition, First Year 1828–1829*, New York: Schocken Books.

Ikram, K. (1980) *Egypt: Economic Mangement in a Period of Transition*, the report of a mission sent to the Arab Republic of Egypt by the World Bank, Baltimore and London: Johns Hopkins University Press, for the World Bank.

Illich, I. (1973a) *Celebration of Awareness*, Harmondsworth: Penguin.

—— (1973b) *Tools for Conviviality*, London: Calder and Boyers.

—— (1977) *Toward a History of Needs*, Berkeley: Heyday Books.

—— (1992) 'Needs', in W. Sachs (ed.) *The Development Dictionary: A Guide to Knowledge as Power*, London: Zed Books.

ILO (International Labor Organization) (1977) *Employment, Growth and Basic Needs*, Geneva: International Labor Organization.

Inikori, J. (ed.) (1982) *Forced Migration: The Impact of the Export Slave Trade on African Societies*, London: Hutchinson.

Innes, D. (1987) 'Privatisation: the solution?' in *South African Review 4*, Johannesburg: Ravan Press.

International Finance Corporation (1983) 'Report and recommendation of the President to the Board of Directors on a proposed investment in Delta Sugar Company S.A.E., Arab Republic of Egypt', Washington D.C.: International Finance Corporation.

IUCN (1980) *The World Conservation Strategy*, Geneva: International Union for Conservation of Nature and Natural Resources, United Nations Environment Programme, World Wide Fund for Nature.

—— (1991) *Caring for the Earth: A Strategy for Sustainable Living*, Geneva: International Union for Conservation of Nature and Natural Resources, United Nations Environment Programme, World Wide Fund for Nature.

Jameson, F. (1983) 'Postmodernism and the consumer society', in H. Foster (ed.) *The Anti-Aesthetic: Essays on Postmodern Culture*, Port Townsend, Washington: Bay Press.

Jelin, E. (1986) 'Otros silencios, otras voces: el tiempo del la democratización en Argentina', in F. Calderón (ed.) *Los Movimientos Sociales ante la Crisis*, Buenos Aires: CLACSO/UNU.

Johnson, P. R. *et al.* (1983) *Egypt: The Egyptian American Rural Improvement Service, A Point Four Project, 1952–63*, AID Project Evaluation No. 43, Washington D.C.: Agency for International Development.

Johnston, H. (1895) 'The British Central Africa Protectorate' *Geographical Journal* 5, 3: 193–217.

Jones, G. (1936) *The Earth Goddess: A Study of Native Farming in the West African Context*, London: Longman, Green and Co.

Juma, C. (1990) *The Gene Hunters*, London: Zed Books.

Kabeer, N. (1991) 'Rethinking development from a gender perspective: some insights from the decade', paper presented at the Conference on Women and Gender in Southern Africa, University of Natal, Durban.

Kandiyoti, D. (1990) 'Women and rural development policies: the changing agenda', *Development and Change* 21, 1: 5–22.

Karp, I. (1991) 'Development and personhood', paper presented to the Institute for Advanced Study and Research in the African Humanities, Northwestern University, Chicago.

Katz, C. and Kirby, A. (1991) 'In the nature of things: the environment in everyday life', *Transactions of Institute of British Geographers* 16: 259–271.

Kay, C. (1993) 'For a renewal of development studies: Latin American theories and neoliberalism in the era of structural adjustment', *Third World Quarterly* 14, 4: 691–702.

Kelley, A.C. (1991) 'The human development index', *Population and Development Review* 17, 2: 315–324.

Kelley, A.C., Khalifa, A.M., and el-Khorazaty, M. Nabil (1982) *Population and Development in Rural Egypt*, Durham, North Carolina: Duke University Press.

Kent, G. (1984) *The Political Economy of Hunger: The Silent Holocaust*, New York: Praeger.

Kenya (1929) *Native Land Tenure in Kikuyu Province*, Report of Committee and Appendix. Chaired by G.V. Maxwell, Nairobi.

Kenya (Fort Hall District) (1948) *Annual Report*, Nairobi.

Kenya (Colony and Protectorate of) (1954) *A Plan to Intensify the Development of African Agriculture in Kenya*, Nairobi.

Kershaw, G. (1975–1976) 'The changing role of men and women in the Kikuyu family by socio-economic strata', *Rural Africana* 19: 173–194.

Kilby, P. and Porter, D. (1992) 'Governance, sustainability and aid', *Development Bulletin* October: 7–10.

Kirby, A. (ed.) (1990) *Nothing to Fear: Risks and Hazards in American Society*, Tuscon: University of Arizona Press.

—— (1994) *Power/Resistance: Local Politics and the Chaotic State*, Bloomington: Indiana University Press.

Kitching, G. (1980) *Class and Economic Change in Kenya: The Making of an African Petite Bourgeoisie*, New Haven, Connecticut: Yale University Press.

—— (1982) *Development and Underdevelopment in Historical Perspective: Populism, Nationalism, Industrialism*, London: Methuen.

Kjekshus, H. (1977) *Ecology Control and Economic Development in East African History: the Case of Tanganyika 1850–1950*, London: Heinemann.

Klitgaard, R. (1990) *Tropical Gangsters*, New York: Basic Books.

Koestler, A. (1969) 'Afterthoughts', in A. Koestler and J.R. Smythies (eds.) *Beyond Reductionism: New Perspectives in the Life Sciences*, London: Hutchinson.

Kohák, E. (1984) *The Embers and the Stars: A Philosophical Inquiry into the Moral Sense of Nature*, Chicago: University of Chicago Press.

Koponen, J. (1986) 'Population growth in historical perspective – the key role of changing fertility', in J. Boesen *et al.* (eds.) *Tanzania: Crisis and Struggle for Survival*, Uppsala: Scandinavian Institute for African Studies.

Kothari, R. (1987) 'On humane governance', *Alternatives* 12, 3: 277–290.

—— (1988) *Rethinking Development: In Search of Humane Alternatives*, Delhi: Ajanta.

Kropotkin, P. (1972) *The Conquest of Bread*, London: Allen Lane.

—— (1974) *Fields, Factories and Workshops Tomorrow*, London: Allen Lane.

Kuczynski, R. (1948, 1949) *Demographic Survey of the British Colonial Empire*, volumes I and II, Oxford: Oxford University Press.

La Touche, S. (1992) 'Standard of living', in W. Sachs (ed.) *The Development Dictionary: A Guide to Knowledge as Power*, London: Zed Books.

Laclau, E. (1977) *Politics and Ideology in Marxist Theory*, London: Verso.

Laclau, E. and Mouffe, C. (1985) *Hegemony and Socialist Strategy*, London: Verso.

Laïdi, Z. (1989) *Enquête sur la Banque Mondiale*, Paris: Fayard.

Lansing, S. (1991) *Priests and Programmers: Technologies of Power and the Engineered Landscape of Bali*, Princeton: Princeton University Press.

Lasaga, R.J. (1973) 'Geography and the geographers in the changing Pacific', in H. C. Brookfield (ed.) *The Pacific in Transition*, London: Arnold.

Lasch, C. (1991) *The True and Only Heaven*, New York: Norton.

Lauren, P.G. (1988) *Power and Prejudice: The Politics and Diplomacy of Racial Discrimination*, Boulder, Colorado and London: Westview Press.

Lavie, S. (1990) *The Poetics of Military Occupation*, Berkeley: University of California Press.

Lazreg, M. (1988) 'Feminism and difference: the perils of writing as a woman on women in Algeria', *Feminist Studies* 14, 1: 81–107.

Lechner, N. (1988) *Los Patios Interiores de la Democracia: Subjectividad y Política*, Santiago: FLACSO.

Lee, E. (1979) 'Egalitarian peasant farming and rural development: the case of South Korea', *World Development* 7: 493–517.

Lehmann, D. (1990) *Democracy and Development in Latin America*, Philadelphia: Temple University Press.

Leinbach, T. (1972) 'The spread of modernization in Malaya 1885–1969', *Tidschrift voor Economische en Sociale Geografie* 63: 263–277.

Leipziger, D.M. and Streeten, P. (eds.) (1981) *Basic Needs to Development*, Cambridge, Mass.: Oelgeschlager, Gunn and Hain Publishers.

Lélé, S. (1991) 'Sustainable development: a critical review,' *World Development* 19: 607–621.

Lenzer, G. (ed.) (1983) *Auguste Comte and Positivism: The Essential Writings*, Chicago: University of Chicago Press.

Leo, C. (1984) *Land and Class in Kenya*, Toronto: Toronto University Press.

Lewis, M. (1992) *Green Delusions: An Environmentalist Critique of Radical Environmentalism*, Durham: Duke University Press.

Lewsen, P. (ed.) (1988) *Voices of Protest: From Segregation to Apartheid 1938–1948*, Craighall, South Africa: A.D. Donker.

Leys, C. (1994) 'Confronting the African tragedy', *New Left Review* 204: 33–47.

—— (forthcoming) *The Rise and Fall of Development Theory*, London: James Currey.

Lipietz, A. (1989) 'The debt problem, European integration and the new phase of the world crisis', *New Left Review* 178: 37–50.

Lipton, M. and Longhurst, R. (1989) *New Seeds and Poor People*, London: Allen and Unwin.

List, F. (1856) *The National System of Political Economy*, Philadelphia: J.B. Lippincott.

—— (1991[1885]) (Sampson S. Lloyd, trans.) *The National System of Political Economy*, New York: Augustus M. Kelly.

Lithgelm, A.A. and Coetzee, S.F. (1984) 'An appropriate development strategy for Southern Africa', *Development Southern Africa* 1, 1: 6–35.

Lithgelm, A.A. and Van Wyk, L. (1985) 'The policy implications of an appropriate development strategy in Southern Africa', *Development Southern Africa* 2, 3: 324–345.

Livingstone, D. (1992) *The Geographical Tradition: Episodes in the History of a Contested Enterprise*, Oxford: Blackwell.

Lloyd, C. (1979) 'The importance of rural development in the defence strategy of South Africa and the need for private sector involvement', paper presented at a workshop on the Urbanisation Process in Natal and KwaZulu and the Need for a Total Development Strategy, Urban Foundation, Durban.

Locke, J. (1960) *Two Treatises of Government*, Cambridge: Cambridge University Press.

Lockwood, M. (1989) 'Fertility and labour in Rufiji District, Tanzania', unpublished Ph.D. thesis, University of Oxford.

Lombard, J.A. (1971) 'Political and adminstrative principles of homeland development', *BANTU*, February.

—— (1978) *Freedom, Welfare and Order: Thoughts on the Principles of Political Co-operation in the Economy of Southern Africa*, Pretoria: BENBO.

Long, N. and Long, A. (eds.) (1992) *Battlefields of Knowledge: The Interlocking of Theory and Practice in Social Research and Development*, London, Routledge.

López Maya, M. (ed.) (1991) *Pensamiento Crítico. Un Diálogo Interregional. 3. Desarollo y Democracia*, Caracas: Universidad Central/Nueva Sociedad.

Lorde, A. (1981) 'The master's tools will never dismantle the master's house', in C. Moraga and G. Anzaldua (eds.) *This Bridge Called My Back*, New York: Kitchen Table Press.

—— (1984) *Sister Outsider*, Freedom, California: Crossing Press.

Louis Berger International Inc. (1985) *Agricultural Mechanization Project: Final Report*, East Orange, New Jersey: Louis Berger International Inc.

Louw, L. (1986) 'Ciskei's economic reforms: correcting the critics', *Indicator South Africa*, Rural and Regional Monitor 3, 4: 18–21.

Love, N. (1989) 'Foucault and Habermas on discourse and democracy', *Polity* 22, 2: 269–293.

Low, D. and Lonsdale, J. (1976) 'Introduction: towards a new order 1945–1963', in D. Low and A. Smith (eds.) *History of East Africa*, Volume III, Oxford: Clarendon Press.

Lowe, L. (1991) *Contested Terrains: British and French Orientalisms*, Berkeley: University of California Press.

Ludden, D. (1992) 'India's development regime', in N. Dirks (ed.) *Colonialism and Culture*, Chicago: University of Chicago Press.

Lummis, D.C. (1991) 'Development against democracy', *Alternatives* 16: 31–66.

Macdonald, J.R.L. (1973) [1897] *Soldiering and Surveying in British East Africa 1891–1894*, Folkestone and London: Dawson of Pall Mall.

Macdonnell, D. (1986) *Theories of Discourse: An Introduction*, Oxford: Basil Blackwell.

Mackenzie, F. (1986) 'Land and labour: women and men in agricultural change, Murang'a District, Kenya, 1880–1984', unpublished Ph.D. thesis, University of Ottawa.

—— (1990) 'Gender and land rights in Murang'a District, Kenya', *Journal of Peasant Studies* 17, 4: 609– 643.

—— (1991) 'Political economy of the environment, gender and resistance under colonialism: Murang'a District, Kenya, 1910–1950', *Canadian Journal of African Studies* 25, 2: 226–256.

Mackenzie, J. (1987) 'Chivalry, social Darwinism and ritualised killing: the hunting ethos in Central Africa up to 1914', in D. Anderson and R. Grove (eds.) *Conservation in Africa: People, Policies and Practice*, Cambridge: Cambridge University Press.

—— (1989) *The Empire of Nature: Hunting, Conservation and British Imperialism*, Manchester: Manchester University Press.

—— (1990) *Imperialism and the Natural World*, Manchester: Manchester University Press.

Maguire, P. (1984) 'Women in development: an alternative analysis', Amherst, Massachussetts: Center for International Education, University of Massachusetts.

Malan, M. (1980) 'Speech by General Malan to the Institute for Strategic Studies'. (trans.), University of Pretoria, 3 September 1980, duplicated by the Institute of Strategic Studies, University of Pretoria.

Malthus, T. (1986[1798]) *An Essay on the Principle of Population*, Harmondsworth: Penguin.

Mandala, E. (1983) 'Capitalism, ecology and society: the Lower Tchiri Valley of Malawi, 1860–1960', unpublished Ph.D. thesis, University of Minnesota.

Mani, L. (1989) 'Multiple mediations: feminist scholarship in the age of multinational reception', *Inscriptions* 5: 1–24.

Mannoni, O. (1956) *Prospero and Caliban: The Psychology of Cannibalism*, London: Methuen.

Manzo, K. (1991) 'Modernist discourse and the crisis of development theory', *Studies in Comparative International Development* 26, 2: 3–36.

—— (1992a) 'Global power and South African politics: a Foucauldian analysis', *Alternatives* 17: 23–66.

—— (1992b) *Domination, Resistance, and Social Change in South Africa: The Local Effects of Global Power*, New York: Praeger.

—— (1992c) 'The limits of Liberalism', *Transition* 55: 115–124.

Marglin, S. (1990) 'Towards the decolonisation of the mind', in F. Appfel Marglin and S. Marglin (eds.) *Dominating Knowledge: Development, Culture and Resistance*, Oxford: Clarendon Press.

Marx, K. (1968[1848]) *The Communist Manifesto*, New York: Washington Square Press.

—— (1973) *Grundrisse*, Baltimore: Penguin.

Marx, K. and Engels, F. (1969) *Selected Works in Three Volumes*, Moscow: Progress Publishers.

Mathur, G. (1989) 'The current impasse in development thinking: the metaphysics of power', *Alternatives* 14: 463–479.

Matless, D. (1991) 'Nature, the modern and the mystic: tales from early twentieth century geography', *Transactions of Institute of British Geographers* 16: 272–286.

Maturana, H. and Varela, F. (1980) *Autopoiesis and Cognition: the Realization of the Living*, Boston, Mass.: D. Reidel.

—— (1987) *The Tree of Knowledge: the Biological Roots of Human Understanding*, Boston, Mass.: Shambhala.

McAllister, P. (1989) 'Resistance to "betterment" in the Transkei: a case study from Willowvale district', *Journal of Southern African Studies* 15, 2: 346–368.

McCarney, P. (1991) 'The life of ideas in donor agencies', paper presented at Special Session on Discourse and Development, Canadian Association of Geographers Annual Meetings, Kingston, Ontario.

McClintock, A. (1992) 'The angel of progress: pitfalls of the term "postcolonialism", *Social Text* 31/32: 84–98.

McCloskey, D. (1985) *The Rhetoric of Economics*, Madison: University of Wisconsin Press.

—— (1990) *If You're So Smart: The Narrative of Economic Expertise*, Chicago and London: University of Chicago Press.

McCormick, J. (1989) *Reclaiming Paradise: The Global Environmental Movement*, Bloomington, Ind.: Indiana University Press.

McGee, T.G. (1967) *The Southeast Asian City*, London: G. Bell & Sons.

—— (1969) 'Malays in Kuala Lumpur City: a geographic study of the process of urbanisation', unpublished Ph.D. thesis, Victoria University of Wellington, New Zealand.

—— (1971) *The Urbanisation Process in the Third World: Explorations in Search of a Theory*, London: G. Bell & Sons.

—— (1974a) *Hawkers in Hong Kong: A Study of Policy and Planning in the Third World City*, Hong Kong: Centre for Asian Studies.

—— (1974b) 'In praise of tradition: towards a geography of anti-development', *Antipode* 6: 30–47.

—— (1978) 'An invitation to the ball, dress "formal" or "informal"?', in P. Rimmer, D. Drakakis-Smith and T.G. McGee (eds.) *Studies in Food, Shelter and Transport in the Third World: Challenging the Conventional Wisdom*, Canberra: Department of Geography, Australian National University.

—— (1991) 'The emergence of Desakota regions in Asia: expanding a hypothesis', in N. Ginsburg, B. Koppel, and T.G. McGee (eds.) *The Extended Metropolis: Settlement Transition in Asia*, Honolulu: University of Hawaii Press.

McGee, T.G. and Yeung, Y.M. (1977) *Hawkers in Southeast Asian Cities: Planning for the Bazaar Economy*, Ottawa: International Development Research Centre.

McNamara, R. (1968, 1969) 'Address[es] to the Board of Governors of the World Bank', Washington D.C.

—— (1981) 'Address to the Board of Governors', World Bank, Nairobi, Kenya, 24 September 1973, in *The McNamara Years at the World Bank: Major Policy Addresses of Robert S. McNamara 1968–1981*, Baltimore: Johns Hopkins University Press for World Bank.

McRae, E. (1990) *A Construção da Igualdade. Identidade Sexual e Política no Brasil da 'Abertura'*, São Paulo: Universidade Estadual de Campinas.

Meadows, D., Meadows, D., and Randers, J. (1992) *Beyond the Limits: Global Collapse or a Sustainable Future*, London: Earthscan.

Meadows, D., Meadows, D., Randers, J., and Behrens, W. (1972) *The Limits to Growth: A Report for the Club of Rome's Project on the Predicament of Mankind*, New York: Universe Books.

Meek, R.L. (1976) *Social Science and the Ignoble Savage*, Cambridge: Cambridge University Press.

Melucci, A. (1988) 'Social movements and the democratization of everyday life', in J. Keane (ed.) *Civil Society and the State: New European Perspectives*, London: Verso.

—— (1989) *Nomads of the Present*, Philadelphia: Temple University Press.

Mies, M. (1986) *Patriarchy and Accumulation on a World Scale: Women in the International Division of Labour*, London: Zed Books.

Mill, J.S. (1942[1831]) *The Spirit of the Age*, Introductory Essay by Frederick von Hayek, Chicago: University of Chicago Press.

—— (1965[1859]) *On Liberty*, in Max Lerner (ed.) *Essential Works of John Stuart Mill*, New York: Bantam.

—— (1968[1858]) *Memorandum of the Improvements in the Administration of India*, Farnborough: Gregg International Publishers.

—— (1974[1843]) *A System of Logic, Ratiocinative and Inductive*, in J.M. Robson (ed.) *Collected Works of John Stuart Mill*, vol. VIII, Toronto: University of Toronto Press.

—— (1985[1848]) *Principles of Political Economy*, Harmondsworth: Penguin.

—— (1989[1873]) *Autobiography*, Harmondsworth: Penguin.

—— (1990) *Writings on India*, in John M. Robson, Martin Moir, and Zawahir Moir (eds.) *Collected Works of John Stuart Mill*, vol. XXX, Toronto: University of Toronto Press.

Minh-Ha, T. (1989) *Woman, Native, Other*, Bloomington: Indiana University Press.

Mires, F. (1987) 'Continuidad y ruptura en el discurso político', *Nueva Sociedad* 91: 129–140.

Mirowski, P. (1984a) 'The role of conservation principles in twentieth century economic theory', *Philosophy of the Social Sciences* 14: 461–473.

—— (1984b) 'Physics and the "marginalist revolution"', *Cambridge Journal of Economics* 8: 361–379.

Missen, G.E. (1972) *Viewpoints on Indonesia: a Geographical Study*, Melbourne: Thomas Nelson.

Mitchell, J.K. (1988) 'Confronting natural disasters: an international decade for natural hazards reduction', *Environment* 30: 25–29.

Mitchell, J.K., Devine, N., and Jagger, K. (1989) 'A contextual model of natural hazard', *Geographical Review* 79, 4: 391–409.

Mitchell, T. (1988) *Colonising Egypt*, Cambridge: Cambridge University Press.

—— (1989) 'The effect of the state', paper presented at the Social Science Research Council Workshop on State Creation and Transformation in the Middle East, Istanbul.

—— (1990) 'The invention and reinvention of the Egyptian peasant', *International Journal of Middle East Studies* 22, 2: 129–150.

—— (1991) 'The representation of rural violence in writings on political development: the case of Nasserist Egypt', in J. Waterbury and F. Kazemi (eds.) *Peasants and Politics in the Modern History of the Middle East*, Gainsville, Florida: University Presses of Florida.

Mitchell, T. and Abu-Lughod, L. (1993) 'Questions of modernity', *Items* 47, 4: 79–84.

Mlia, J.R. Ngoleka (1987) 'History of soil conservation in Malawi', in *History of Soil Conservation in the SADCC Region*, Report No. 8, Maseru: SADCC, Soil and Water Conservation Programme.

Mohanty, C. (1988) 'Under western eyes: feminist scholarship and colonial discourses', *Feminist Review* 30: 61–88.

—— (1991) 'Under western eyes', in C. Mohanty, A. Russo, and L. Torres (eds.) *Third World Women and the Politics of Feminism*, Bloomington: Indiana University Press.

Mohanty, C., Russo, A., and Torres L. (eds.) (1991) *Third World Women and the Politics of Feminism*, Bloomington: Indiana University Press.

Momsen, J. and Kinnaird, V. (eds.) (1993) *Different Places, Different Voices: Gender and Development in Africa, Asia and Latin America*, London: Routledge.

Moodie, T.D. (1975) *The Rise of Afrikanerdom: Power, Apartheid, and the Afrikaner Civil Religion*, Berkeley and Los Angeles: University of California Press.

Moodley, K. (1992) 'The continued impact of black consciousness', in B. Pityana *et al.* (eds.) *Bounds of Possibility: The Legacy of Steve Biko and Black Consciousness*, London and New Jersey: Zed Books.

Moore, D. (1992) 'The dynamics of development discourse', unpublished report, Ottawa: International Development Research Centre.

Moore, D. and Schmitz, G. (forthcoming) *Crisis and Renewal in Development Discourse: Global and Regional Perspectives on Democracy, Sustainability and Equity*, London: Macmillan.

Moore, M. (1989) 'The ideological history of the Sri Lankan peasantry', *Modern Asian Studies* 23, 1: 179–207.

Morren, G. (1983) 'A general approach to the identification of hazards', in K. Hewitt (ed.) *Interpretations of Calamity: From the Viewpoint of Human Ecology*, London: Allen and Unwin.

Mortimore, M. (1971) 'Population densities and systems of agricultural land use in northern Nigeria', *Nigerian Geographical Journal* 42.

—— (1989) *Adapting to Drought: Farmers, Famines and Desertification in West Africa*, Cambridge: Cambridge University Press.

Morton, D. and Zavarzadeh, M. (eds.) (1991) *Theory/Pedagogy/Politics: Texts for Change*, Urbana: University of Illinois Press.

Moser, C. (1989) 'Gender planning in the Third World: meeting practical and strategic gender needs', *World Development* 17, 11: 1799–1825.

—— (1993) *Gender, Planning and Development: Theory, Practice and Training*, London: Routledge.

Moss, L. (1978) 'Implementing site and services: the institutional environment of comprehensive development projects', unpublished Ph.D. thesis, University of California, Berkeley.

Mudimbe, V.Y. (1988) *The Invention of Africa: Gnosis, Philosophy and the Order of Knowledge*, Bloomington: Indiana University Press and London: James Currey.

Mueller, A. (1987a) 'Peasants and professionals: the social organization of women in development knowledge', unpublished Ph.D. thesis, Ontario Institute for Studies in Education, Toronto.

—— (1987b) 'Peasants and professionals: the production of knowledge about women in the Third World', paper presented at the Association for Women in Development, Washington, D.C.

Mzala (1988) *Gatsha Buthelezi: Chief with a Double Agenda*, London: Zed Books.

Mzamane, M.V. (1992) 'The impact of black consciousness on culture', in B. Pityana *et al.* (eds.) *Bounds of Possibility: The Legacy of Steve Biko and Black Consciousness*, London: Zed Books.

Nandy, A. (1987) *Traditions, Tyranny and Utopias: Essays in the Politics of Awareness*, Delhi: Oxford University Press.

—— (1989) 'Shamans, savages and the wilderness: on the audibility of dissent and the future of civilisations', *Alternatives* 14, 3: 263–278.

—— (1991) *The Intimate Enemy: Loss and Recovery of Self Under Colonialism*, Delhi: Oxford University Press.

—— (1992) 'State', in W. Sachs (ed.) *The Development Dictionary*, London: Zed Books.

Nash, R. (1983) *Wilderness and the American Mind*, New Haven: Yale University Press.

National Research Council (1987) *Confronting Natural Disasters: An International Decade for Natural Hazards Reduction*, Washington D.C.: National Academy.

Nature (1948) 'Aspects of colonial development', *Nature* 162: 547–550.

Netting, R. (1968) *Hill Farmers of Nigeria: Cultural Ecology of the Koyfars of the Jos Plateau*, Seattle: University of Washington Press.

Newman, J. (1992[1845]) 'An essay on the development of Christian doctrine', in *Conscience, Consensus and the Development of Doctrine*, New York: Doubleday.

Nisbet, R. (1969) *Social Change and History: Aspects of the Western Theory of Development*, New York: Oxford University Press.

Nixon, R. (1992) 'The collapse of the communist-anticommunist condominium: the repercussions for South Africa', *Social Text* 31/32: 235–251.

—— (1993) 'Of Balkans and Bantustans', *Transition* 60: 4–26.

Nockrashy, A.S., Galal, O., and Davenport, J. (1987) *More and Better Food: An Egyptian Demonstration Project*, Washington D.C.: National Research Council.

Norgaard, R. (1992) *Development Reportrayed*, London: Routledge.

Norris, C. (1992) *Uncritical Theory: Postmodernism, Intellectuals and the Gulf War*, London: Lawrence & Wishart.

Nyerere, J. (1973) *Freedom and Development*, Oxford: Oxford University Press.

O'Hanlon, R. and Washbrook, D. (1992) 'After Orientalism: culture, criticism and the politics of the Third World', *Comparative Studies in Society and History* 34, 1: 141–167.

O'Keefe, P., Westgate, K., and Wisner, B. (1976) 'Taking the naturalness out of natural disaster', *Nature*, 260.

Okoth-Ogendo, H.W.O. (1989) 'Some issues of theory in the study of tenure relations in African agriculture', *Africa* 59, 1: 6–17.

Oliver-Smith, A.S. (1986) *The Martyred City: Death and Rebirth in the Andes*, Prospect Heights, Illinois: Waveland Press.

Olivier, G.C. (1978) 'Political aspects of national security', in M. H. Louw (ed.) *National Security – A Modern Approach*, Pretoria: Institute for Strategic Studies, University of Pretoria.

Ong, A. (1988) 'Colonialism and modernity: feminist representations of women in non-Western societies', *Inscriptions* 3/4, 2: 79–93.

—— (1990) 'State versus Islam: Malay families, women's bodies, and the body politic in Malaysia', *American Ethnologist* 17, 2: 258–276.

O'Riordan, T. (1981) *Environmentalism*, London: Pion.

—— (1988) 'The politics of sustainability', in R. Turner (ed.) *Sustainable Environmental Management: Principles and Practice*, Boulder, Colorado: Westview Press.

O'Riordan, T. and Turner, R. (1983) *An Annotated Reader in Environmental Planning and Management*, Oxford: Pergamon Press.

Osuji, G.E. (1989) 'Splash erosion under sole and mixed cropping systems in southwestern Nigeria', *Journal of Environmental Management* 28: 1–9.

Packard, R. (1989) *White Plague, Black Labor: Tuberculosis and the Political Economy of Health and Disease in South Africa*, Berkeley: University of California Press.

—— (1994) 'Historicizing development', summary of a Workshop at Emory University, December 1993.

Pain, A. (1986) 'Agricultural research in Sri Lanka: an historical account', *Modern Asian Studies*, 20, 1: 755–778.

Palmer, B. (1990) *Descent into Discourse: The Reification of Language and the Writing of Social History*, Philadelphia: Temple University Press.

Pantojas-Garcia, E. (1990) *Development Strategies as Ideology: Puerto Rico's Export-Led Industrialization Experience*, Boulder, Colorado: Westview Press.

Papanek, H. (1984) 'Coming out of the niche: intellectual consequences of segregating advocacy research on women and development', mimeo.

Parajuli, P. (1991) 'Power and knowledge in development discourse: new social movements and the state in India', *International Social Science Journal* 127: 173–190.

Parpart, J. (1989) *Women and Development in Africa*, Lanham, Maryland: University Press of America.

Pearce, D., Markyanda, A., and Barbier, E. (1988) *Blueprint for a Green Economy*, London: Earthscan.

Peet, R. (1990) *Global Capitalism: Theories of Societal Development*, London: Routledge.

Peet, R. and Watts, M. (1993) 'Development theory and environment in an age of market triumphalism', *Economic Geography* 69, 3: 227–253.

Pepper, D. (1984) *The Roots of Modern Environmentalism*, London: Croom Helm.

Phimister, I. (1986) 'Discourse and the discipline of historical context: conservationism and ideas about development in Southern Rhodesia 1930–1950', *Journal of Southern African Studies* 12, 2: 263–275.

Pickles, J. (1992) 'Texts, hermeneutics and propaganda maps', in T. Barnes and J. Duncan (eds.) *Writing Worlds: Discourse, Text and Metaphor in the Representation of Landscape*, London: Routledge.

Pieterse, J.N. (1991) 'Dilemmas of development discourse: the crisis of developmentalism and comparative method', *Development and Change* 22: 5–29.

Pigg, S. (1992) 'Inventing social category through place: social representations and development in Nepal', *Comparative Studies in Society and History* 34, 3: 491–513.

Pike, A.H. (1938) 'Soil conservation amongst the Matengo tribe', *Tanganyika Notes and Records*, 6: 79–81.

Pityana, B., Ramphele, M., Mpumlwana M., and Wilson, L. (1992) *Bounds of Possibility: The Legacy of Steve Biko and Black Consciousness*, London: Zed Books.

Plant, C. and Plant, J. (1990) *Turtle Talk: Voices for a Sustainable Future*, Philadelphia: New Society Publishers.

Pletsch, C. (1981) 'The three worlds, or the division of social scientific labor, ca 1950–1975', *Comparative Studies in Society and History*, 23: 565–590.

Polier, N. and Roseberry, W. (1989) 'Tristes tropes: post-modern anthropologists encounter the other and discover themselves', *Economy and Society* 18: 245–264.

Porter, D. (1993) 'The limits and beyond: global collapse or a sustainable future', *Sustainable Development: People, Economy and Environment* 1, 2: 53–67.

—— (1991) *Haunted Journeys: Desire and Transgression in European Travel Writing*, Princeton: Princeton University Press.

302

Porter, D., Allen, B., and Thompson, G. (1991) *Development in Practice: Paved with Good Intentions*, London: Routledge.

Portes, A. and Kincaid, D. (1989) 'Sociology and development in the 1990's: critical challenges and empirical trends', *Sociological Forum* 4, 4: 479–503.

Posel, D. (1987) 'The language of domination', in S. Marks and S. Trapido (eds.) *The Politics of Race, Class and Nationalism in Twentieth Century South Africa*, London: Longman.

Pottenger, J. (1989) *The Political Theory of Liberation Theology*, Albany, New York: SUNY Press.

Pottier, J. (ed.) (1992) *Practising Development*, London: Routledge.

Prakash, G. (1992) 'Postcolonial criticism and Indian historiography', *Social Text* 31/32: 8–19.

Pratt, M.L. (1992) *Imperial Eyes: Travel Writing and Transculturation*, London: Routledge.

Pred, A. and Watts, M. (1992) *Reworking Modernity: Capitalisms and Symbolic Discontent*, New York: Rutgers University Press.

Preston, P. and Simpson-Housley, P. (eds.) (1994) *Writing the City: Eden, Babylon and the New Jerusalem*, London: Routledge.

Procacci, G. (1991) 'Social economy and the government of poverty', in G. Burchell, C. Gordon and P. Miller (eds.) *The Foucault Effect: Studies in Governmentality*, London: Harvester Wheatsheaf.

PRODDER (Programme for Development Research) (1988) *Development Research*, PRODDER newsletter, November.

Quijano, A. (1988) *Modernidad, Identidad y Utopia en América Latina*, Lima: Sociedad y Política Ediciones.

Rabinow, P. (1990) *French Modern*, Boston: MIT Press.

Rahnema, M. (1988a) 'A new variety of AIDS and its pathogens: homo economicus, development and aid', *Alternatives* 13, 1: 117–136.

—— (1988b) 'Power and regenerative processes in micro-spaces', *International Social Sciences Journal* 117: 361–375.

—— (1990) 'Participatory action research: the "Last Temptation of Saint" Development', *Alternatives* 15: 199–226.

—— (1992) 'Poverty', in W. Sachs (ed.) *The Development Dictionary: A Guide to Knowledge as Power*, London: Zed Books.

Raikes, P. (1981) *Livestock Development and Policy in East Africa*, Uppsala: Scandinavian Institute of African Studies.

—— (1985) *Modernising Hunger: Famine, Food Surplus and Farm Policy in the EEC and Africa*, London: CIIR with James Currey.

Ramphal, S. (1990) 'Endangered earth', in D. Angell, J. Comer, and M. Wilkinson (eds.) *Sustaining Earth: Response to Environmental Threats*, Basingstoke: Macmillan.

Ramphele, M. (1992) 'Empowerment and symbols of hope: black consciousness and community development', in B. Pityana *et al.* (eds.) *Bounds of Possibility: The Legacy of Steve Biko and Black Consciousness*, London: Zed Books.

Ranger, T. (1989) 'Whose heritage? The case of Matobo National Park', *Journal of Southern African Studies* 15, 2: 217–249.

Rathgeber, E. (1990) 'WID, WAD, Gad: trends in research and practice', *Journal of Developing Areas* 24: 489–502.

Redclift, M. (1984) *Development and the Environmental Crisis: Red or Green Alternatives?*, London: Methuen.

—— (1987) *Sustainable Development: Exploring the Contradictions*, London: Methuen.

Republic of Transkei (1982) *Debates of the National Assembly*, 13 May.

Rich, B. (1991) 'World Bank – green Frankenstein', *ECOS: A Review of Conservation* 12, 1: 82–83.

Richards, A. (1980) 'Egypt's agriculture in trouble', *MERIP Reports*, No. 84.

Richards, A. and Baker, R. (1992) 'Political economy review of Egypt', prepared for USAID Governance and Democracy Program, Washington D.C.: Management Systems International.

Richards, A. and Waterbury, J. (1993) *A Political Economy of the Middle East: State, Class, and Economic Development*, Boulder, Colorado: Westview Press.

Richards, P. (1983) 'Ecological change and the politics of African land use', *African Studies Review* 21, 2: 1–72.

—— (1985) *Indigenous Agricultural Revolution*, London: Hutchinson.

—— (1986) *Coping with Hunger: Hazard and Experiment in an African Rice-Farming System*, London: Allen and Unwin.

—— (1990) 'Indigenous approaches to rural development: the agrarian populist tradition in West Africa', in M. Altieri and S. Hecht (eds.) *Agroecology and Small Farm Development*, Boca Raton: CRC Press.

Richards, T. (1993) *The Imperial Archive*, London: Verso.

Rist, G. (1990) '"Development" as part of the modern myth', *European Journal of Development Research*, 2, 1: 10–21.

Robequain, C. (1954) *Malaya, Indonesia, Borneo and the Philippines*, London: Longman, Green and Co.

Robertson, A.F. (1984) *The People and the State: An Anthropology of Planned Development*, Cambridge: Cambridge University Press.

Robertson, C. and Berger, I. (eds.) (1986) *Women and Class in Africa*, New York: Africana.

Rocheleau, D. (1991) 'Gender, ecology, and the science of survival: stories and lessons for Kenya', *Agriculture and Human Values*, Winter-Spring: 158–165.

Roe, E. (1989) 'Folktale development', *American Journal of Semiotics* 6, 2/3: 277–90.

—— (1991) 'Development narratives, or making the best of blueprint development', *World Development* 19, 4: 287–300.

Rogers, B. (1980) *The Domestication of Women: Discrimination in Developing Societies*, New York: St. Martin's Press.

Rogers, P. (1979) 'The British and the Kikuyu 1890–1905: a reassessment', *Journal of African History* 20: 255–269.

Rosen, G. (1985) *Western Economists and Eastern Societies*, Baltimore: Johns Hopkins University Press.

Rostow, W. (1960) *The Stages of Economic Growth*, Cambridge: Cambridge University Press.

Roszak, T. (1979) *Person/Planet: The Creative Disintegration of Industrial Society*, London: Victor Gollancz.

RSA (Republic of South Africa) (1974) *Multi-National Development in South Africa: The Reality*, Pretoria: Department of Information.

Ruf, T. (1988) *Histoire contemporaine de l'agriculture egyptienne: Essai de synthèse*, Bondy, France: Editions de l'Orstom.

Runte, A. (1987) *National Parks: The American Experience*, Lincoln: University of Nebraska Press.

Sachs, W. (1990) 'Interview' on 'The Age of Ecology', Canadian Broadcasting Corporation Ideas Program, 18 June 1990.

Sachs, W. (ed.) (1992) *The Development Dictionary: A Guide to Knowledge as Power*, London: Zed Books.

Said, E. (1978) *Orientalism*, New York: Pantheon.

—— (1983) *The World, the Text, the Critic*, Cambridge, Mass.: Harvard University Press.

—— (1985) 'Orientalism reconsidered', in F. Baker *et al.* (eds.) *Europe and its Others*, volume 1, Colchester: University of Essex.

—— (1993) *Culture and Imperialism*, New York: Knopf.

Santos, M. (1975) *L'Espace Partage, les Deux Circuits de l'Economie Urbaine des Sous Développés*, Paris: Éditions, M. Th. Genin-Libraries Techniques.

—— (1976) 'Articulation of modes of production and two circuits of urban economy: wholesalers in Lima, Peru', *Pacific Viewpoint* 17, 1: 23–36.

Sardar, Z., Nandy, A., and Davies, M. (1993) *Barbaric Others: A Manifesto on Western Racism*, London: Pluto Press.

Scaff, L. (1991) *Fleeing the Iron Cage*, Berkeley: University of California Press.

Scheper-Hughes, N. (1992) *Death Without Weeping: The Violence of Everyday life in Brazil*, Berkeley: University of California Press.

Schick, I. C. (1990) 'Representing Middle Eastern women: feminism and colonial discourse', *Feminist Studies* 16, 2: 345–380.

Schon D. (1982) *The Reflective Practitioner: How Professionals Think in Action*, London: Maurice Temple Smith.

Schultz, T.W. (1964) *Transforming Traditional Agriculture*, New Haven: Yale University Press.

Schumpeter, J. (1952) *Capitalism, Socialism and Democracy*, London: Thames.

Schuurman, F. (ed.) (1993) *Beyond the Impasse: New Directions in Development Theory*, London: Zed Books.

Scott, A. (1988) 'Flexible production systems and regional development: the rise of new industrial spaces in North America and Western Europe', *International Journal of Urban and Regional Research* 12, 1.

Scott, C. (1994) *Gender and Development: Rethinking Modernization and Dependency Theory*, Boulder, Colorado: Lynne Rienner.

Scott, J. (1985) *Weapons of the Weak: Everyday Forms of Peasant Resistance*, New Haven, Connecticut: Yale University Press.

—— (1990) *Domination and the Arts of Resistance*, New Haven, Connecticut: Yale University Press.

Sen, A. (1981) *Poverty and Famines: An Essay on Entitlement and Deprivation*, Oxford: Clarendon Press.

Sen, G. and Grown, C. (1987) *Development, Crises, and Alternative Visions*, New York: Monthly Review Press.

Sessions, G. (1985) 'Western process metaphysics' in B. Devall and G. Sessions (eds.) *Deep Ecology: Living as if Nature Mattered*, Salt Lake City: Gibbs Smith Publisher.

Sewell, W.H. (1980) *Work and Revolution in France*, Cambridge: Cambridge University Press.

Shabalala, S. (1991) 'Economic emancipation: a Pan Africanist Congress view', in IDASA, *Economy: Growth and Redistribution*, Cape Town: IDASA.

Shava, P.V. (1989) *A People's Voice: Black South African Writing in the Twentieth Century*, London: Zed Books.

Sheail, J. (1976) *Nature in Trust: The History of Nature Conservation in Britain*, Glasgow: Blackie.

Shet, D.L. (1987) 'Alternative development as political practice', *Alternatives* 12, 2: 155–171.

Shiva, V. (1988) *Staying Alive: Women, Ecology and Development*, London: Zed Books.

—— (1991) *The Violence of the Green Revolution*, London: Zed Books.

—— (1992) 'Resources', in W. Sachs (ed.) *The Development Dictionary: A Guide to Knowledge as Power*, London: Zed Books.

Shohat, E. (1992) 'Notes on the post colonial', *Social Text* 31/32: 99–113.

Showers, K. (1989) 'Soil erosion in the Kingdom of Lesotho: origins and colonial response, 1830s to 1950s', *Journal of Southern African Studies* 15, 2: 263–286.

Sikkink, K. (1991) *Ideas and Institutions: Developmentalism in Brazil and Argentina*, Ithaca, New York: Cornell University Press.

Singleton, J.A. (1983) 'The training of local Development Fund officials and the decentralization process in Egypt', paper presented to the Conference on Organizational Policy and Development, University of Louisville, Louisville, Kentucky.

Skinner, A. (1982) 'A Scottish contribution to Marxist sociology', in I. Bradley and M. Howard (eds.) *Classical and Marxian Political Economy: Essays in Honour of R.L. Meek*, London: Macmillan.

Skinner, G.W. (1964–65) 'Marketing and social structure in rural China', *Journal of Asian Studies* 24: 3–43, 195–228, 363–399.

Slater, D. (1975) 'The poverty of modern geographic enquiry', *Pacific Viewpoint* 17, 1: 23–36.

—— (1976) 'Anglo-Saxon geography and the study of underdevelopment', *Antipode* 8, 3: 88–93.

—— (ed.) (1985) *New Social Movements and the State in Latin America*, Amsterdam: CEDLA.

—— (1990) 'Fading paradigms and new agendas: crisis and controversy in development studies', *European Review of Latin American and Caribbean Studies* 49: 25–32.

—— (1992a) 'Theories of development and the politics of the postmodern, *Development and Change* 23: 283–319.

—— (1992b) 'On the borders of social theory learning from other regions', *Society and Space* 10: 307–327.

—— (1993) 'The geopolitical imagination and the enframing of development theory', *Transactions of Institute of British Geographers* 18: 419–437.

Sloterdjik, P. (1987) *Critique of Cynical Reason*, Minneapolis: University of Minnesota Press.

Smith, A. (1937[1776]) *Inquiry into the Nature and Causes of the Wealth of Nations*, New York: Modern Library.

Smith, K. (1992) *Environmental Hazards: Assessing Risk and Reducing Disaster*, London: Routledge.

Sondheimer, S. (ed.) (1991) *Women and the Environment: A Reader*, New York: Monthly Review Press.

Sorj, B. (1990) 'Modernity and social disintegration: crisis of society and crisis of the social sciences in Brazil and Latin America', *European Journal of Development Research* 21, 1: 108–120.

Sorrenson, M. (1967) *Land Reform in Kikuyu Country*, Nairobi: Oxford University Press.

Spate, O.H.K. (1954) *India and Pakistan: A General and Regional Geography*, London: Methuen.

Spelman, E. (1990) *Inessential Woman*, London: The Woman's Press.

Spencer, J.E. (1954) *Asia, East by South: A Cultural Geography*, New York: Wiley.

Spies, P. (1983) 'Community development in black rural areas', in P. Spies (ed.) *Urban-Rural Interaction in South Africa*, Stellenbosch: Unit for Futures Research, University of Stellenbosch.

Spivak, G.C. (1987) *In Other Worlds: Essays in Cultural Politics*, London: Methuen.

—— (1990a) 'Gayatri Spivak speaks on the politics of the postcolonial subject', *Socialist Review* 20: 81–90.

—— (1990b) *The Post-Colonial Critic: Interviews, Strategies, Dialogue*, London: Routledge.

Springborg, R. (1989) *Mubarak's Egypt: Fragmentation of the Political Order*, Boulder, Colorado: Westview Press.

—— (1990) 'Rolling back Egypt's agrarian reform', *Middle East Report* 166: 28–30, 38.

Sprinker, M. (ed.) (1992) *Edward Said: A Critical Reader*, Oxford: Blackwell.

Spurr, D. (1993) *The Rhetoric of Empire*, Durham: Duke University Press.

Stamp, L.D. (1938) 'Land utilisation and soil erosion in Nigeria', *Geographical Review* 28: 32–45.

Stamp, P. (1989) *Technology, Gender, and Power in Africa*, Ottawa: International Development Research Centre.

Statman, J. (1992) 'In the heroic vanguard of the normalization: professional elites and the creation of the new South Africa', paper presented at the Third Annual Conference of the South African Azanian Student Movement (SAASM), Dover, Delaware, 8 August.

Staudt, K. (1985) *Women, Foreign Assistance and Advocacy Administration*, New York: St. Martin's Press.

—— (1991) *Managing Development: State, Society, and International Contexts*, Newbury Park, California: Sage.

Stauffer, B. (1990) 'After socialism: capitalism, development and the search for critical alternatives', *Alternatives* 15: 401–430.

Stauth, G. (1989) 'Capitalist farming and small peasant households in Egypt', in K. Glavanis and P. Glavanis (eds.) *The Rural Middle East: Peasant Lives and Modes of Production*, London: Zed Books.

Stenhouse, A.S. (1944) 'Agriculture in the Matengo Highlands', *East African Agricultural Journal* July: 22–24.

Stichter, S. and Parpart, J. (eds.) (1988) *Patriarchy and Class: African Women in the Home and the Workforce*, Boulder, Colorado: Westview Press.

Stiglitz, J. (1989) *The Economic Role of the State*, Oxford: Blackwell.

Stocking, M. (1985) 'Soil conservation policy in colonial Africa', *Agricultural History* 56, 2: 148–161.

Stoler, A. (1992) 'Sexual affronts and racial frontiers', *Comparative Studies in Society and History* 34, 3: 514–551.

Stone, R. (1991) *The Nature of Development: A Report from the Tropics on the Quest for Sustainable Economic Growth*, New York: Knopf.

Storper, M. (1991) *Industrialisation, Economic Development and the Regional Question in the Third World*, London: Pion.

Streeten, P. (1982) 'A cool look at outward-looking strategies for development', *The World Economy* 5, 2: 159–169.

Studies in Family Planning (1990) 'Egypt 1988: results from the demographic and health survey', *Studies in Family Planning* 21, 6: 347–351.

Susman, P., O'Keefe, P., and Wisner, B. (1983) 'Global disasters, a radical interpretation', in K. Hewitt (ed.) *Interpretations of Calamity*, London: Allen and Unwin.

Sutton, F. (1989) 'Development ideology: its emergence and decline', *Daedalus* 118, 1: 35–60.

Sutton, K. L. and Lawless, R. (1987) 'Progress in human geography of the Maghreb', *Progress in Human Geography* 11, 1: 60–105.

Swart, H. (1985) ''n Gesonder wetenskaplike basis as vertrekpunt vir saamwerking in die ontwikkelingsveld', in L. A. Van Wyk (ed.) *Development Perspectives in Southern Africa* ABEN: Research Paper 85–1, Potchefstroom: Potchefstroom University, for Christian Higher Education.

Tarrow, S. (1988) 'National politics and collective action: recent theory and research in Western Europe and the United States', *Annual Review of Sociology* 14: 421–440.

Tax, S. (1953) *Penny Capitalism*, Chicago: University of Chicago Press.

Tennekoon, S. (1988) 'Rituals of development: the accelerated Mahavali development program in Sri Lanka', *American Ethnologist* 15, 2: 294–310.

Therborn, G. (1982) *The Ideology of Power and the Power of Ideology*, London: Verso.

Thiong'o, Ngugi wa (1986) *Decolonising the Mind*, London: James Currey.

Thomas, A. (1992) 'Introduction', in T. Allen and A. Thomas (eds), *Poverty and Development in the 1990s*, Oxford: Oxford University Press.

Thomas, K. (1983) *Man and the Natural World: Changing Attitudes in England 1500–1800*, London: Allen Lane.

Thompson, N. (1988) *The Market and Its Critics: Socialist Political Economy in Nineteenth Century Britain*, London: Routledge.

Thomson, J. (1885) *Through Masailand*, London: Sampson Low, Marston, Searle and Rivington.

Thoreau, H.D. (1977) *The Portable Thoreau*, New York: Penguin Books.

Tiffen, M., Mortimore, M., and Gichuki, F. (eds.) (1994) *More People Less Erosion*, Chichester: John Wiley.

Tinker, I. (ed.) (1990) *Persistent Inequalities*, Oxford: Oxford University Press.

Tobey, R. (1981) *Saving the Prairies: The Life of the Founding School of American Plant Ecology*, Berkeley: University of California Press.

Torgovnick, M. (1990) *Gone Primitive: Savage Intellects, Modern Lives*, Chicago: University of Chicago Press.

Torry, W.I. (1978) 'Natural disasters, social structure and changes in traditional societies', *Journal of Asian and African Studies*, 13: 167–183.

—— (1986) 'Morality and harm: Hindu peasant adjustments to famines', *Social Science Information*.

Touraine, A. (1981) *The Voice and the Eye: An Analysis of Social Movements*, Cambridge: Cambridge University Press.

—— (1988) *The Return of the Actor*, Minneapolis: University of Minnesota Press.

Toye, J. (1987) *Dilemmas of Development*, Oxford: Blackwell.

Truman, H. (1967) [1949] 'Inaugural Address', Washington D.C., 20 January, in *Documents on American Foreign Relations*, New York: Simon and Schuster.

Turner, R. (ed.) (1988a) *Sustainable Environmental Management: Principles and Practice*, Boulder: Westview Press.

—— (1989b) 'Sustainability, resource conservation and pollution control: an overview', in R. Turner (ed.) *Sustainable Environmental Management: Principles and Practice*, Boulder: Westview Press.

UK (United Kingdom) (1934) *Kenya Land Commission: Evidence and Memoranda*, 3 volumes, London: HMSO.

Ullrich, O. (1992) 'Technology', in W. Sachs (ed.) *The Development Dictionary: A Guide to Knowledge as Power*, London: Zed Books.

U.N. Centre for Human Settlements (1987) *Global Report on Human Settlements*, Oxford: Oxford University Press.

UNDRO (1990) 'World launches International Decade for Natural Disaster Reduction', *UNDRO News*, Special Issue, Jan/Feb., United Nations Disaster Relief Organization, Geneva.

U.S. Congress (House Committee on Foreign Affairs, Subcommittee on Europe and the Middle East) (1984) *Foreign Assistance Legislation for Fiscal Year 1985 (Part 3): Economic and Military Aid Programs in Europe and the Middle East*, 98th Congress, second session, Feb.-March, Washington D.C.

—— (House Committee on Foreign Affairs, Subcommittee on Europe and the Middle East) (1987) *Hearings on Agency for International Development Policy on the Use of Cash Transfer: The Case of Egypt*, 100th Congress, second session, 10 December, Washington D.C.

USAID (United States Agency for International Development) (1973) *The Logical Framework: Modifications Based on Experience*, Washington D.C.
—— (1982) 'Women in development', USAID Policy Paper, Washington D.C.
USAID/c (United States Agency for International Development, Cairo) (1989a) *Agricultural Data Base*, Cairo: Office of Agricultural Credit and Economics.
—— (1989b) 'Common misconceptions about USAID in Egypt/Mafâhîm khâti'a 'an barnâmij al-musâ'idât al-amrîkîya li-misr', pamphlet, Cairo: Public Affairs Office.
—— (1989c) *Status Report: United States Economic Assistance to Egypt*, Cairo: Public Affairs Office.
USAID/w (United States Agency for International Development, Washington) (1988) *Annual Budget Submission, FY 1990: Egypt*, Washington D.C.
—— (1989) *Congressional Presentation, FY 1990*, main volume, and Annexe II: *Asia and the Near East*, Washington D.C.
USDA (United States Department of Agriculture) (1976) *Egypt: Major Constraints to Increasing Agricultural Productivity*, Foreign Economic Report No. 120, Washington D.C.: Department of Agriculture.
—— (1989) *Agricultural Outlook*, October, Washington D.C.: Department of Agriculture.
Van der Berg S. and Van der Kooy, R. (1980) 'Dimensions of interdependence: a review', *Development Studies Southern Africa* 2, 4: 517–522.
Van der Kooy, R. (1979) 'In search of a new economic development paradigm for Southern Africa: an introduction', *Development Studies Southern Africa* 2, 1: 3–34.
—— (ed.) (1985) *An Introduction to Economic Development in Southern Africa and the Role of the DBSA*, Position Paper No. 4, Sandton: Development Bank of Southern Africa.
—— (1988) 'A development perspective of southern Africa', in R. Van der Kooy (ed.) *Prodder Development Annual*, Pretoria: Programme of Development Research.
Van der Merwe, P. J. (1983) 'Unemployment and employment creation', *Development Studies Southern Africa* 5, 2: 146–59.
Van der Merwe, S. (1980) 'A response to Cleary', in S. Cleary and S. Van der Merwe 'The homelands policy – a neo-colonial solution to South Africa's future', Occasional Papers, South African Institute of International Relations.
Van der Waal, K. and Sharp, J. (1988) 'The informal sector: a new resource', in E. Boonzaier and J. Sharp (eds.) *South African Keywords: The Uses and Abuses of Political Concepts*, Cape Town: David Philip.
Vandergeest, P. and Buttel, F. (1988) 'Marx, Weber and development sociology', *World Development* 16: 683–695.
Van Maanen, J. (1988) *Tales of the Field: On Writing Ethnography*, Chicago: University of Chicago Press.
Van Niekerk, A.S. (1986) 'African religion and development', *Development Southern Africa* 3, 1: 50–66.
Viljoen, F. (1984) 'Investigation into the small business sector in the RSA, with specific reference to factors that may retard the growth of development thereof', *Development Southern Africa* 1, 2: 223–228.
Viola, E. (1987) 'O movimento ecológico no Brasil (1974–1986)', *Revista Brasileria de Ciencias Sociais* 1, 3: 5–25.
von Höhnel, L. (1894) *Discovery of Lakes Rudolf and Stefanie*, London: Longman, Green and Co.
Waddell, E. (1977) 'The hazards of scientism: a review article', *Human Ecology* 5, 1: 67–76.
—— (1983) 'Coping with frosts, governments and disaster experts: some relevant reflections based on a New Guinea experience and a perusal of the relevant literature', in K. Hewitt (ed.) *Interpretations of Calamity: From the Viewpoint of Human Ecology*, London: Allen and Unwin.

Wager, W. (1967) 'Modern views of the origins of the idea of progress', *Journal of the History of Ideas* 28 1: 22

Wallerstein, I. (1991) *Unthinking Social Science*, Cambridge: Polity Press.

Wamba, E. Wamba dia (1991) 'Some remarks on culture, development and revolution in Africa', *Journal of Historical Sociology* 4, 3: 219–235.

Watnick, M. (1952) 'The appeal of Communism to the peoples of underdeveloped areas', *Economic Development and Cultural Change* 1: 22–36.

Watts, M.J. (1983) *Silent Violence: Food, Famine and Peasantry in Northern Nigeria*, Berkeley: University of California Press.

—— (1989) 'The agrarian crisis in Africa: debating the crisis', *Progress in Human Geography* 13, 1: 1–41.

—— (1991a) 'Mapping meaning, denoting difference, imagining identity', *Geografiska Annaler* 73: 7–16.

—— (1991b) 'Visions of excess: African development in an age of market idolatry', *Transition* 51: 126–141.

Watts, M.J. and Bohle, H-G. (1993) 'The space of vulnerability: the causal structure of hunger and famine', *Progress in Human Geography* 17, 1: 43–87.

Weiskel, T. (1988) 'Toward an archaeology of colonialism: elements in the transformation of the Ivory Coast', in D. Worster (ed.) *The Ends of the Earth: Perspectives on Modern Environmental History*, Cambridge: Cambridge University Press.

Welsh, B. and Butorin, P. (1990) *Dictionary of Development*, New York: Garland.

Wessel, J. (1983) *Trading the Future: Farm Exports and the Concentration of Economic Power in Our Food System*, San Francisco: Institute for Food and Development Policy.

Weurleusse, J. (1946) *Paysans de Syrie du Proche-Orient*, Paris: Gallimard.

White, G.F. (1961) 'The choice of use in resource management', *Natural Resources Journal* 1: 23–40.

White, H. (1987) *The Content of the Form: Narrative Discourse and Historical Representation*, Baltimore: Johns Hopkins University Press.

Wignaraja, P., Hussain, A., Sethi, H., and Wignaraja, G. (eds.) (1992) *Participatory Development: Learning from South Asia*, Tokyo: Oxford University Press.

Wijkman, A. and Timberlake, L. (1987) *Natural Disasters: Acts of God or Acts of Man?*, London: Earthscan.

Williams, G. (1981) 'The World Bank and the peasant problem', in J. Heyer, P. Roberts and G. Williams (eds.) *Rural Development in Tropical Africa*, London: Macmillan.

—— (1982) 'Equity, growth and the state', *Africa* 52, 3: 114–120.

—— (1984) 'Review of the Berg Report', *Review of African Political Economy*, 27–28.

—— (1988a) 'The World Bank in Northern Nigeria revisited: a review of the World Bank's Agricultural Sector Report, 1987', *Review of African Political Economy* 43: 42–67.

—— (1988b) 'Why is there no agrarian capitalism in Nigeria?' *Journal of Historical Sociology* 1, 4: 345–398.

Williams, R. (1976) *Keywords: A Dictionary of Culture and Society*, Oxford: Oxford University Press.

Willis, P. (1990) *Common Culture*, Boulder, Colorado: Westview Press.

Wilson, K.B. (1989) 'Trees in fields in Southern Zimbabwe', *Journal of Southern African Studies* 15, 2: 369–383.

Wilson, L. (1992) 'Bantu Stephen Biko: a life', in B. Pityana *et al.* (eds.) *Bounds of Possibility: The Legacy of Steve Biko and Black Consciousness*, London: Zed Books.

Wiltshire, R. (1988) 'Indigenisation issues in women and development studies in the Caribbean: towards a holistic approach', paper presented at Canadian Research Institute for the Advancement of Women, Annual Conference, Quebec City, 11–13 November.

Winrock International (1986) *Policy Guidelines for Agricultural Mechanization in Egypt*, Winrock International.

Wisner, B. (1988) *Power and Need in Africa*, London: Earthscan.

Wood, R. (1985) 'The politics of development policy labelling', *Development and Change* 16: 347–373.

Woods, G. (1968) 'Address of the President of the World Bank before UNCTAD', New Delhi.

World Bank (1975a) *Assault on World Poverty*, Baltimore: Johns Hopkins University Press.

—— (1975b) *Lesotho: A Development Challenge*, Washington D.C.

—— (1979–93) *World Bank Development Reports*, Washington, D.C.

—— (1981) *Accelerated Development in Sub-Saharan Africa* (Berg report), Washington D.C.

—— (1986) *Population Growth and Policies in Sub-Saharan Africa*, Washington D.C.

—— (1989a) *Sub-Saharan Africa: From Crisis to Sustainable Growth*, Washington D.C.

—— (1989b) *Kenya: The Role of Women in Economic Development*, Washington D.C.

—— (1989c) *World Development Report 1989*, New York: Oxford University Press.

—— (1989d) *Trends in Developing Economies 1989*, Washington D.C.

—— (1990a) *World Development Report 1990*, New York: Oxford University Press.

—— (1990b) *The Population, Agriculture and Environment Nexus in Sub-Saharan Africa*, draft for discussion, The World Bank, Africa Region, Washington D.C., 29 May 1990.

—— (1990c) 'Executive summary of *The Population, Agriculture and Environment Nexus in Sub-Saharan Africa* (World Bank 1990b).

—— (1991a) *World Development Report 1991*, New York: Oxford University Press.

—— (1991b) *Egypt: Alleviating Poverty*, Washington D.C.

—— (1992) *World Development Report 1992*, New York: Oxford University Press.

—— (1993) *World Development Report 1993*, New York: Oxford University Press.

Worster, D. (1985) *Nature's Economy: A History of Ecological Ideas*, Cambridge: Cambridge University Press.

—— (1990a) 'The ecology of order and chaos', *Environmental History Review* 14, 1/2: 1–18.

—— (1990b) 'Interview' on *The Age of Ecology*, Canadian Broadcasting Corporation Ideas Program, 18 June 1990.

Worthington, E. (1938) *Science in Africa: A Review of Scientific Research Relating to Tropical and Southern Africa*, London: Royal Institute of International Affairs.

—— (1983) *The Ecological Century: A Personal Appraisal*, Cambridge: Cambridge University Press.

Young, D. (1990) *Post-Environmentalism*, London: Belhaven Press.

Young, K., Walkowitz, C., and McCallogh, R. (eds.) (1981) *Of Marriage and the Market*, Berkeley: University of California.

Young, R. (1990) *White Mythologies: Writing History and the West*, London: Routledge.

Zaytoun, M.A. (1982) 'Income distribution in Egyptian agriculture and its main determinants', in G. Abdel-Khalek and R. Tignor (eds.) *The Political Economy of Income Distribution in Egypt*, New York: Holmes and Meier.

Zille, H. (1983) 'Restructuring the industrial decentralisation strategy', in Southern African Research Service, *South African Review 1*, Johannesburg: Ravan Press.

311

INDEX

Abu-Lughod, J. 192, 196
Acton, J.E.E.D. 30
Adams, William *see* conservation;
environmentalism; sustainable
development
Africa: agricultural ecology 91;
agriculture 101–5; big game hunting
92–3; British rule in 1–2, 10, 15, 16;
colonial 1–2, 69, 92, 100–12; famines
115; food imports 163; resistance
movements 103–4; soil erosion/
conservation *see* soil conservation;
soil erosion; technology imports 103;
World Bank and 158–75, *see also*
Kenya; South Africa
African National Congress (ANC) xi,
xii, xiii, 19, 229, 230, 251
Afshar, H. 257
agribusiness 77, 79, 80
agriculture: in Africa 101–5, *see also*
Egypt; Kenya; mechanization of
139–42, *see also* soil conservation;
soil erosion
Allen, R. 89
Alonso, W. 197
alternative development 19, 20, 45, 58,
61; DAWN 259–60, 263; populism
and 20; women and 258–60
alternatives to development 19, 20, 58,
215–16; alternative forms of
knowledge *see* knowledge; anti-
developmentalism as 45; post-
modern 47; Watts on 19, 20, 45, 47,
58, *see also* new social movements
Alvares, Claude 5, 51
Alvarez, S. 222
Amin, Samir; alternative development
45, 258; Eurocentrism 45, 58–9, 194;

green ideas 96
Anderson, Ben 45
Anderson, B.R. 195
Anderson, D. 91, 102, 103
Angelou, Maya 266
anti-developmentalism 19, 20, 45, *see
also* new social movements
apartheid xi, 13, 178, 182, 229;
bantustans xii, 176, 177–81, 187, 189;
collaborators 245; parent/child
metaphor and 12, 236, 237, 239;
responsibility for 234; 'separate
development' 16, 177–81, 182, 191,
237; *see also* South Africa
Appadurai, A. 57
appropriate technology 79
Apthorpe, R. 56, 76
Arato, A. 59
archaeology of development 8, 11, 46,
56, *see also* Watts, Michael J.
Arditi, B. 224
Aristotle 67
Armstrong, W.R. 201
Arndt, H. 69
Aseniero, G. 90
Ashcraft, R. 235
Ashforth, A.: the common good 174–5;
experts 190
Ashley, R.K. 232, 233
Asia 199–202; Eurocentric creation of
195–7; geographical texts on 18;
urban theory 202–5; Western
Geography and 196–8
authoritarianism 64–5, 84

Bacon, Francis 232
Bagehot, Walter 48
Baker, R. 115